Network Science

Network Science

Analysis and Optimization Algorithms for Real-World Applications

Carlos Andre Reis Pinheiro
SAS Institute Inc., USA

Registered Offices
John Wiley & Sons, Inc., 111 River Street, Hoboken, NJ 07030, USA

Editorial Office
111 River Street, Hoboken, NJ 07030, USA

For details of our global editorial offices, customer services, and more information about Wiley products visit us at www.wiley.com.
Wiley also publishes its books in a variety of electronic formats and by print-on-demand. Some content that appears in standard print versions of this book may not be available in other formats.

Library of Congress Cataloging-in-Publication Data Applied for:
Hardback ISBN: 9781119898917

Cover Design: Wiley
Cover Image: © GarryKillian/Shutterstock

Set in 9.5/12.5pt STIXTwoText by Straive, Pondicherry, India

To Dani, Lucas, and Maitê.

Contents

Preface

I was first exposed to the term network science in 2008 when I moved to Dublin to conduct a post-doctoral research term at the Dublin City University in partnership with Eircom. By that time, there was much more network analysis, or social network analysis. At first glance, I thought it was related to social media, such as Facebook, Twitter, LinkedIn, and Instagram, along with many others. Many people likely have this misunderstanding. When we mention social network analysis most people are directly pointed to social media, or the analysis of social interactions.

Network science involves several disciplines like social sciences, graph theory, mathematical modeling, statistics analysis, and optimization, to name a few. Assuming that everything is connected, the main goal is to solve complex problems or to understand a scenario in a unique perspective. When I say everything is connected, I mean that most of the real-world problems can be analyzed in a network perspective, where descriptive attributes are linked to constraints or restrictions, which are linked to possible outcomes or targets, which are linked to goals and solutions, which ultimately are linked to the problems. Even traditional approaches such as predictive modeling can have a network understanding, where input or independent variables are linked to output, target, or dependent variables. How strongly or weakly are they connected to each other? How strongly or weakly are they connected to the target? Surrogate variables can be connected to the original input variables. Network metrics or centralities can turn out to be the most important and relevant predictors in supervised modeling. It can definitely be a more complex approach, but certainly give us opportunities for a more comprehensive understanding of the problem and the possible optimal solutions.

For example, in association rules, items are correlated to each other based on some specific transactions. These correlations create rules, which are symmetric. Similar correlations can be visualized as a network, where items are connected to each other upon a very particular frequency, defining therefore distinct weights for the links among them. The weighted links between items define the importance of the relationships, similar to how confidence and support define the importance of the association rules. This is a very straightforward network analysis that can be conducted in a similar approach as association rules. As in sequence analysis, if we have the time identifier, or the information about the sequence of the transactions, we can also define the direction of the links between the items, and then we can produce a similar analysis as we do in sequence association rules, but again, in a network perspective. Something that we cannot see in association rules or in sequence analysis is the strength of the weak links or the missing links. Imagine that from the association rules, we find two strong rules, Coke associated to Lays, and Pepsi associated to Lays. There is no association between Coke and Pepsi. In network analysis, we would see a "triangle" with a missing link, with a link between Coke and Lays, a link between Pepsi and Lays, but no link between Coke and Pepsi. Perhaps, this missing link indicates they are surrogate products. Speaking about Coke and Pepsi it is easy to figure that out. But what about a grocery store with dozens of thousands of items?

In business problems there is always a question to be answered, an operational task to be improved, a challenge to be overcome, an insight to be produced. Are the customers willing to purchase this specific product? How likely are they to purchase it? How do subscribers consume this specific service? Are they willing to increase or decrease their usage? How likely? How can we improve a telecommunications network due to the customers' usage? How can we improve our supply chain across all stores based on past purchases? Can we describe a specific economic scenario considering different countries, states, or cities? Can we explain cause and effect of complex events in politics, international trade relations, or immigration? Network science works nicely to describe social relationships. However, by social here you can understand almost anything. Social can be people, employees, companies, countries, equipment, products, services, governments, or a combination of them. Network science also works well as an exploratory analysis tool, clearly describing complex scenarios, particularly when relations between entities play a key role.

How do we answer business questions? We often answer questions, solve problems, or explain business scenarios by working on the data that describe that problem or scenario. Machine learning models adjust mathematical and statistical equations according to the data available, based on the data distribution, the types of the variables, their variation, and many other data aspects. Different models work better for some specific type of data, but all models require data to create, or better, to find the right correlations between the problem and the solution. The nice thing about network science is that we can create different networks upon the same set of data, which creates distinct exploratory models and then different outcomes. The way we translate the data into the nodes and links, the way we use the data to define the relationships between them, the way we weight the nodes and links, everything changes the input network and therefore the results. For instance, based on the CDRs – Call Detail Records – we can define a network where the mobile phone is a node, a household is a node, or a switch is a node. The way these nodes are related also depends on the way we want to envision the network. The link can be calls and messages, physical connections of the telecommunications network, and people moving around and being handled by different cell towers. This flexibility in building different "inputs" and performing multiple exploratory models gives us more possibilities to describe and understand the problem, and for sure, more options to find viable solutions.

The first real problem I experienced in terms of network science was working in Dublin at Dublin City University in partnership with Eircom. We had a wonderful challenge to better understand the churn event at Eircom, a major telecommunications company in Ireland. We were asking ourselves if the churn event could occur as a viral event. When subscribers decide to leave, would they influence other subscribers to leave as well? In order to answer this question, we should understand the subscribers' relationships. At the end, this is what communications providers do right? They allow people to get connected to each other. Then, we got the fundamental data from carriers, calls, and text messages. This data describes in detail when and how one subscriber gets connected to another. Based on this data, we built the network, considering all subscribers, and all relationships. In addition to that, we considered the churn event over time. By doing this, we could monitor what happened when a subscriber decided to leave. What happened with their friends, relatives, co-workers, and so on? Did they leave afterwards? We investigated a substantial amount of data, considering a reasonable timeframe, to understand the overall viral effect some subscribers could exert over the others. In addition to the traditional transactional and demographic data about the customers, we described them in terms of their network centralities (what valued them as nodes within a network), and in terms of communities (how they were grouped together based on their relationships). At the end, a classification model was trained to estimate the likelihood of each subscriber's behavior as an influencer in terms of the churn event. We also noticed that some subscribers can be influencers in one specific business event and not be for another. That means, the characteristics of the influencer subscribers differ from one business event to another. The patterns of influence in churn were different from purchasing, consuming, or product adoption.

After that experience, I had the privilege and luck to work on many projects involving network science, looking at vastly different type of problems in a variety of industries. Even though each project is unique in terms of the particular problem, or the best possible solution, or based on the timely data available, most of them search for solutions in the same space. The solutions go from traditional business demands such as avoiding churn and boosting product adoption in communications and retail, to detecting fraud in finance, insurance, communications, taxpayers, and consumer goods. This required, a search for the optimal learning path not based on the content or subject of the courses but upon the relationships of the courses created by the student enrollments to evaluate and understand the players in economic trading among government agencies. Also, to search the main actors in illicit trade to find the best delivery routing in wholesales – from depot to stores, to optimize a work force scheduling in restaurants and hotels, to find optimal routes using the public transportation systems, to understand the virus spreading and predict new outbreaks, to foresee population movements and the impact in society, to evaluate the urban mobility and create solutions for specific big events, and unexpected situations, among others. The more I work in the field of network science in different industries with distinct customers, the more I believe it is part of an overall solution for complex problems.

After many projects in the field of network science, sometimes having network analysis and network optimization as a unique solution for the problem, sometimes having these techniques as part of the solution, either by identifying new inputs to supervised or unsupervised machine learning models, or by just revealing new insights about the problem, I have decided to enhance the work I started back in 2010 when I wrote *Social Network Analysis in Telecommunications*, also published by Wiley. This book describes in detail most of the important algorithms in subnetwork analysis, centrality metrics, and network optimization. In addition to covering the fundamental math behind the algorithms, I tried to emphasize a holistic explanation about each algorithm, why it is used, how it can be used, the benefits of using it, and so on. Each algorithm within the books is somehow associated to a business description and a possible application to solve complex real-world problems. Throughout the book, code examples for each algorithm, and most important, the analysis of their outcomes, are also added to complete a step-by-step approach that in my view makes it easier to understand each technique, how to deploy it, and some possible ways to interpret the results. At the end of the book, I shared some real-world case studies to demonstrate how beneficial network science can be in industry, government, and society.

All code examples in this book were based on SAS Viya, a new in-memory distributed engine for analytics. This environment is perfect for network science. Most of the network analysis and network algorithms can benefit from a distributed computing environment to execute the procedures faster and better. I have been using SAS, mostly as a customer, since 2002. I have been using the network procedures at SAS (formerly OptGraph and OptNet) since 2010. Most of the projects I have been involved with in the field of network science was before joining SAS at the end of 2015. I was a customer, a partner, or a researcher. Since the release of Viya, around 2016, I am truly amazed about how fast and reliable the environment is, and how I can benefit from it when running network analysis and network optimization algorithms. The new procedures for the Viya platform are now named as Network and OptNetwork. We are going to use them a lot throughout this book.

For me, at the end of the day, it is hard to think of a problem that cannot benefit from network science, either as a straightforward solution or as part of a more overall approach in solving complex problems. Even when I engage in an initially traditional problem, where machine learning or statistical modeling can be the solution, I always envision a network, and always search for the nodes and links. In my opinion, everything is connected. Finding the dots and connecting them, can be a great analytical approach to solving real-world problems.

I hope you enjoy reading this book, and I anticipate that it will be somehow helpful to you.

Carlos Andre Reis Pinheiro
March 2022

Acknowledgments

To all my colleagues and customers who have worked with me during the past 14 years in the field of network science in different companies, countries, and industries. It is hard to name all of them, but I would like to give a special thanks to some individuals who were extremely important throughout this journey.

James Davey, Gavin Kiernan, and Brian Buckley from Eircom; Karl Langan from SAS Ireland; Bruno Martins, Carlos Pamplona, and João Pedro Sant'Anna from Oi; Ozlem Demirboga and Tamer Cagatay from Turkcel; Pelin Özbozkurt, Ömer Nadiler, and Onuralp Öztürk from SAS Turkey; Marcelo Pimenta and Paula Fadul from Telefonica Vivo; Mirella Micheloni, Anderson Martins, and Elliton Matos from TIM; Jay Laramore from SAS US; Mike McCarthy from Sainsbury's; Mihai Paunescu from Austrian Ministry of Finance; Hans de Wit from Telenor; Pamela Morales-Cabudoy from Globe Telecom; and Mustapha Ramli from Philip Morris International.

A special thanks to Manoj Chari, Ed Hughes, Rob Pratt, Matthew Galati, Brandon Reese, and Natalia Summerville from SAS US; and Falko Schulz from SAS Australia. My thanks is especially for those who created such amazing network science tool, and others, working with me on such awesome projects in the network science field. I give a double thanks to Matt for doing both.

As everything is connected, and different knowledge can always be combined, I feel compelled to thank my colleagues from SAS Education, where I learned so much in many other fields of analytics. I thank you all. However, I would like to name a few who I have worked with closely in teaching and course development. Thanks to Cat Truxillo, Chip Wells, Jeff Thompson, Robert Blanchard, Jay Laramore, Terry Woodfield, Mike Patetta, Peter Christie, and Danny Modlin.

I am grateful to have been advised and supervised by the brilliant minds in academia. A very special thanks to Luciana Ferraz Thomé from Universidade Federal Fluminense; Alexandre G. Evsukoff and Nelson F. F. Ebecken from Universidade Federal do Rio de Janeiro; Vladas Sidoravicius from Instituto Nacional de Matemática Pura e Aplicada; Markus Helfert from Dublin City University; Kavé Salamatian from Université de Savoie; Bart Baesens from KU Leuven; and Moacyr Silva from Fundação Getúlio Vargas.

About the Author

Carlos Andre Reis Pinheiro holds a D.Sc. in Engineering from Universidade Federal do Rio de Janeiro (UFRJ) where he focused on new methods and topologies of artificial neural networks to prevent fraud and detect voluntary bad debt in telecommunications. He holds a M.Sc. in Computer Science from Universidade Federal Fluminense (UFF) and a B.Sc. in Mathematics and Computer Science, from UFF and Universidade Carioca.

Dr. Pinheiro has successfully accomplished a series of post-Doctoral research terms over the past years. He started off with Optimization at Institute of Pure and Applied Mathematics (IMPA), Brazil, during 2006 and 2007, where he focused on genetic algorithms to perform survival analysis. During 2008 and 2009, he worked on Social Network Analysis at Dublin City University (DCU), Ireland, focusing on the use of network analysis to identify the viral effects in telecommunications business events. After a few years back in the industry, in 2012 he worked on Urban Mobility at Université de Savoie, France, focusing on the study of the population movements, particularly on the gravity and radiation models to predict massive displacements. Immediately after in 2013 and 2014, he worked on Human Mobility and Dynamic Social Network Analysis at KU Leuven, Belgium, focusing on the pattern recognition of the human mobility and how it is correlated to social networks. In 2014 and 2015, he worked on Human Mobility and Multimodal Traffic Behavior at Fundação Getúlio Vargas (FGV), Brazil, continuing his work from KU Leuven and focusing on the estimation of commuting trips based on the availability of transportation options in major cities.

Dr. Pinheiro began his career in analytics in 1996, working for a Brazilian shipping company called Libra/CSAV. He worked for some of the largest telecommunications providers in Brazil, at Embratel during 1999 and 2000, at Brasil Telecom from 2000 to 2010, and at Oi from 2010 to 2012. He also worked as a Senior Data Scientist for EMC in 2014 and as a Lead Data Scientist for Teradata in 2015. He joined SAS US at the end of 2015 as a Principal Analytical Training Consultant to work in the Advanced Analytics group in Education. He currently works as a Distinguished Data Scientist at the Global Technology Practice.

Dr. Pinheiro has authored *Social Network Analysis in Telecommunications*, published by Wiley in 2011, and coauthored with his colleague Fiona McNeill *Heuristics in Analytics: A Practical Perspective of What Influences Our Analytical World*, also published by Wiley in 2014. With his friend, Mike Patetta, he also coauthored *Introduction to Statistical and Machine Learning Methods for Data Science*, published by SAS in 2021.

Dr. Pinheiro has published several chapters and papers in international journals and conferences, and he is a frequent speaker in academic and professional seminars, workshops, and tutorials.

About the Book

Network science is the study of connected things. Things can be represented by people, devices, companies, governments, agencies, bank accounts, etc. Connected can be represented by calls, messages, likes, money transferring, references, geo-positioning, contracts, etc.

The study of networks has emerged in diverse disciplines as a means of analyzing complex relational data. Every industry, in every domain, has information that can be analyzed in terms of linked data. Network science can be applied to understand spatiotemporal events like virus spread or traditional business events like churn and product adoption in telecommunications and entertainment, or service consumption in retail, fraud in insurance, money laundering in banking, among many other business cases.

Network analysis includes graph theory algorithms that can augment statistical and machine learning modeling. In many practical applications, pairwise interaction between the entities of interest in the model often plays an important role. This role can be used as input or independent variable in supervised models and often they turn out to be one of the best predictors. Network analysis goes beyond traditional unsupervised modeling like clustering and supervised modeling like predictive models. Both supervised and unsupervised models are frequently used to identify hidden patterns in data. However, these models are based on the attributes describing the observations, normally entities. Network science commonly reveals hidden patterns on the relationship of these entities, rather than on their individual attributes.

Network analysis can be used to understand customers' behavior, highlighting possible influencers within the network. Therefore avoiding churn, diffusing products and services, detecting fraud and abuse, identifying anomalies, and many other business and life science applications. It is hard to think of an application in any industry that cannot benefit from network analysis, from industries such as communications and media, banking, insurance, retail, utilities, travel, hospitality, transportation, education, healthcare, and life science, etc.

Network optimization includes graph theory algorithms that can augment more generic mathematical optimization approaches. Many practical applications of optimization depend on an underlying network. Networks also appear explicitly and implicitly in many other application contexts. Networks are often constructed from certain natural co-occurrence types of relationships, such as relationships among researchers who coauthor articles, actors who appear in the same movie, words or topics that occur in the same document, items that appear together in a shopping basket, terrorism suspects who travel together or are seen in the same location, and so on. In these types of relationships, the strength or frequency of interaction is modeled as weights on the links of the resulting network. Particularly in network optimization, nodes can represent computers, stores, locations, cell towers, bank accounts, vehicles, trains, airplanes, or even a person walking. The occurrence of a transaction involving two nodes can represent a link between them. For example, a truck delivering goods to multiple stores with the stores representing the nodes. The truck going through these multiple stores represent a link between the stores, or the links between the nodes. The weight of these links can be defined by the frequency of the trips the truck performs between any two stores, or the number of goods delivered, or a combination of both. Analogously to the traditional network analysis approach, the network optimization methods can represent nodes and links in the same way but use them differently when distinct optimization algorithms are applied. For example, in the finance industry, bank accounts, individual persons, and organizations can represent nodes. All transactions between bank accounts, persons, and organizations can represent links. Money transfers, withdraws, payments, wires, etc., can all represent varied connections between these nodes, which may have different importance, and thus, different weights. In traditional network analysis, components, cores, or node communities can be detected, revealing groups of accounts, persons, or organizations that are strongly related. Highly weighted links can highlight connections between account, persons, and organizations that are not usual. In network optimization, the same network, based on the same graph representation (bank accounts, individual persons, and organizations as nodes, and transactions as links) can be used differently under distinct optimization algorithms. Cycles, clique, paths, and shortest paths can reveal connection patterns that cannot be highlighted by traditional network analysis. Pattern matching can search for a specific group of nodes based on particular types of links among them throughout the entire network, identifying target connections or entities. In the previous example, where stores are nodes and trucks delivering goods are links, network optimization algorithms can search for optimal routes to minimize the time or the distance when trucks are delivering the goods. By adding a depot as a node and a vehicle capacity for the truck, network optimization algorithms can also search for the optimal trips that trucks need to perform when delivering the goods from the depot throughout all the stores.

1

Concepts in Network Science

1.1 Introduction

Network science is the study of connected things. Things can be represented by people, devices, companies, governments, agencies, bank accounts, etc. Connected can be represented by calls, messages, likes, money transferring, references, geo-positioning, contracts, etc.

The study of networks has emerged in diverse disciplines as a means of analyzing complex relational data. Every industry, in every domain, has information that can be analyzed in terms of linked data. Network science can be applied to understand different types of spatiotemporal events, from (virus spread to traditional business events such as churn and product adoption in telecommunications and entertainment, service and product consumption in retail industry, fraud or exaggeration in insurance, or money laundering and fraudulent transactions in banking, among many others business scenarios.

Network analysis includes graph theory algorithms that can augment statistical and machine learning modeling. In many practical applications, pairwise interaction between the entities of interest in the model often plays an important role and can be used as input or independent variables in supervised models. Very often they turn out to be one of the best predictors. Network analysis goes beyond traditional unsupervised modeling like clustering and supervised modeling like predictive models. Both supervised and unsupervised models are frequently used to identify hidden patterns in data. However, these models are based on the attributes describing the observations, normally entities. Network science often reveals hidden patterns on the relationship of these entities, rather than on their individual attributes.

Network analysis can be used to understand customers' behavior, highlighting possible influencers within the network, and therefore avoid churn, diffuse products and services, detect fraud and abuse, identify anomalies, and many other business and life science applications. Network analysis applies to a wide range of industries such as communications and media, banking, insurance, retail, utilities, travel, hospitality, transportation, education, healthcare, life science, among many others. It is hard to think about any application, in any industry, that cannot benefit from network analysis.

Network optimization includes graph theory algorithms that can augment more generic mathematical optimization approaches. Many practical applications of optimization depend on an underlying network. Networks also appear explicitly and implicitly in many other application contexts. Networks are often constructed from certain natural co-occurrence types of relationships, such as among researchers who coauthor articles, actors who appear in the same movie, words or topics that occur in the same document, items that appear together in a shopping basket, terrorism suspects who travel together or are seen in the same location, and so on. In these types of relationships, the strength or frequency of interaction is modeled as weights on the links of the resulting network. Particularly in network optimization, nodes can represent computers, stores, locations, cell towers, bank accounts, vehicles, trains, airplanes, or even a person walking. The occurrence of a transaction involving two nodes can represent a link between them. For example, a truck delivering goods to multiple stores. The stores represent the nodes. The truck going through these multiple stores represents a link between the stores, or the links between the nodes. The weight of these links can be defined by the frequency of the truck's trips between any two stores, the number of goods delivered, or a combination of both. Analogously to the traditional network analysis approach, the network optimization methods can represent nodes and link in the same way but use them differently when distinct optimization algorithms are applied. For example, in the finance industry, bank accounts, individual persons and organizations can represent nodes. All transactions between bank accounts, persons, and organizations can represent link. Money transfers, withdrawals, payments, wires, etc., all represent different types of connections between these nodes, which may have different importance, and thus, different weights. In traditional network analysis, components, cores, or communities of nodes can be detected, revealing groups of accounts, persons, or organizations that are strongly related. Highly weighted

Network Science: Analysis and Optimization Algorithms for Real-World Applications, First Edition. Carlos Andre Reis Pinheiro.
© 2023 John Wiley & Sons, Inc. Published 2023 by John Wiley & Sons, Inc.

links can highlight connections between account, persons, and organizations that are not usual. In network optimization, the same network, based on the same graph representation (bank accounts, individual persons, and organizations as nodes, and transactions as links) can be used differently under distinct optimization algorithms. Cycles, clique, paths, and shortest paths can reveal patterns on the connections that cannot be highlighted by traditional network analysis. Pattern matching can search for a specific group of nodes based on particular types of links among them throughout the entire network, identifying target connections or entities. In the previous example, where stores are nodes and trucks delivering goods are links, network optimization algorithms can search for optimal routes to minimize the time or the distance when trucks are delivering the goods. By adding a depot as a node and a vehicle capacity for the truck, network optimization algorithms can also search for the optimal trips that truck needs to perform when delivering the goods from the depot throughout all the stores.

Somehow, when looking at the problem through a network's perspective, anything can be represented as nodes, and anything can be represented as links. It is just a matter of how the problem will be described by a graph, using nodes and links to represent entities and relationships. Based on this, any problem can be envisioned as a network problem. Perhaps, the network analysis or the network optimization methods will not actually solve the problem, but definitely these methods will help with the solution by highlighting distinct aspects of the entities relationships that cannot be revealed by traditional approaches.

1.2 The Connector

Tensions ran high between American colonists and British soldiers in the spring of 1775. On the morning of April 19th, a few hundred British soldiers set-off from Boston to capture a cache of arms and arrest some rebel leaders. Marching north into the small towns of Lexington and Concord, the British were astonished to encounter fierce and well-organized resistance. Soundly beaten, they beat a hasty retreat back to Boston under constant harassment from colonial militia. The nascent rebellion was given a huge boost in confidence. Thus began the war for American Independence from Great Brittan.

How the colonial militia was notified in time and able to assemble ahead of the advancing British troops is a story well known to American schoolchildren, many of whom have read or perhaps even memorized the Henry Wadsworth Longfellow poem called "The Midnight Ride of Paul Revere."

On the afternoon of April 18, 1775, a boy who worked at a livery stable in Boston overheard one British army officer say to another something about "hell to pay tomorrow." The stable boy ran with the news to Boston's North End to the home of a silversmith named Paul Revere. Revere listened to the news very seriously, as it was not the first rumor to come his way that day. Earlier, he had been told of an unusual number of British officers gathered on Boston's Long Wharf, talking in hushed tones. British crewmen had been spotted moving quickly in the boats tethered beneath the HMS Somerset and the HMS Hoyne in Boston Harbor. Several other sailors were seen on shore that morning, running what appeared to be last-minute errands. Revere and his close friend, Joseph Warren, became more and more convinced that the British were about to make the major move on the long-rumored march to the town of Lexington, northwest of Boston, to arrest the colonial leaders John Hancock and Samuel Adams, and then on to the town of Concord to seize the stores of guns and ammunition that some of the local colonial militia had stored there.

Revere and Warren decided they had to warn the communities surrounding Boston that the British were on their way, so that local militia could be roused to meet them. Revere was spirited across Boston Harbor to the ferry landing at Charlestown. He jumped on a horse and began his "midnight ride" to Lexington. In two hours, he covered 13 miles. Along the way, in every town he passed through, he knocked on doors and spread the word, telling local colonial leaders of the oncoming British, and telling them to spread the word to others. The news spread like a virus as those informed by Paul Revere sent out riders on their own, until alarms were going off throughout the entire region. The word was in Lincoln, Massachusetts, by 1 A.M., in Sudbury by 3 A.M., in Andover, 40 miles northwest of Boston, by 5 A.M., and by nine in the morning had reached as far west as Ashby, near Worcester. When the British finally began their march toward Lexington on the morning of April 19, 1775, their attack into the countryside faced an organized and fierce resistance. In Concord that day, the British were confronted and soundly beaten by the colonial militia, and from that exchange came the war known as the American Revolutionary War.

Revere rode his horse through the night alerting militia men of the approaching army. Hundreds quickly turned up ready to fight the heavily armed British troops. However, a young man named William Dawes also rode out of Boston the same night with the same message. Starting earlier than Revere, and riding through more populous towns along a more westerly route, Dawes could have been expected to arouse even more rebels. But he didn't. The few militia men who responded to Dawes turned up too late to be of much use.

Paul Revere is an essential figure in American History. William Dawes is but a footnote. Both men carried the same message into towns where the people had equal motivations. One started a word-of-mouth epidemic; the other was mostly ignored. But why?

Malcolm Gladwell, proposes an answer to this question in his 1999 book named *The Tipping Point, How Little Things Can Make a Big Difference*. According to Gladwell, Paul Revere was a special type of person, a connector. William Dawes was a 26-year-old shoemaker. He rode through towns knocking on random doors and shouting that the British were coming. But people in these towns didn't know who he was, didn't have confidence that his message represented an immediate threat. They didn't see their neighbors doing anything so they went back to bed, figuring they would check it out in the morning. Paul Revere on the other hand was well known. He was a successful businessman who at age 40 had for years been at the center of events in and around Boston. When Paul Revere rode into a town he didn't knock on a random door but went straight to the local militia commanders. Opening their doors in the middle of the night and recognizing Paul Revere on the doorstep, they quickly acted, alerted their neighbors, and spread the word. Revere alerted enough of the right people, so the alarm reached the tipping point. The answer is that the success of any kind of social epidemic is heavily dependent on the involvement of people with a particular and rare set of social gifts. Revere's news tipped and Dawes's didn't because of the differences between the two men.

This story is a classic example of how connectors are essential to get an idea to tip. But what can this history lesson from 235 years ago tell us about today's fast-moving, high-technology marketplace? Instead of riding around on horses shouting to each other, we use mobile phones, e-mail, and social network sites. But these special people, the connectors, are still with us. The technology may have changed but connectors still influence our behavior, and they are worth paying attention to.

Connectors know lots of people, many times the number of people the average person knows. But they also know essential facts about people. Chances are you know someone like this, someone that seems to know everyone and stops to talk wherever they go, but it isn't all small talk. Connectors know what to talk about with the people they meet. If you tell a connector about your hobby, they can give you the names of several people who share that interest. Tell them about a problem you have, and they will give you the names of people who can be helpful. Tell them about a great new product and a connector will spread the word. Research has shown that most people have at one time found a job because a connector put them in contact with someone they did not know who helped them get the job. Connectors play a vital role in our social networks.

The impact of connectors to a small neighborhood business is quite obvious. A local restaurant owner would give his best table to a guy who is very influential in the neighborhood while some average person will likely find himself seated near the kitchen. This local restaurant owner may realize that there are more customers on nights when this influential guy shows up and may also observe other customers saying hello to the influential guy and asking him to join them. People stay longer and spend more money when he is around. Owning a local restaurant may be a high-risk business and the presence of one person like that influential guy can determine if the business will tip toward profitability.

However, big corporations, for example communications service providers, are not a local restaurant. Business managers can't personally know millions of customers. Yet the impact of connectors can be vital clues to what others will be doing in the future. Losing a connector to a competitor can represent an increase in churn numbers for the next quarter. On the other hand, seduce a connector away from your competitor and other customers may follow. Have a new service? Get the word to a connector and see how fast it spreads throughout his social network.

But how does a communications service provider know that someone is a connector? Companies can't just put a check box on the service application that asks: "Are you a connector: yes or no?" Unlike gender, connector is not a binary status. Connectors could be expected to be heavy service users, but just accumulating call minutes and numbers of messages is not a good indicator, because these calls and messages may be only assigned to a small number of people. Looking at the number of people someone calls won't help either. A taxi driver may talk with hundreds of people a month, but if he's just arranging pick-ups, he's not a connector. That is when network analysis comes into play, computing centrality measures that can reveal the different roles people have within the social network and highlighting the well-connected customers. Different centralities describe different behaviors, representing something different to the company. Most likely, a combination of different centralities like closeness, betweenness, influence, degree, PageRank, etc., can better reveal the connectors within the communications network, allowing the company to avoid influential people to move away to the competitors, which eventually will be followed by other customers.

1.3 History

Social network analysis considers social life as relations between individuals. The different behaviors within a network are critical to understanding social connections, and more importantly, the implications of those behaviors on communities. One of the most impactful implications within social relations is how someone can influence other people to do similar actions. For example, in the fashion industry, influencers affect how followers dress. In the corporate world, business and thought leaders influence how people manage teams or even companies. Sports idols influence fans to not just support their clubs but also consume products they advertise. In academia, respectful authors influence other scholars on their teaching and research assignments. The list of examples of influence in different environments is virtually endless.

Traditional methods of data analysis and analytical modeling usually consider individual attributes from all observations when exploring the data and searching for hidden patterns. For example, what is the customers' average purchase behavior? How many customers use certain products or services? How long do they remain customers? Do they delay their payments? How long on average? All these questions are about the individuals, looking at their attributes or characteristics. Network analysis focuses on the attributes and characteristics of the relations rather than of the individuals. Individuals' attributes can also be important, and they are also considered in the analysis. However, most often, the attributes of the relations between individuals reveal more hidden patterns. Even when individuals' attributes are considered, they are analyzed from a relationships' perspective. For example, what do the relations of specific entities look like? Specific entities are defined by entities attributes. But the focus of the analysis is on the relationships of those entities.

1.3.1 A History in Social Studies

In the 19th century, Émile Durkheim, a French sociologist, published *The Rules of Sociological Method*. Durkheim describes a theory of sociology as the science of social facts. The article explores the phenomena that is created by interactions between individuals. These relations describe a reality that is independent of any individual actor. He defined the social fact as any way of acting, whether fixed or not, capable of exerting an external constraint over the individual. The social fact is a phenomenon, which is general over the whole of a given society whilst having an existence of its own, independent of its individual manifestations.

In the 20th century, George Simmel, a German sociologist and philosopher, was one of the first scholars to think about explicitly describing social network terms. He examined how third parties could affect the relationship between two individuals. He examined how organizational structures were needed to coordinate interactions in large groups.

One of the first examples of network research was described in 1922. John Almack, a professor at Stanford, published a paper that anticipated the development of the sociometric instrument. He wrote "The Influence of Intelligence on the Selection of Associates." In this study, he asked California elementary school children, grades 4–7, to identify the classmates that they wanted as playmates. One question for example was, "If you had a party, which boy from your class would you invite?" Almack then correlated the IQs for the choosers and the chosen and examined the hypothesis that choices were homophilous.

In 1926, Beth Wellman, an American psychologist, was also focused on homophilous choices among pairs of individuals. Wellman recorded pairs of individuals who were observed as being together frequently. She noted additional attributes such as students' height, grades, gender, age, IQ, score in physical coordination tests, and degree of introversion versus extroversion (based on teachers' ratings). She studied 63 boys and 50 girls who were enrolled in Lincoln Junior High. She then studied homophily with respect to all these traits. Wellman then examined whether interactions were homophilous or not.

In 1928, Helen Bott, a Canadian developmental psychologist, performed an ethnographic approach to examine the behavior of the play activities among preschool children at the nursery attached to the University of Toronto. She identified five types of interactions: talking to one another, interfering with one another, watching one another, imitating one another, and cooperating with one another. She then used the focal sampling to observe one child each day. Bott organized her data into matrices and discussed the results in terms of linkages between individuals. It was a brand-new approach, largely based on the network analyses that are used today.

In 1933, Elizabeth Hagman, an American psychologist, observed interactions within a particular timeframe and interviewed children to measure their recollections of their interactions earlier in that term. This study was called "The Companionships of Preschool Children." Those studies established an important approach for network analysis, particularly in defining how to link individual attributes to interactions, such as IQ, gender, social class, and so on. She highlighted the difference in the observational approaches and the individuals' own accounts of their patterns of interactions. Hagman emphasized the ability to take into account the many different types of interactions between individuals, and the fact that sometimes the linkage attributes might say more about the individuals than the individuals' attributes.

In 1933, *The New York Times* reported on the new science of psychological geography, which "aims to chart the emotional currents, cross currents, and under currents of human relationships in a community." Jacob Moreno, a Romanian American psychiatrist and psychosociologist, is considered by many as one of the fathers of social network analysis, with the graphical social network of interactions between school children in *The New York Times*. He developed the sociogram and presented it to the public at an April 1933 convention of medical scholars. Moreno claimed that "Before the advent of sociometry no one knew what the interpersonal structure of a group 'precisely' looked like." The sociogram was a representation of the social structure of a group of elementary school students. Moreno analyzed interconnections among 500 girls in the State Training School for Girls and also interconnections of students in two New York City schools. He observed that many relationships were non-reciprocal, and as a result, many individuals were isolated. The boys were friends of boys, and the girls were friends of girls, with the exception of one boy who said he liked a single girl. This network representation of social structure was so intriguing that it was printed in *The New York Times* on April 3, 1933. The sociogram has found many applications and has grown into the field of social network analysis. Moreno initially referred to his research as psychological geography, but later changed the name to sociometry. In 1938 he started the *Sociometry* journal. His work was largely theoretical, and his social networks were more illustrative than mathematical representations. Moreno recognized that, and began working with Paul Lazarsfeld, a mathematical sociologist from Columbia University. Lazarsfeld developed a probabilistic model of a social networks, which was published in the first volume of *Sociometry*.

Between 1935 and 1939, Theodore Newcomb studied the women of Bennington College. He observed that when women were exposed to the relatively liberal referent group of fellow students and faculty, they became more liberal. "Becoming radical meant thinking for myself and, figuratively, thumbing my nose at my family. It also meant intellectual identification with the faculty and students that I most wanted to be like."

In 1950, Festinger studied the influence of dorm room location and found that individuals were more likely to associate with those who were similar to them. In this particular case, similar means proximity. Festinger proposed that those who were physically close to each other were more likely to form positive associations. The study showed that the dorm room arrangements could influence the formation of both weak and strong relationships between the students.

After the 1950s, networks were less evident in social psychology and more evident in sociology. Developments in the last few decades included much attention to concepts such as the strength of weak ties and the small worlds. These findings were mainly important to social behavior studies. In the last decade several business applications, including a huge database, were developed based on the "social" study of networks.

In 1998, David Krackhardt and Kathleen Carley introduced the idea of a meta-network with the PCANS Model. They suggested that "All organizations are structured along these three domains, individuals, tasks, and resources." Their paper introduced the concept that networks are interrelated and occur across multiple domains. This field has grown into another sub-discipline of network science called dynamic network analysis.

More recently other network science efforts have focused on mathematically describing different network topologies. Duncan Watts reconciled empirical data on networks with mathematical representation, describing the small-world network. Albert-László Barabási and Reka Albert developed the scale-free network, which is a loosely defined network topology that contains hub vertices with many connections, and grow in a way to maintain a constant ratio in the number of the connections, versus all other nodes. Although many networks, such as the internet, appear to maintain this aspect, other networks have long tailed distributions of nodes that only approximate scale free ratios.

1.4 Concepts

In mathematics, specifically in graph theory, a graph is a structure represented by a set of objects, where these objects are connected in pairs. Objects can describe any type of entity, like a person, a company, a bank account, or a country. The connections between these pairs of objects can describe any type of relationship, like text messages, contracts, money transfers, or international trades. These objects can be named differently, according to the domain knowledge, or even different authors. The same for those connections. Objects can be referred to as nodes, vertices, or points. Connections can be referred to as links, edges, arcs, or lines. In this book, the nomenclature used will be nodes and links. A node will be an individual unit in the graph, and the link will be an interaction between a pair of nodes.

A network, or a graph G, is then defined as a pair $G = (N, L)$, where N is a set of nodes and L is a set of links. The nodes i and j of a link (i, j) are also called endpoints of the link. The link is commonly referred to as join nodes i and j or to be incident on i and j. This pair of nodes i and j is also referred to as adjacent if there is a link (i, j) connecting them. A node can be isolated in the graph, which means, it is not joined to any other node, or has no link incident to it.

A node can be connected to itself, representing an auto relationship. A graph can be undirected or directed. A directed graph means that all the links within the graph have a source node (origin) and a sink node (destination). The link (i, j) can be represented by $i \rightarrow j$. An undirected graph has links with no directions, which means a link (i, j) can represent $i \rightarrow j$ or $j \rightarrow i$. In undirected networks, the association between two nodes is reciprocal. That is, if node i is connected to node j, it is implied that node j is also connected to node i. In other words, a link between two nodes implies a two-way reciprocal association. Friendship is a good example of an undirected network, where two folks are friends to each other, reciprocally. In directed networks, this reciprocal association does not exist. If a link exists pointing from node i to node j, it does not mean that node j also points to node i. If node j also points to node i, another link would need to represent this relationship. Authorship is a good example of a directed network, where an author of a paper refers to many other papers, and therefore, to many other authors. That reference doesn't imply that the referred authors also refer to the original author of the paper.

Finally, both nodes and links can have attributes to represent some particular characteristics of the entity or the relationship. For example, in a telecommunications network, the nodes can represent customers, and one important characteristic is the average revenue per user (ARPU). The ARPU will be the node weight. The link can represent a call between two customers or a text message, or even a multimedia message. Each of these connection types has a different importance for the carrier, and then can be weighted accordingly. For example, the call can be weighted by the number of minutes, the text message can be weighted by the number of characters, and the multimedia message can be weighted by the megabytes. Each one of these measures will be the link weight.

Figure 1.1 shows directed and undirected links, as well as nodes and links with different weights.

The link weight is a crucial measure in most of the network analysis and network optimization algorithms. The link weight can represent the importance or the frequency of the connection between two nodes in network analysis, affecting the way to calculate some centrality metrics like influence, eigenvector, PageRank, closeness and betweenness. The link weight also can represent the cost, the demand, or the capacity of the connection between two nodes in network optimization. Cost, demand, or capacity affect the calculation of some network optimization algorithms like shortest path, traveling salesman problem, vehicle routing problem, minimum spanning tree, minimum-cost network flow and maximum flow.

Nodes and links can be represented as input data sets, in a sequential data fashion, or as matrices. For example, Figure 1.2 shows a graphical representation of a small network.

The directed graph presented before can be represented by a matrix, as shown in Figure 1.3. In this representation, all nodes can be connected to all other nodes, including the self-relation. When a link occurs connecting a pair of nodes, the matrix's cell assigned to these two nodes is flagged. The flag can represent the frequency, the link weight, or any information that represents the importance of that connection.

In addition to the matrix representation, the previous graph can be also described as a table. This representation only shows the pairs of nodes that are connected by a link. This type of representation may save a substantial amount of space, either in memory or disk, particularly when representing large networks. For example, the matrix representation of the directed graph, containing 8 nodes and 10 links uses 64 cells (8 nodes × 8 nodes). The table representation uses on 10 cells or observations, exactly the 10 links connecting the 8 nodes throughout the network. Figure 1.4 shows the table representation for that directed graph.

Matrices to represent networks can frequently be very sparse, especially in large networks, as not all nodes are connected to all other nodes. A large storage may be required to represent a network, even though the majority of the possible combinations of source and sink nodes will be empty. Tables represent networks by only using the existing relations, which can not only save storage but can also be highly effective to process, read and write.

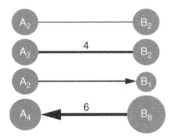

Figure 1.1 Directed and undirected links with weighted nodes.

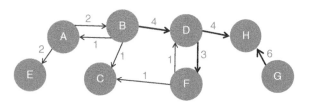

Figure 1.2 Representation of a directed graph with weighted nodes and links.

1.4.1 Characteristics of Networks

Two factors have a great impact on the way the information flows between the nodes within a network.

- Topology: the shape of the network, where one node cannot pass information to another unless they are connected.
- Time: the timing for the connections, where a node cannot pass information before it has received it from another node.

Two key features of the network topology consider the way the nodes are connected throughout the network.

- Connectivity: Describes how nodes in one part of the network are connected to other nodes in another part of the network. It is the possible paths between one node to another perhaps taking intermediates along the way.
- Centrality: Describes how some nodes are connected to other nodes in particular locations of the network. It describes how central the node is within the network.

Connectivity and centrality are mostly translated by the connections between the nodes within the network. Neighborhood and degree are two metrics that describe how connected a node is.

- A neighborhood N for a node is the set of its immediate connected nodes.
- Degree is the network metric k_i of a node representing the number of other nodes in its neighborhood. It is, in practice, the number of distinct connected nodes that a particular node n has.
  - Degree-out represents the number of nodes that a node n points out (outbound interactions).
  - Degree-in represents the number of nodes that a node n is pointed to (inbound interactions).

	A	B	C	D	E	F	G	H
A		2			2			
B	1		1	4				
C								
D						1		4
E								
F			1	3				
G								6
H								

Figure 1.3 A matrix representation for a graph.

Figure 1.4 A table representation for a graph.

1.4.2 Properties of Networks

Some of the network metrics can be used to describe the topology of the network. For example, the metric density describes how connected the network is considering the total number of existing links and the number of possible links that network could have. The number of possible links in the network derives from the total number of nodes. It accounts for all possible connections that all the nodes within the network could have. The size of the network is calculated based simply on the number of existing nodes. The average degree indicates how balanced or well distributed the connections are throughout all nodes within the network. Most of the networks tend to present a power law distribution, where a few nodes have a great number of connections, while the majority of the nodes present only a few connections. The simple network metrics can initially describe the main characteristics of the network and its topology. The average path length describes how close the nodes within the network are to each other. The average path length recalls the concept of six degrees of separation, where there are up to six people between any two people in the world. Some dense networks may have low average path length, indicating nodes are close to each other in terms of reachability, and some networks may present higher average path length, showing that nodes may require multiple steps to reach out to other nodes within the network. Diameter is another important measure to describe the topology of the network. It measures the longest shortest path between all pairs of nodes within the network. This measure describes the possible longest distance between a pair of nodes within the network. Clustering coefficient is a measure that describes how connected a node's neighborhood is. It considers the number of existing connections between a node's neighbors and the possible number of connections these neighbors could have. A well-connected neighborhood means that the information flowing through it may be more persistent than an information flowing throughout a sparse or low-connected neighborhood.

The density of a network is defined as the ratio of the number of the existing links to the number of possible links. L is the number of existing links within the network and N is the number of existing nodes.

$$density = \frac{L}{N(N-1)}$$

The size of a network is simply the number of nodes N.

$$size = N$$

The average degree is the sum of degree k for all nodes divided by the number of existing nodes N within the network.

$$average\ degree = \frac{\sum_{i=1}^{N} k_i}{N}$$

The average path length is calculated by finding the shortest path between all pairs of nodes divided by the total number of pairs.

$$average\ path\ length = \frac{\sum_{i,j=1}^{N} sp_{ij}}{l_{ij}}$$

The diameter of a network is the longest path considering all shortest paths within the network, where u and v represent a pair of nodes, and $d(u,v)$ is the path that connects nodes u and v, no matter the number of steps it takes, or the number of nodes in between them. For example, node u is connected to node v passing through nodes i, j, and k. The path connecting nodes u and v is $u \rightarrow i \rightarrow j \rightarrow k \rightarrow v$. The distance $d(u,v)$ between nodes u and v is 4.

$$diameter = \max_{u,v} d(u,v)$$

Both metrics, average path length, and diameter are metrics to describe how wide and spanned the network is.

The clustering coefficient of a node is the ratio of the existing links connecting a node's neighbors to each other to the maximum possible number of links between those node's neighbors.

$$clustering\ coefficient = \frac{2l_{jk}}{k_i(k_i-1)}$$

The clustering coefficient C_i for a node n_i considers all existing links l_{jk} between the set of neighbors N_i for the node i, and all possible links these neighbors can have, $k_i(k_i-1)$. In a directed graph, the link l_{jk} differs from the link l_{kj}. Then, considering the set of neighbors N_i of node i and the set of existing links L between those neighbors, the clustering coefficient C_i of node i is:

$$C_i = \frac{\left| l_{jk} :: n_j, n_k \in N_i, l_{jk} \in N \right|}{k_i(k_i-1)}$$

In an undirected graph, the links l_{jk} and l_{kj} are identical. Then, for a particular node n_i with k_i neighbors, the number of possible links is $\frac{k_i(k_i-1)}{2}$. The clustering coefficient of node i in an undirected graph is:

$$C_i = \frac{2\left| l_{jk} :: n_j, n_k \in N_i, l_{jk} \in N \right|}{k_i(k_i-1)}$$

Those basic network metrics give an overall description of the network's topology and how nodes are connected and close to each other. Chapter 3 presents a full spectrum of network centrality metrics to better describe the main characteristics of the network, its nodes, and its links.

1.4.3 Small World

Mark Sanford Granovetter is an American sociologist and professor at Sanford University. In 1973, he published a paper in the American Journal of Sociology called "The Strength of Weak Ties." The idea of the strength of the weak ties is probably one of the most important concepts in social network analysis, and that study is one of the top sociology papers published in recent decades.

Granovetter argues that those weak ties observed within social networks actually might be more advantageous in politics or in seeking employment than the expected strong ties. It happens because the weak ties enable individuals to reach more individuals than the strong ties. In fact, by having more weak ties, a person can reach out to more people than by having a few strong ties.

In social network analysis, this is one of the most important concepts. The idea of weak ties is about reachability. It states that, in most of the cases, it is more important to have lots of weak connections than a few strong ones. The reachability from having lots of connections, even though it is based on weak ties, is greater than the reachability from having strong links but only a small number of them.

Some of Granovetter's observations include: the presence of weak ties often reduces path lengths or distances between any two individuals within the network. Short paths enable the information diffusion to speed up throughout the network. This is actually the concept of closeness and betweenness. Weak ties shorten the distances between nodes within the network. The presence of weak ties in the network creates multiple paths between nodes, gathering them closer to each other.

One of the most famous concepts in network analysis is the one assigned to small worlds. The concept basically argues that the world is small because there are a few nodes separating any pair of two distinct nodes. In social networks, it means that a person can reach out to any other person in the world by contacting a fixed number of other persons in the path.

Two possible graphs, almost at opposite ends of a spectrum, are random graphs and regular graphs. A small world can be thought of as existing between a random graph and a regular graph.

The small world concept is assigned to the concept of six degrees of separation. It states that there is an average number of people (or nodes) in the world (or the network) separating any two persons. This number changes according to the network under analysis.

The concept of small worlds is associated with the concepts of random graphs and regular graphs, as well as with clustering coefficients and network diameter. There are basically three types of graphs: regular, small-world, and random. They differ based on the probability that any two nodes are connected within the network. Figure 1.5 shows representations of these three distinct types of graphs.

In random graphs, each pair of nodes i, j has a link with an independent probability p. For example, Figure 1.6 shows a graph with 16 nodes and 19 links.

With 16 nodes, this graph has 120 possible links, which is represented by how many combinations are possible from 16 nodes taken 2 at a time. The possible number of combinations is described by the following formula:

$$C_{n,r} = \frac{n!}{r!(n-r)!}$$

where n represents the number of items, here 16 nodes, and r represents the number of items being chosen at a time, here 2 nodes to create a link. The total number of possible links is then:

$$C_{16,2} = \frac{16!}{2! \times (16-2)!} = \frac{16!}{2! \times 14!} = \frac{16 \times 15 \times 14!}{2!14!} = \frac{16 \times 15}{2} = \frac{240}{2} = 120$$

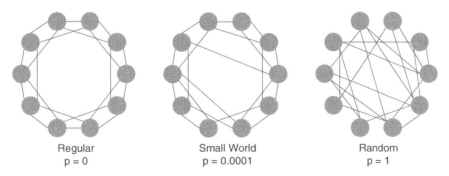

| Regular | Small World | Random |
| p = 0 | p = 0.0001 | p = 1 |

Figure 1.5 Representation of the three types of graphs, regular, small world, and random.

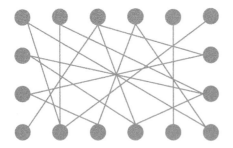

Figure 1.6 Random graph.

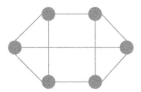

Figure 1.7 Regular graph.

Figure 1.8 Graph representing four legislators working in committees.

However, this graph has only 19 existing links. The probability of a link, or a node being connected to another node, is 0.16 (19 existing links divided by the 120 possible links). In other words, the probability of a node being connected to another node is 0.16.

In random graphs, the presence of a link between A and B and a link between B and C will not influence the probability of a link between A and C. Therefore, in a random graph, nodes have an independent probability of being connected.

In regular graphs, each node has the same number k of neighbors, which is represented by the degree. Figure 1.7 shows a graph with 6 nodes and 9 links. With 6 nodes, this graph has 15 possible links.

$$C_{6,2} = \frac{6!}{2! \times (6-2)!} = \frac{6!}{2! \times 4!} = \frac{6 \times 5 \times 4!}{2!4!} = \frac{6 \times 5}{2} = \frac{30}{2} = 15$$

This graph has only 9 existing links. The probability of any two nodes being connected to each other is 0.6 (9 links by 15 possible).

In this regular graph, all nodes have a degree k of 3, which means, every node within the network is connected to exactly three other nodes. They all have the same number of neighbors ($k = 3$). In regular graphs, each node has the same number of connections.

In 1998, Duncan Watts, an American sociologist and professor at University of Pennsylvania, and Steve Strogatz, an American mathematician and professor at Cornell University, introduced a network metric called clustering coefficient. Clustering coefficients is a measure of how close a node and its neighbors are from being a clique or a complete graph within a larger network. The clustering coefficient of a node is the number of actual connections across its neighbors as a percentage of the possible connections between them. The concept of small worlds is associated with the clustering coefficient network measure. It measures how connected a node's neighborhood is.

Suppose that four legislators are working in committees. It is an undirected graph, which means that if legislator A serves with B in a committee, then legislator B serves with A. Figure 1.8 describes this small network of legislators.

Following the previous concept, four legislators (nodes) in a network creates six possible connections. Legislator A has three neighbors. They have 1 link out of 3 possible (1/3). Legislator B has two neighbors. They have 1 link out of 1 possible (1/1). Legislator C has two neighbors. They have 1 link out of 1 possible (1/1). Legislator D has one neighbor. There is no possible link in this case (0/0). The clustering coefficient is computed for each node by dividing the number of existing connections among its neighbors by the number of possible connections.

Graph diameter is the longest shortest path between any two nodes within the network. Figure 1.9 shows four graphs with the same number of nodes, but a different number of links.

The first graph has a diameter of 3, the second graph has a diameter of 4, the third graph has a diameter of 5, and finally the fourth and last graph has a diameter of 7. The last graph on the right has a larger diameter because it takes, at most, seven links to travel from one node to another. The two highlighted nodes at the bottom in the graph are actually not well-connected, which makes the diameter for this graph the largest among the four graphs presented.

As the longest shortest distance between any two nodes in the network, the graph diameter is associated with the degrees of separation. It shows how many nodes or steps are in between any pair of nodes in the network.

The small world is basically the idea of six degrees of separation. It is the number of steps or links needed to connect any node to another one within the network. The main idea behind the concept of the small world is that the number separating any pair of nodes is often small. That means, observed networks present lower diameters than expected.

In 1967, Stanley Milgram, an American social psychologist and former professor at Yale and Harvard Universities, published an important study, "The Small-World Problem," which generated great publicity. This publicity was supported primarily by several research projects that were conducted simultaneously and that focused on the concept that the world was becoming highly interconnected. Notice that this experiment took place in 1967. It is not difficult to imagine now the

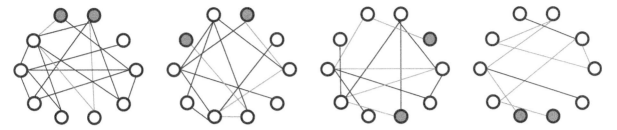

Figure 1.9 Four graphs with same number of nodes and different number of links.

impact this concept had, viewing the possible types of relationships people could establish with the use of different devices, technologies, and geographies.

In this experiment, individuals were asked to reach a particular individual target by passing a message along a chain of acquaintances. For successful chains, the average number of acquaintances was five people, which means, six steps. However, in that study, many of those chains were not actually completed. The idea of the small world is that most networks have smaller diameters than expected, which means that there are fewer people than expected between any two people in the network.

Stanley Milgram conducted some experiments to examine the average length of the path for particular networks. The main objective of these experiments was to suggest that the human communities were not too big. One individual could reach any other one by some maximum number of steps or, in other words, by a certain number of intermediate individuals between them. Indeed, the research outcomes suggested that the networks were characterized by short paths among the nodes considering the number of those intermediate steps. One experiment was associated with the famous phrase "six degrees of separation." However, Milgram himself had never used this term to reference his experiment.

Milgram's experiment intended to highlight the likelihood that two randomly selected people would know each other, regardless of the length of the path, or in other words, the number of steps (nodes and links) between them. This is certainly one way to solve the small-world problem. A straight method to solve the small-world problem is by calculating the average length of the path connecting any two distinct nodes inside the network. The Milgram experiment ultimately created a particular algorithm to calculate the average number of links that connected any two nodes.

The idea behind finding out the number of degrees that separate people in a network has triggered further concepts about how people behave inside the network rather than the number of connections that they have. In terms of business, it is more important to know how each individual behaves inside the network than it is to know how many connections there are among people – that is, of prime importance is their correlated nodes through their connected links.

Georg Simmel, a German sociologist, and philosopher was influential in the field of sociology. According to Simmel, the premise that social ties are primarily based on viewing components as isolated units, those components are better understood as being at the intersection of particular relations and deriving their defining characteristics from the intersection of these relations. He argues that society itself is nothing more than a web of relations.

He wrote: "The significance of these interactions among men lies in the fact that it is because of them that the individuals, in whom these driving impulses and purposes are lodged, form a unity, that is, a society. For unity in the empirical sense of the word is nothing but the interaction of elements. An organic body is a unity because its organs maintain a more intimate exchange of their energies with each other than with any other organism; a state is a unity because its citizens show similar mutual effects."

This statement is an argument against the premise that a society is just a bunch of individuals who react individually and independently to particular circumstances according to their personal desires. Based on his belief that the social world is found in interactions rather than in an aggregation of individuals, Simmel argued that the primary work of sociologists is to study patterns among these interactions rather than to study the individual motives.

Simmel wrote in the same study: "A collection of human beings does not become a society because each of them has an objectively determined or subjectively impelling life-content. It becomes a society only when the vitality of these contents attains the form of reciprocal influence; only when one individual has an effect, immediate or mediate, upon another, is mere spatial aggregation or temporal succession transformed into society."

1.4.4 Random Graphs

Let's examine four different graphs. They all have the same number of nodes, 12, but different a number of links. Let's assume that N represents the number of nodes, p represents the probability that a pair of nodes is connected, and k represents the average degree for the network, or the number of neighbors of a node. In this analysis, let's examine the size of the

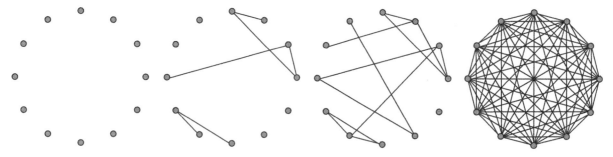

Figure 1.10 Four graphs with same number of nodes but different number of links.

Size of the largest component			
1	5	11	12
Diameter of the largest component			
0	4	7	1
Average path length between connected nodes			
0.0	2.0	4.2	1.0

Figure 1.11 Largest component, diameter, and average path for the four graphs.

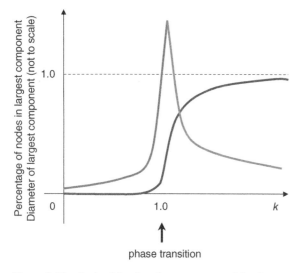

phase transition

Figure 1.12 Ratio of the size of components and the diameter.

largest connected cluster within the four graphs, the diameter (longest shortest path between nodes) of the largest cluster, and the average path length between all nodes within the network. Figure 1.10 shows these four different graphs.

All four graphs have $N = 12$. The first graph has $p = 0$ and $k = 0$, which means no probability to have nodes connected and then no neighbors for all nodes. The second graph has $p = 0.09$ (6 existing links divided by 66 possible links) and $k = 0.9$ (11 total degree divided by 12 nodes). The third graph has $p = 0.18$ (12 existing links divided by 66 possible links) and $k = 2$ (24 total degree divided by 12 nodes). Finally, the four graph has $p = 1$ (as a complete graph, 66 existing links divided by 66 possible links) and $k = 11$ ($k \approx N$).

The size of the largest component in graph 1 is 1, in graph 2 is 5, in graph 3 is 11, and in graph 4 is 12. The diameter for graph 1 is 0, in graph 2 is 4, in graph 3 is 7, and in graph 4 is 1. Finally, the average path length between connected nodes in graph 1 is 0, in graph 2 is 2, in graph 3 is about 4, and in graph 4 is 1. Figure 1.11 summarizes these measures.

When $k < 1$, there are small, isolated clusters, the diameter is small, and there are short path lengths. That means, If the average number of connections is small, the network is shaped by isolated nodes. When $k = 1$, giant components may appear, there is a peak on the diameter, and path lengths can be very high. That means, If the average number of connections is equal to one, the network topology can be assigned to a huge single group. When $k > 1$, almost all nodes are connected into a single cluster, diameter shrinks to almost 1 step, and path lengths are short. That means, finally, if the average number of connections is greater than one, the network is shaped by high density, where almost all nodes are connected. Figure 1.12 shows the relation between k to the size of the clusters and diameters.

If connections between people can be modeled as a random graph, and because an average person can easily know more than one person ($k \gg 1$), we live indeed in a small world. The path length between connected nodes would be calculated as $\frac{\ln N}{\ln K}$.

That brings us back to the concept of six degree of separation. I know someone, who knows someone, who knows someone, who knows someone, who knows someone, who knows someone, who knows you!

1.5 Network Analytics

This book covers network analytics in basically three different approaches. The technical aspects of network analytics are covered in Chapters 2–4. Chapter 5 shows some case studies where network analytics were crucial to solve business problems.

Chapter 2 focuses on the subgraphs, emphasizing a set of algorithms to identify subgraphs and compute some measures about them. For example, it begins by describing how to identify connected components and biconnected components from a major network. Then it moves to how to detect communities by using different algorithms and distinct approaches. It covers how to compute k-cores, an alternative from community detection. Reach network, an especially useful subnetwork, mostly used in marketing campaigns, is also covered. Network projection is described as a way to identify multiple types of subgraphs from a main graph considering different scenarios. Node similarity is not exactly a subgraph detection, but it is included in Chapter 2 as this algorithm may result in a set of nodes, which can be considered as a subgraph. Chapter 2 finishes with pattern matching, a critical network algorithm in detecting fraud and suspicious transactions from a main graph based on a query graph.

Chapter 3 focuses on the network metrics as a method to evaluate and identify the different roles the nodes can play within a network. Network metrics, or centralities, is also a method to measure the importance of the nodes within the network considering different aspects, like number of connections, strength of the relationships, how central the nodes are in the network, how they may control the information flow, or how influential they can be with other nodes within the network. Degree, influence, clustering coefficient, closeness, betweenness, eigenvector, PageRank, hub, and authority are the centrality measures covered in Chapter 3.

Chapter 4 emphasizes the network optimization area, describing algorithms that aim to minimize or maximize objective functions considering a set of resources, demands, constraints, supplies, or costs. This chapter covers algorithms to find cliques and cycles, compute linear assignment, calculate minimum-cost network flow, find minimum cuts, identify minimum spanning tree, find paths and shortest paths, compute transitive closure, solve traveling salesman problems, vehicle routing problems, and identify topological sort.

All examples in this book are created using two main procedures in SAS Viya. NETWORK and OPTNETWORK procedures provide a substantial set of network analysis and network optimization algorithms. All algorithms presented in Chapter 2 are available in the NETWORK procedure. Some of them are also available in the OPTNETWORK procedure. All centralities presented in Chapter 3 are available in the NETWORK procedure as well. Finally, all the optimization algorithms presented in Chapter 4 are available in the OPTNETWORK procedure, and some of them are also available in the NETWORK procedure.

There are plenty of materials describing the SAS Viya architecture and how the in-memory distributed engine works. However, this is not part of the scope of this book. What is important to mention here is that both NETWORK and OPTNETWORK procedures can run in single machines, multiple machines, and in multithreaded executions. Some algorithms run only in single machine mode. Some can use multiple machines, even though each machine requires a global view of the input data, which means, links and nodes, subset of nodes, query graphs, etc. Some algorithms in multiple machines require only a portion of the input data. In this case, the input data is shuffled between machines and then it is randomly distributed across multiple cores. Some algorithms require no data movement, and then the original input data is directly distributed across multiple machines.

In both NETWORK and OPTNETWORK procedures, some algorithms are sensitive to the order in which the data is loaded. Due to the in-memory distributed engine, just reading input data tables in the same order at each algorithm invocation is not guaranteed. When the order of the nodes or links is different in different executions, the final result may change. By default, both NETWORK and OPTNETWORK procedures ensure that each invocation, considering the same machine configuration and parameter settings, will produce the same final result. However, there is a performance cost for that. The option DETERMINISTIC=FALSE allows both procedures to improve performance, but the final results might differ. Sometimes the difference is only a permutation of identifiers, sometimes it may be an alternative local solution.

1.5.1 Data Structure for Network Analysis and Network Optimization

Relationships in network are mutually exclusive. That means, considering a pair of nodes i and j, a link may or may not connect them. As a result, storing every single relationship among all nodes in a network produces redundant information that the network procedures do not need to perform network analysis and network optimization. In this case, both network analysis and network optimization procedures do not use an adjacency matrix as an input data table. Instead, it uses only the unique elements to represent nodes and links. One example is, if there are nodes with no links, or nodes have no weights, the input data table to represent the nodes is not required for network analysis and network optimization. The data input table to represent the links suffices. The nodes representation can be derived from the links representation.

The NODES= option in both NETWORK and OPTNETWORK procedures defines the data table that contains the list of nodes in the graph. This data table is used to assign node attributes, for example, its labels and weights. The nodes dataset may contain the following variables:

- node: the node label (can be numeric or character).
- weight: the node weight (must be numeric).

The LINKS= option in both NETWORK and OPTNETWORK procedures defines the data table that contains the list of links in the graph. A link is represented by pair of nodes, which can be seen as from source or origin node, and to sink or destination node in directed graphs, or simply from and to nodes in undirected graphs with no differentiation on the direction of the link. The links dataset may contain the following variables:

- from: the from node (can be numeric or character).
- to: the to node (can be numeric or character).
- weight: the link weight (must be numeric).
- auxweight: the auxiliary link weight (must be numeric).

For both nodes and links, additional columns can be read in using the VARS= option. For example, if the nodes in the graph represent customers, an additional column describing their average revenue may be defined and can be read into the node's dataset by the network analysis and network optimization procedures. Analogously, if the links in the graph represent several types of connections between customers like calls, text messages, or multimedia messages, the type of connection can be specified in the links dataset and be read into the network procedure. An example would be VARS=(revenue) or VARS=(type). The VARS= option is used to define additional attributes to be read into the network procedures. Similarly, the option VARSOUT= is used to write additional attributes for both nodes and links into the final results. If this option is not used, all columns specified in the option VARS= are written to the final datasets.

Both NETWORK and OPTNETWORK procedures assume some reserved words for nodes and links attributes. For example, if the nodes dataset has a column named *node* to identify the nodes and a column named *weight* to identify the node weight, there is no need to use the NODESVAR= statement to define NODE=node and WEIGHT=weight. The same concept applies to the links dataset. If there is a column named *from* to define the source node, a column named *to* to define the sink node, and a column named *weight* to identify the link weight, there is no need to use the LINKSVAR= statement to define FROM=from, TO=to and WEIGHT=weight.

1.5.1.1 Multilink and Self-Link

A multigraph is a graph that allows multiple links between nodes. Usually, the links describe different types of relationships or distinct spatiotemporal representation. For example, customers can be represented by nodes in a graph, and multiple types of connections between customers like calls, texts, and multimedia texts can be represented by multiple links. If all types of connections between nodes are summarized into a single link, or there are no multiple types of connections between pairs of nodes, the graph is called a simple graph, or simply a graph. Both NETWORK and OPTNETWORK procedures can execute on simple graphs or multigraphs. In order to run on multiple links between the same pair of nodes, the option MULTILINKS= must be specified. If the option is TRUE, multiple links between a single pair of nodes will be processed individually. If the option is FALSE, the multiple links between the same pair of nodes will be aggregated into one single link that uses the minimum of each attribute value. In that case, the multigraph is transformed into a simple graph. By default, both network procedures work with multigraphs (MULTILINKS=TRUE).

A self-link is a link for which the *from* node and *to* node are the same. It represents an auto relationship. Sometimes the self-link is confusing or simply does not make any sense. But in some cases, it does. A network representing relationships between employees working in multiple projects with multiple roles might be the case. Employees are represented by nodes. Links represent the hierarchy on the report. If employee (node) A reports to the manager (node) B, there is a link from A to B. However, it might be the case where B also work in the same project, so, B reports to itself in that graph representation. Both NETWORK and OPTNETWORK procedures can work with self-links. The option SELFLINKS= specifies is the network procedure must remove the self-links or not before invoking any algorithm. If the option is TRUE, self-links are allowed. If the option is FALSE, self-links are removed. By default, both network procedures work with self-links (SELFLINKS=TRUE).

1.5.1.2 Loading and Unloading the Graph

Some of the network tasks can be time consuming, particularly when large graphs are involved. Substantial time can be spent simply loading the graph into memory before the network procedure can actually invoke any analysis or optimization algorithm. Particularly for large graphs, where loading the input data tables to represent nodes and links can be computationally expensive, and a series of analysis will be performed upon the same network, loading this graph into memory can be beneficial. The LOADGRAPH statement reads the input data tables and stores a standard representation of the graph in memory. The graph identifier is saved in a macro variable called _NETWORK_, or in an output data table created by using the option OUTGRAPHLIST=. A variable named *graph* stores the graph identifier. This in-memory graph can be referenced in the network procedure by using the GRAPH= option, by specifying the graph identifier stored in the macro variable or in the output table.

When referencing an in-memory graph in the network procedure by using the option GRAPH=, sometimes additional data structures are required depending on the algorithm invoked. When that happens, there is an additional one-time expense on the first execution of this algorithm. However, subsequent calls can then use these data structures directly. For this reason, a first call that includes the GRAPH= option to some algorithms can sometimes take longer than subsequent calls.

Multiple graphs can be loaded into memory at the same time. In order to better manage graphs loaded into memory, there is also an option to unload them from memory when needed. The UNLOADGRAPH statement deletes an in-memory graph and its persistent data structures.

1.5.2 Options for Network Analysis and Network Optimization Procedures

The PROC NETWORK statement invokes the network analysis procedure, and the PROC OPTNETWORK statement invokes the network optimization procedure. The following options work most of the time for both procedures and specify how both procedures will process the graphs, identify subgraphs, compute centralities, and solve optimization problems.

The option DETERMINISTIC= specifies whether to enforce determinism when running the network algorithms. When the option is TRUE, each invocation produces the same final result. If the option is FALSE, the same final results are not guaranteed.

The option DISTRIBUTED= specifies whether to use a distributed graph when running network analysis and network optimization algorithms. When the option is FALSE, a distributed graph is not used. When the option is TRUE, the graph can be distributed throughout multiple machines and some algorithms can use that distribution to reduce the running time.

The option GRAPH= specifies the in-memory graph to use when the input graph is loaded into memory to be used in multiple executions.

The option DIRECTION= specifies the direction of the input graph. When the option equals to DIRECTED, the input graph is processed considering directed links, where each link (i, j) has a direction defined in the links input data table. It determines how the information flows through the nodes over that link. For example, considering the link (i, j), the information can flow only from node i to node j, and not node j to node i. Node i is called the source node, and node j is called the sink node. When the option equals to UNDIRECTED, the input graph is processed considering undirected links, where each link (i, j) has no determined direction. In this case, the information can flow from node i to node j, and from node j to node i.

The option LINKS= specifies the input data table that contains the graph link information. The option NODES= specifies the input data table that contains the graph node information. The option NODESSUBSET= specifies the input data table that contains the graph node subset information.

The option OUTLINKS= specifies the output data table to contain the graph link information along with any results from the algorithms that calculate metrics on links. The option OUTNODES= specifies the output data table to contain the graph node information along with any results from the algorithms that calculate metrics on nodes.

The option LOGFREQUENCYTIME= controls the frequency in number of seconds for displaying iteration logs for some algorithms. This option is useful for computationally intensive algorithms as it can show how the algorithms progress over time.

The option LOGLEVEL= controls the amount of information that is displayed in the log. The option NONE turns off all messages to the log. BASIC displays a brief summary of the algorithmic processing. MODERATE displays a moderately detailed summary of the input, output, and algorithmic processing. AGGRESSIVE displays a more detailed summary of the input, output, and algorithmic processing.

The option NTHREADS= specifies the maximum number of threads to use for multithreaded processing. Some of the algorithms can take advantage of multicore machines and can run faster when the number is greater than 1. Algorithms that cannot take advantage of this option use only one thread even if the number is greater than 1.

The option STANDARDIZEDLABELS specifies that the input graph data are in a standardized format. The option STANDARDIZEDLABELSOUT specifies that the output graph data include standardized format.

The option TIMETYPE= specifies whether CPU time or real time is used for each algorithm's MAXTIME= option when applicable. The value CPU specifies units of CPU time. The value REAL specifies units of real time.

1.5.3 Summary Statistics

In both NETWORK and OPTNETWORK procedures, it is possible to compute some important summary statistics about the graph before executing the network analysis or the network optimization algorithms. The SUMMARY statement calculates various summary statistics for the graph and its nodes. Two main categories of summary statistics are created. The first one is based on the entire graph. The second one is based on the nodes and links of the graph. The summary statistics about nodes and links are appended to the output nodes and the output links data tables specified in the options OUTNODES= and OUTLINKS=. The summary statistics about the graph are reported in the data table specified in the OUT= option in the SUMMARY statement. The output columns may differ according to the direction of the input graph.

The option OUT= specifies the output data table in the SUMMARY statement that reports the summary statistics for the graph. This output table contains the following columns:

- nodes: the number of nodes in the graph.
- links: the number of links in the graph.
- avg_links_per_node: the average number of links per node.
- density: the number of existing links in the graph divided by the number of possible links in a complete graph.
- self_links_ignored: the number of self-links that are ignored.
- dup_links_ignored: the number of links removed in multilink aggregation.
- leaf_nodes: the number of leaf nodes.
- singleton_nodes: the number of singleton nodes.

The summary statistics also reports information about the connectedness of the graph by using the CONNECTEDCOMPONENTS and BICONNECTEDCOMPONENTS options within the NETWORK and OPTNETWORK procedures. When using the options CONNECTEDCOMPONENTS or BICONNECTEDCOMPONENTS, additional statistics are reported to the summary output data table. For an undirected graph, the following columns are added:

- concomp: the number of connected components in the graph.
- biconcomp: the number of biconnected components in the graph.
- artpoints: the number of articulation points in the graph.
- isolated_pairs: the number of isolated pairs of nodes.
- isolated_stars: the number of isolated stars.

For a directed graph, the following columns are added:

- concomp: the number of strongly connected components in the graph.
- isolated_pairs: the number of isolated pairs of nodes.
- isolated_stars_out: the number of isolated outward stars.
- isolated_stars_in: the number of isolated inward stars.

Similarly, the summary statistics can also produce statistics about the shortest paths in the graph by using the SHORTESTPATH= option within the NETWORK and OPTNETWORK procedures. When using the option SHORTESTPATH=, the following columns are added to the summary output data table:

- diameter_wt: the longest weighted shortest path distance in the graph.
- diameter_unwt: the longest unweighted shortest path distance in the graph.
- avg_shortpath_wt: the average weighted shortest path distance in the graph.
- avg_shortpath_unwt: the average unweighted shortest path distance in the graph.

When using the option CLUSTERINGCOEFFICIENT, the summary statistics can add the number of triangles of an undirected graph. A triangle is identified by a set of three distinct nodes where each node is a neighbor of the other two. The CLUSTERINGCOEFFICIENT option adds the following column to the summary output data table:

- triangles: the total triangle count of the graph.

The OUTNODES= option specifies the output table containing the summary statistics for the nodes in the graph. The following columns are appended to the output data table:

- sum_in_and_out_wt: the sum of the link weights from and to the node.
- leaf_node: 1, if the node is a leaf node; otherwise, 0.
- singleton_node: 1, if the node is a singleton node; otherwise, 0.
- isolated_pair: the identifier, if the node is in an isolated pair; otherwise, missing (.).
- neighbor_leaf_nodes: the number of leaf nodes connected to the node.

When the options CONNECTEDCOMPONENTS and BICONNECTEDCOMPONENTS are used, the following columns are added to the output data table. For an undirected graph:

- isolated_star: the identifier, if the node is in an isolated star; otherwise, missing (.).

For a directed graph:

- isolated_star_out: the identifier, if the node is in an isolated outward star; otherwise, missing (.).
- isolated_star_in: the identifier, if the node is in an isolated inward star; otherwise, missing (.).

When the option SHORTESTPATH= is used, the summary statistics include the calculation of the eccentricity of a node. This measure reports the longest of all possible shortest path distances between a particular node and any other node in the graph. The following columns are added to the output data table. For an undirected graph:

- eccentr_out_wt: the longest weighted shortest path distance from the node.
- eccentr_out_unwt: the longest unweighted shortest path distance from the node.

For a directed graph

- eccentr_in_wt: the longest weighted shortest path distance to the node.
- eccentr_in_unwt: the longest unweighted shortest path distance to the node.

The OUTLINKS= option specifies the output table containing the summary statistics for the links in the graph. When the options CONNECTEDCOMPONENTS and BICONNECTEDCOMPONENTS are used, the following columns are added to the output data table. For an undirected graph:

- isolated_pair: the identifier, if the link is in an isolated pair; otherwise, missing (.).
- isolated_star: the identifier, if the link is in an isolated star; otherwise, missing (.).

For a directed graph:

- isolated_star_out: the identifier, if the link is in an isolated outward star; otherwise, missing (.).
- isolated_star_in: the identifier, if the link is in an isolated inward star; otherwise, missing (.).

The summary statistics are always a good approach to better understand the main characteristics of the input graph before beginning a deep analysis of the main graph and invoking network analysis and network optimization algorithms.

1.5.3.1 Analyzing the Summary Statistics for the Les Misérables Network

As a first step in evaluating the network topology in some networks, and in order to better understand the output from the summary statistics, let's take the Les Misérables network presented in Figure 1.13.

The Les Misérables network is an undirected graph, where it represents all the actors who play scenes together throughout the move. As the actors play multiple scenes together during the movie, the number of these scenes is used to create the link weight. The network is small, with 77 nodes (actors) and 254 links (distinct pairs of actors playing scenes together).

The following code invokes proc. network to execute the summary statistics on the Les Misérables network. The SUMMARY statement considers all possible summary statistics, calculating the biconnected components, the connect components, the clustering coefficient, the diameter of the network, the finite path, and the shortest path.

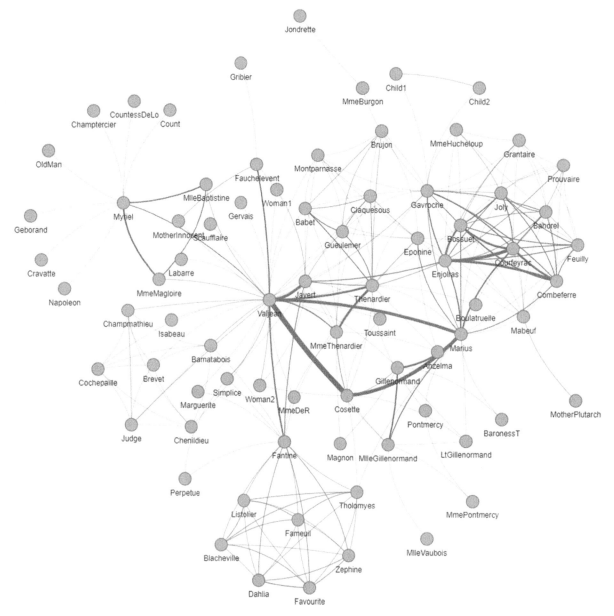

Figure 1.13 Les Misérables network.

```
proc network
   direction = undirected
   nodes = mycas.lmnodes
   links = mycas.lmlinks
   outnodes = mycas.outnodessummary
   outlinks = mycas.outlinkssummary
   ;
   nodesvar
      node = node
      vars = (label)
      varsout = (label)
   ;
```

```
linksvar
   from = from
   to = to
   weight = weight
;
summary
   biconnectedcomponents
   connectedcomponents
   clusteringcoefficient
   diameterapprox = both
   finitepath
   shortestpath = both
   out     = mycas.summary;
run;
```

Figure 1.14 shows the summary results reported by proc. network when running the summary statistics. It shows the size of the network, with 77 nodes and 254 links. The summary statistics are computed considering the graph with undirected links. Three output tables are produced by the summary statement, one for the nodes, one for the links, and the last one for the summary metrics.

Figure 1.15 shows the SUMMARY output table, consisting of summary metrics for the network, such as the number of nodes and links, the average number of links per node, the number of links divided by the number of nodes, the number of self-links within the network that were ignored, the number of links that were removed by the multilink aggregation, the number of leaf nodes, and the number of singleton nodes. Leaf nodes are simply external nodes, sometimes called outer or terminal. Simply put, are the nodes with degree of 1. Singleton nodes, sometimes referred to as singleton graphs, are nodes with no links, or nodes isolated in the network.

The NETWORK Procedure

Problem Summary

Number of Nodes	77
Number of Links	254
Graph Direction	Undirected

The NETWORK Procedure

Solution Summary

Problem Type	Summary
Solution Status	OK
CPU Time	0.00
Real Time	0.01

Output CAS Tables

CAS Library	Name	Number of Rows	Number of Columns
CASUSER(Carlos.Pinheiro@sas.com)	OUTLINKSSUMMARY	254	5
CASUSER(Carlos.Pinheiro@sas.com)	OUTNODESSUMMARY	77	10
CASUSER(Carlos.Pinheiro@sas.com)	SUMMARY	1	20

Figure 1.14 Summary statistics results.

Figure 1.15 Summary statistics for the Les Misérables network.

	node	label	leaf_node	singleton_node	neighbor_leaf_nodes	
1	0	Myriel	0	0	7	
2	1	Napoleon	1	0	0	
3	2	MlleBaptistine	0	0	0	
4	3	MmeMagloire	0	0	0	
5	4	CountessDeLo	1	0	0	
6	5	Geborand	1	0	0	
7	6	Champtercier	1	0	0	
8	7	Cravatte	1	0	0	

OUTNODESSUMMARY — Table rows: 77 — Columns: 10 of 10 — Rows 1 to 77

	isolated_star	isolated_pair	sum_in_and_out_wt	eccentr_wt_out	eccentr_unwt_out
1			31	12	4
2			1	13	5
3			17	10	4
4			19	10	4
5			1	13	5
6			1	13	5
7			1	13	5
8			1	13	5

Figure 1.16 Output data for the nodes.

Figure 1.16 shows the OUTNODESSUMMARY output table consisting of information about all the nodes within the network. Each node in the network has an indicator informing if the node is a leaf node and a singleton node. It also shows how many leaf node neighbors the node has, if the node is an isolated star node (connected to two or more nodes where all of them have degree 1), if the node is an isolated pair node (connected to one single node, and that node it is connected only to it), the sum of the link weights from and to the node, the longest weighted shortest path distance from the node, and the longest unweighted shortest path distance from the node.

	⊕ from	⊕ to	⊕ weight	⊕ isolated_star	⊕ isolated_pair
1	2	0	8		
2	5	0	1		
3	9	0	1		
4	11	0	5		
5	15	11	1		
6	19	16	4		
7	20	17	3		
8	21	17	3		

Figure 1.17 Output data for the links.

Figure 1.17 shows the OUTLINKSSUMMARY output table consisting of information about all the links within the network. It shows the original origin and destination nodes, the weight of the link, if the link partakes an isolated star, and if the link partakes an isolated pair.

The Les Misérables network is a small network is very dense, with 77 nodes consisting of one single connected component. That means, all actors can reach out to all other actors throughout a particular path. And that path is not long, with a diameter of 5, which means there four other actors on top between any two. But on average, there are only 2.6 actors in between any pair of actors. There are only 17 nodes with one single connection. Most of the actors play scenes with more than one actor along the movie. On average, they play with 6.6 other actors along the movie.

1.6 Summary

The study of social networks is formally defined as a set of nodes, which consist of network members. Those nodes are connected by different types of relations, which are formally defined as links. Network analysis takes all those connections as the primary building blocks of the social world. Traditional analysis considers individual attributes. Social network analysis considers the attributes of relations in a unique perspective than that adopted by individualist or attribute-based social science. Traditional methods of data analysis usually consider individual attributes from all observations. What is the average characteristic from a particular population of companies, employees, customers, or markets? Beyond individual attributes, social network analysis considers all information about the relationships (links) among the network members (nodes). In fact, the information about the relations among the individuals within a social network is usually more relevant than the individual attributes of the members of the network. Ultimately, the relations between the individuals within a network can reveal more hidden information about the network members than their individual attributes. This is the basic difference between data analysis and social network analysis.

Graphs can reveal valuable information about social interactions, and therefore, characteristics of communities, groups of people, or even individuals. Graph algorithms can also be used to compute network measures that identify roles and the importance of the network members. There is a wide range of network measures that can be used to describe the importance of nodes and links. Graph algorithms can also detect and identify groups of members in the network based on their relations, whether considering the frequency, the strength, or any combination of relations attributes.

Network analysis is frequently used to better understand customers behavior, highlight influencers in networks, avoid churn, diffuse products, detect fraud, and search for anomalies, among many other applications. Network optimization includes algorithms to solve multiple practical real-world problems.

Connectors are essential roles within networks. They may know lots of people, many times the number of people the average person knows. They also know essential facts about people. The impact of connectors to business is obvious.

With its origins in social science, there is a long history of network analysis studies to evaluate how people relate to each other and how these relations shape not just how groups of people are created, but also how individuals behave in those groups.

A mathematical formalism is used to describe the network in terms of a graph, and then builds up a set of algorithms to analyze the graph, compute centrality metrics, identify subgraphs, and search for optimal solutions in optimization problems.

One of the most important concepts in network analysis is the small world idea. The concept implies that the world is small because there are a few nodes separating any pair of nodes. In social networks, it means that a person can reach out to any other person in the world by contacting a fixed number of persons in the path. The small world concept is assigned to the concept of six degrees of separation. It states that there is an average number of people (or nodes) in the world (or the network) separating any two persons. This number changes according to the network analyzed.

The following chapters will describe in great details how to identify subgraphs from main network, how to compute network centralities to describe nodes and links, and how to solve multiple optimization problems that can be represented as a network.

All examples presented in this book use SAS procedures in Viya, a multi distributed in-memory engine. Two procedures are explored. First is the NETWORK procedure, which covers all the subgraphs algorithms shown in Chapter 2 and all the centrality metrics algorithms shown in Chapter 3. Second is the OPTNETWORK procedure, which covers all the optimization algorithms shown in Chapter 4. Chapter 5 presents some case studies to better exemplify how valuable network analysis and network optimization are.

2

Subnetwork Analysis

2.1 Introduction

An important concept in graph theory is the concept assigned to subgraphs. A subgraph of a graph is a smaller portion of that original graph. For example, if H is a subgraph of G and i and j are nodes of H, by the definition of a subgraph, the nodes i and j are also nodes of the original graph G. However, a critical concept in subgraphs needs to be highlighted. If the nodes i and j are adjacent in G, which means that they are connected by a link between them, then the definition of a subgraph does not necessarily require that the link joining the nodes i and j in G is also a link of H. If the subgraph H has the property that whenever two of its nodes are connected by a link in G, this link is also a link in H, then the subgraph H is called an induced subgraph.

For example, the three graphs shown in Figure 2.1 describe (a) a graph, (b) a subgraph, and (c) an induced subgraph.

Notice that the subgraphs in (b) and (c) have the same subset of nodes. However, the subgraph (b) misses all the links between their nodes. Subgraph (c) has the original links between the subset of nodes from the original graph. For that reason, the subgraph (c) is an induced subgraph.

The idea of working with subgraphs is simple and straightforward. If all the nodes and links we are interested in analyzing are in the graph, we have the whole network to perform the analysis. However, from the entire data collected to perform the network analysis, we may often be interested in just some subset of the nodes, or a subset of the links, or in most cases, in both sets of nodes and links. In those cases, we don't need to work on the entire network. We can focus only on the subnetwork we are interested in, considering just the subset of nodes and links we want to analyze. This makes the entire analysis easier, faster, and more interpretable, as the outcomes are smaller and more concise.

We can think of a graph as a set of two sets. A set of nodes and a set of links, or sometimes, just a set of nodes and links. Based on the set theory, it is always possible to consider only a subset of the original set. Since graphs are sets of nodes and links, we can do the same we consider only a subset of its members. A subset of the original nodes of a graph is still a graph, and we can call it a subgraph. Based on that, if $G = (N, L)$ is the original graph, the subgraph $G' = (N', L')$ is a subset of G, The subset of nodes and links G must contain only elements of the original set of nodes and links G'. This property is written as $G' \subset G$. We read this equation as G' is a subset of G. If every element of the subset of nodes and links is also an element of the original set of nodes and links, then we can also write the following equations, $N' \subset N$ and $L' \subset L$.

In Figure 2.1, the graph represented by (b) is also sometimes called a disconnected subgraph of the original graph, and the graph represented by (c) is called a connected subgraph of the original graph. The subgraph in (c) is well connected, where all nodes are a link to each other, even if is not through a directed connection. This subgraph captures well all relationships between the subset of nodes in the subgraph. It misses just the links to the nodes that are not included in the subgraph. Conversely, the subgraph in (b) has two pairs of nodes isolated. The nodes within the two pairs are linked to each other, but the pairs of nodes are not. That means a node from one pair cannot reach out to a node in the other pair. The absence of these existing connections in the original graph compromise the information power of the subgraph and makes it a disconnected subgraph.

It is important to note that we can create a subgraph from a subset of nodes from the original graph, but also from a subset of links from the original graph. When selecting a subset of links from the original graph, by consequence, we also select the nodes connected by those links. Therefore, we also select a subset of nodes from the original set. For example, by selecting three particular links from the previous graph, we also select the four nodes connected by those links. This scenario is shown in Figure 2.2.

Network Science: Analysis and Optimization Algorithms for Real-World Applications, First Edition. Carlos Andre Reis Pinheiro.
© 2023 John Wiley & Sons, Inc. Published 2023 by John Wiley & Sons, Inc.

(a) (b) (c)

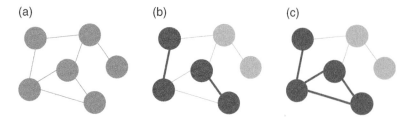

Figure 2.1 (a) Graph (b) Subgraph (c) Induced graph.

(a) (b) (c)

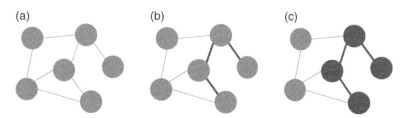

Figure 2.2 (a) Graph (b) Links selection (c) Subgraph by links selection.

Notice that the three links in (b) selected as part of the subgraph from the original graph in (a) also brings the fours nodes connected by them, creating the subgraph in (c).

When a subgraph is defined based on a subset of nodes, which is quite common, it is called a node induced subgraph of the original graph. When a subgraph is defined by selecting a subset of links, it is called a link induced subgraph of the original graph.

In addition to reducing the size of the original network, one of the most common reasons we use subgraphs when analyzing network is to understand the impact of removing nodes and links from the original graph; how the subnetwork would look if some actors or relationships are removed from the original network. This approach is quite important to better understand how significant a particular actor is, or how important specific relationships are. Eventually, by removing an actor, or a relationship, the original network is split into multiple subnetworks, disconnected from each other. That means that actor or relationship is absolutely crucial in the original network. Eventually, the network will keep a similar interconnectivity, still allowing the other actors to reach out to each other through other actors or other relationships. It means removing that actor or that relationship from the original network is not too important and doesn't much change the overall network connectivity.

For example, in a star graph, the central node plays a very important role holding the network together. If the central node is removed from the original network, all satellite nodes will be isolated from each other. This case is shown in Figure 2.3.

Notice that the central node in the star graph connects all other nodes in the graph. If this node is removed from the original network, all other nodes will be isolated.

A specific connection can be also very important in a network. The following example shows that, if the central connection highlighted in the graph (a) is removed from the original graph, it will split the graph into two disconnected subgraphs in (b). In that case, one single connection can completely change the shape and the properties of the original network. This can be seen in Figure 2.4.

Removing nodes and links from the original network can therefore help us to understand the properties of the graph and the outcomes produced by these deletions. For example, how the final topology and structure of the subnetworks will look after removing particular nodes and links. Do these resulting subnetworks create communities, cores, cliques, or other different types of groups from the original network? What would be the importance of the removed nodes? What would be the relevance of the removed links? All these questions guide us to better exploring all the properties of the original network and identifying hidden patterns within the nodes and most important, within their relationships.

(a) (b)

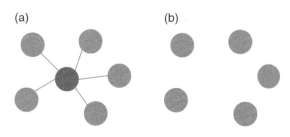

Figure 2.3 (a) Graph (b) Subgraph by removing nodes.

Figure 2.4 (a) Graph (b) Subgraphs by removing links.

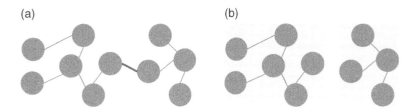

2.1.1 Isomorphism

A frequent problem in network analysis and network optimization is known as the subgraph isomorphism problem. A subgraph isomorphism problem is a computational task that involves two graphs G and H as input. The problem is to determine if G contains a subgraph that is isomorphic to H. Two graphs are isomorphic to each other if they have the same number of nodes and links, and the link connectivity is the same. The following graphs and subgraphs exemplify the subgraph isomorphism problem. The subgraph G' of the graph G is isomorphic to the graph H. They have the same number of nodes and the same number of links. Also, the link connectivity is the same, which means the existing links connect the existing nodes within the graph and the subgraph in the same way. Figure 2.5 shows a case of isomorphism.

The subgraph isomorphism problem can be reduced to find a fixed graph as a subgraph in a given graph. The answer to the subgraph isomorphism question attracts many researchers and practitioners because many graph properties are hereditary for subgraphs. Ultimately, this means that a graph has a particular property if and only if all its subgraphs have the same property. However, the answer to this question can be computationally intensive. Finding maximal subgraphs of a certain type is usually a NP-complete problem. The subgraph isomorphism problem is a generalization of the maximum clique problem or the problem of testing whether a graph contains a cycle, or a Hamiltonian cycle. Both problems are NP-complete.

NP is a set of decision problems that can be solved by a non-deterministic Turing Machine in polynomial time. NP is often a set of problems for which the correctness of each solution can be verified quickly, and a brute-force search algorithm can find a solution by trying all workable solutions. Some heuristics are often used to reduce computational time here. The problem can be used to simulate every other problem for which a solution can be quickly verified whether it is correct or not. NP-complete problems are the most difficult problems in the NP set. The problem, particularly in large networks, can be very computationally intensive. For instance, in the previous example, the very small graph G has another five other different subgraphs that are isomorphic to graph H. Graph G is a very small graph with just eight nodes and contains a total of six subgraphs isomorphic to graph H. In a very large network, the search for all possible subgraphs can be computationally exhaustive.

Analyzing subgraphs instead of the entire original graph becomes more and more interesting, especially in applications like social network analysis. Finding patterns in groups of people instead of in the entire network can help data scientists, data analysts, and marketing analysts to better understand customers' relationships and the impact of these relations in viral churn or influential product and service adoption. Particularly for large networks, we can find multiple distinct patterns throughout the original network when looking deeper into groups of actors. Each group might have a specific pattern, and it can be vastly different from the other groups. Understanding these differences may be a key factor not to just interpret the network properties but mostly to apply network analysis outcomes to drive business decisions and solve complex problems. The subgraph approach allows us to focus on specific groups of actors within the network, reducing the time to process the data and making the outcomes easier to interpret. The sequence of steps to divide the network into subnetworks, analyze less data, and more quickly interpret the outcomes can be automatically replicated to all groups within the network.

Figure 2.5 Subgraph G' isomorphic to graph H.

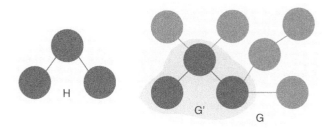

Many network problems rely on the concepts behind subgraphs. In this chapter we will focus on some of the most important subgraphs in network analysis and network optimization. This chapter covers, therefore, the concepts on connected components, biconnected components, and the role of the articulation points, communities, cores, reach networks (or ego networks), network projection, nodes similarity, and pattern match. Some other concepts on subgraphs like clique, cycle, minimum cut, and minimum spanning tree, for example, will be covered in the network optimization chapter.

2.2 Connected Components

A connected component is a set of nodes that are all reachable from each other. For a directed graph, there are two types of components. The first type is a strongly connected component that has a directed path between any two nodes within the graph. The second type is a weakly connected component, which ignores the direction of the graph and requires only that a path exist between any two nodes within the graph. Connected components describe that if two nodes are in the same component, then there is a path connecting these two nodes. The path connecting these two nodes may have multiple steps going through other nodes within the graph. All nodes in the connected component are somehow linked, no matter the number of nodes in between any pair.

Connected components can be calculated for both directed and undirected graphs. A graph can have multiple connected components. Therefore, a connected component is an induced graph where any two nodes are connected to each other by a path, but they are not connected to any other node outside the connected component within the graph.

Connected components are commonly referred to as just components of a graph. A common question in graph theory relies on the fact of whether or not a single node can be considered as a connected component. By definition, the answer would be yes. A graph can be formed by a single node with no link incident to it. A graph is connected if every pair of nodes in the graph can be connected by a path. Theoretically, a single node is connected to itself by the trivial path. Based on that, a single node in a graph would be also considered a component, or a connected component in that graph.

In proc network, the CONNECTEDCOMPONENTS statement invokes the algorithm that solves the connected component problem, or the algorithm that identifies all connected components within a given graph. The connected components can be identified for both directed and undirected graphs. Proc network uses three different types of algorithms to identify the connected components within a graph. The option ALGORITHM = has four options to either specify one of the three algorithms available or allow the procedure to automatically identify the best option. The option ALGORITHM = AUTOMATIC uses the union-find or the afforest algorithms in case the network is defined as an undirected graph. If the network is defined as a directed graph, the option ALGORITHM = AUTOMATIC uses the depth-first search algorithm to identify all connected components within the network. The option ALGORITHM = AFFOREST forces proc network to use the algorithm afforest to find all connected components within an undirected graph. Similarly, the option ALGORITHM = DFS forces proc network to use the algorithm depth-first in order to find all connected components within a directed graph or an undirected graph. Finally, the option ALGORITHM = UNIONFIND forces proc network to use the union-find algorithm to find all connected components within an undirected graph. The algorithm union-find can also be executed in a distributed manner across multiple server nodes by using the proc network option DISTRIBUTED = TRUE. This option specifies whether or not proc network will use a distributed graph. By default, proc network doesn't use a distributed graph. The afforest algorithm usually scales well for exceptionally large graphs.

The final results for the connected component algorithm in proc network are reported in the output table defined by the options OUT =, OUTNODES =, and OUTLINKS=.

The option OUT = within the CONNECTEDCOMPONENTS statement produces an output containing all the connected components and the number of nodes in components.

- concomp: the connected component identifier.
- nodes: the number of nodes contained in the connected component.

The option OUTNODES = within the PROC NETWORK statement produces an output containing all the nodes and the connected component to which each one of them belongs.

- node: the node identifier.
- comcomp: the connected component identifier for the node.

The option OUTLINKS = within the PROC NETWORK statement produces an output containing all the links and the connected component each link belongs to.

- from: the origin node identifier.
- to: the destination node identifier.
- concomp: the connected component identifier for the link.

Let's see some examples of connected components in both directed and undirected graphs.

2.2.1 Finding the Connected Components

Let's consider a simple graph to demonstrate the connected components problem using proc network. Consider the graph presented in Figure 2.6. It shows a network with just 8 nodes and only 10 links. Weight links are not relevant when searching for connected components, just the existence of a possible path between any two nodes. As in many graph algorithms, when the input graph grows too large, the process can be increasingly exhaustive, and the output table can be extremely big.

The following code describes how to create the input dataset for the connected component problem. First, let's create only the links dataset, defining the connections between all nodes within the network. Later, we will see the impact of defining both links and nodes datasets to represent the input graph. The links dataset has only the nodes identification, the from and to variables specifying the origin and the destination nodes. When the network is specified as undirected, the origin and destination are not considered in the search. The direction doesn't matter here. Another way to think of the undirected graph is that every link is duplicated to have both directions. For example, a unique link A→B turns into two links A→B and B→A. This link definition doesn't provide a link weight as it is not considered when searching for the connected components.

```
data mycas.links;
    input from $ to $ @@;
datalines;
A B   B C   C A   B D   D H   H E   E F   E G   G F   F D
;
run;
```

Once the input dataset is defined, we can invoke the connected component algorithm using proc network. The following code describes how this is done. Notice the links definition using the LINKSVAR statement. The variables from and to receive the origin and the destination nodes, respectively. As the name of the variables in the links definition are exactly the same names required by the LINKSVAR statement, the use of the LINKSVAR statement can be suppressed. We are going to keep the definition in the code (and in all of the codes throughout this book) as in most of the cases, the name of the nodes can be different than to and from, like caller and callee, origin and destination, departure and arrival, sender and receiver, among many others.

```
proc network
    direction = undirected
    links = mycas.links
    outlinks = mycas.outlinks
    outnodes = mycas.outnodes
    ;
    linksvar
        from = from
        to = to
    ;
    connectedcomponents
        out = mycas.concompout
    ;
run;
```

Figure 2.7 shows the outputs for the connected component algorithm. The first one reports a summary of the execution

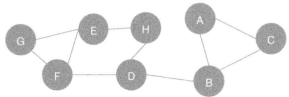

Figure 2.6 Input graph with undirected links.

The NETWORK Procedure

Problem Summary

Number of Nodes	8
Number of Links	10
Graph Direction	Undirected

The NETWORK Procedure

Solution Summary

Problem Type	Connected Components
Solution Status	OK
Number of Components	1
CPU Time	0.00
Real Time	0.00

Output CAS Tables

CAS Library	Name	Number of Rows	Number of Columns
CASUSER(Carlos.Pinheiro@sas.com)	OUTLINKS	10	3
CASUSER(Carlos.Pinheiro@sas.com)	OUTNODES	8	2
CASUSER(Carlos.Pinheiro@sas.com)	CONCOMPOUT	1	2

Figure 2.7 Output results by proc network running the connected components algorithm upon an undirected input graph.

process. It provides a summary of the network, with 8 nodes and 10 links. Notice that we didn't provide a nodes dataset, even though proc network can derive the set of nodes from the links dataset. It also says the graph direction as undirected. The solution summary says that a solution was found, and only one connected component was identified. Finally, the output describes the three output tables created, the links output table, the nodes output table, and the connected components output table.

The following figures show the connected components results. Figure 2.8 presents the OUTNODES table, which describes the nodes and the connected components which each node belongs to.

Notice that there is one single connected component. The undirected graph in this example allows for every node to be reached by any other within the network, based on some particular path.

Figure 2.9 shows the OUTLINKS table, which presents the links and the connected components they belong to. The output table describes the original links variables, from and to, and the connected component assigned to the link.

Finally, the output table CONCOMPOUT shows the summary of the connected component algorithm, presented in Figure 2.10. This table presents the connected components identifiers and the number of nodes in each component. As this undirected graph has one single connected component, this output table has one single row, showing the unique connected component 1 with all the eight nodes from the input graph.

As we can see, the original input graph, considering undirected links, presents one single connected component. What happens if we remove a few links from the original network? Let's recreate the links dataset without the three original links, H→E, D→H, and B→D. The following code shows how to recreate this new network based on that modified links dataset.

```
data mycas.links;
   input from $ to $ @@;
datalines;
A B  B C  C A  E F  E G  G F  F D
;
run;
```

Figure 2.8 Output nodes table for the connected components.

	from	to	concomp
1	A	B	1
2	D	H	1
3	G	F	1
4	B	C	1
5	H	E	1
6	F	D	1
7	C	A	1
8	E	F	1
9	B	D	1
10	E	G	1

Figure 2.9 Output links table for the connected components.

In order to maintain the same set of nodes from the original network, we need to define a nodes dataset. Proc network assumes the set of nodes based on the provided set of links. As we remove the incident links to node H (H→E and D→H), node H wouldn't be in the links dataset and therefore, it wouldn't be in the presumed nodes dataset either. If we want to keep the same set of nodes, including node H from the original network, we need to provide the nodes dataset and force the presence of node H in the input graph. The following code creates the nodes dataset.

```
data mycas.nodes;
   input node $ @@;
datalines;
A B C D E F G H
;
run;
```

Figure 2.10 Output summary table for the connected components.

Once both links and nodes dataset s are created, we can rerun proc network and see how many connected components will be identified upon that new set of links and the declared set of nodes. The following code reruns proc network on the new links dataset and also considers the nodes dataset.

```
proc network
    direction = undirected
    links = mycas.links
    nodes = mycas.nodes
    outlinks = mycas.outlinks
    outnodes = mycas.outnodes
    ;
    linksvar
        from = from
        to = to
    ;
    nodesvar
        node = node
    ;
    connectedcomponents
        out = mycas.concompout
    ;
run;
```

Without the three original links H→E, D→H, and B→D, the new network will present three connected components. One of the components will contain a single node.

Figure 2.11 shows the new network without those three links and the declared set of nodes.

Proc network produces the output summary, stating that the input graph has eight nodes and now only seven links. A solution was found and three connected components were identified. The output tables OUTLINKS, OUTNODES, and CONCOMPOUT report all results. Figure 2.12 shows the summary results for the procedure.

The output table OUTLINKS shows all seven links associated with the input graph, and the connected component each link belongs to. Figure 2.13 shows the output table for the links.

The output table OUTNODES shows all the eight nodes associated with the input graph, and the connected component each one belongs to. Notice that the connected component 1 has three nodes (A, B, and C), the connected component 2 has four nodes (E, F, G, and D), and the connected component 3 has one single node (H). Figure 2.14 shows the output table for the nodes.

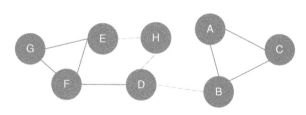

Figure 2.11 Input graph with undirected links.

The output table CONCOMPOUT shows the connected components identified and the number of nodes of each component. The connected component 1 has three nodes, the connected component 2 has four nodes, and finally the connected component 3 has one single node. Figure 2.15 shows the output table for the connected components.

Figure 2.16 shows the connected components highlighted within the original input graph.

The NETWORK Procedure

Problem Summary

Number of Nodes	8
Number of Links	7
Graph Direction	Undirected

The NETWORK Procedure

Solution Summary

Problem Type	Connected Components
Solution Status	OK
Number of Components	3
CPU Time	0.00
Real Time	0.00

Output CAS Tables

CAS Library	Name	Number of Rows	Number of Columns
CASUSER(Carlos.Pinheiro@sas.com)	OUTLINKS	7	3
CASUSER(Carlos.Pinheiro@sas.com)	OUTNODES	8	2
CASUSER(Carlos.Pinheiro@sas.com)	CONCOMPOUT	3	2

Figure 2.12 Output results by proc network running connected components on undirected graph.

≪ OUTLINKS   Table Rows: 7 | Columns: 3 of 3 | Rows 1 to 7

	from	to	concomp
1	A	B	1
2	E	G	2
3	B	C	1
4	G	F	2
5	C	A	1
6	F	D	2
7	E	F	2

Figure 2.13 Output links table for the connected components.

Now, let's see how the connected components algorithm works upon a directed graph. In this example, we are going to use the exact same input graph defined at the beginning of this section, but the direction in proc network will be set as directed. The direction of each link will follow the links dataset definition.

The following code defines the set of links, and the *from* variable here actually specifies the origin or source node, and the *to* variable actually specifies the destination or sink node.

Figure 2.14 Output nodes table for the connected components.

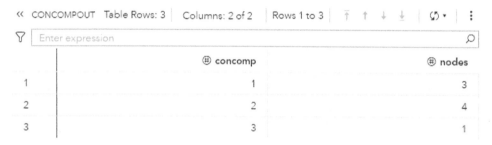

Figure 2.15 Output summary table for the connected components.

```
data mycas.links;
   input from $ to $ @@;
datalines;
A B   B C   C A   B D   D H   H E   E F   E G   G F   F D
;
run;
```

The following code invokes proc network to identify the connected components within a directed graph. Notice the DIRECTION = DIRECTED statement in proc network.

```
proc network
   direction = directed
   links = mycas.links
   outlinks = mycas.outlinks
   outnodes = mycas.outnodes
   ;
   linksvar
      from = from
      to = to
   ;
   connectedcomponents
      out = mycas.concompout
   ;
run;
```

Figure 2.17 shows the original input graph with the directed links. Note that the graph is exactly the same as the one presented previously, except for the direction of the links between the nodes.

The output summary produced by proc network shows the results of the connected components upon a directed graph. It has the same number of nodes and links, but now instead of one single component, two components are identified. Figure 2.18 shows the summary results for the connected components algorithm.

Figure 2.19 presents the output links for the connected components. The output table OUTLINKS shows all ten links associated with the input graph and the connected component each link belongs to. Notice that the link B→D does not belong to any connected component. This particular link is the one that breaks down the original directed input graph into two distinct components.

Figure 2.20 shows the output nodes for the connected components. The output table OUTNODES shows all the eight nodes associated with the input graph, and the connected component

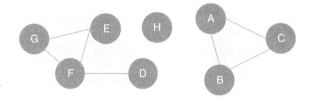

Figure 2.16 Input graph with undirected links and the identified connected components.

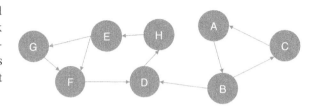

Figure 2.17 Input graph with directed links.

each one of them belongs to. Notice that the connected component 1 has three nodes (B, A, and C), and the connected component 2 has five nodes (D, E, G, H, and F).

Figure 2.21 shows the output connected components. The output table CONCOMPOUT shows the connected components identified and the number of nodes of each component. The connected component 1 has three nodes and the connected component 2 has five nodes.

Figure 2.22 shows the connected components highlighted within the original directed input graph.

When we analyze graphs, sometimes the focus is too much on the role of the nodes. In the next section, we clearly see this while we discuss the concept of biconnected components. However, the last example here also shows the importance of analyzing the roles of the links. The link B→D plays a fundamental role in splitting the original input graph into two disjoined components. In many practical cases and business scenarios, when analyzing graphs created by social or corporate interactions, we focus on the roles of the nodes. Nevertheless, it is important to keep in mind the significant role of the links.

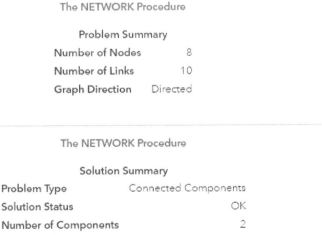

The NETWORK Procedure

Problem Summary

Number of Nodes	8
Number of Links	10
Graph Direction	Directed

The NETWORK Procedure

Solution Summary

Problem Type	Connected Components
Solution Status	OK
Number of Components	2
CPU Time	0.00
Real Time	0.00

Figure 2.18 Output results by proc network running connected components on a directed graph.

« OUTLINKS   Table Rows: 10 | Columns: 3 of 3 | Rows 1 to 10 | ⫪ ↑ ↓ ⫧ | ↻ ▾ | ⋮

▽ | Enter expression   ○

	△ from	△ to	⊕ concomp
1	B	D	.
2	E	G	2
3	A	B	1
4	D	H	2
5	G	F	2
6	B	C	1
7	H	E	2
8	F	D	2
9	C	A	1
10	E	F	2

Figure 2.19   Output links table for the connected components.

« OUTNODES   Table Rows: 8 | Columns: 2 of 2 | Rows 1 to 8 | ⫪ ↑ ↓ ⫧ | ↻ ▾ | ⋮

▽ | Enter expression   ○

	△ node	⊕ concomp
1	B	1
2	D	2
3	E	2
4	G	2
5	A	1
6	H	2
7	F	2
8	C	1

Figure 2.20   Output nodes table for the connected components.

« CONCOMPOUT   Table Rows: 2 | Columns: 2 of 2 | Rows 1 to 2 | ⫪ ↑ ↓ ⫧ | ↻ ▾ | ⋮

▽ | Enter expression   ○

	⊕ concomp	⊕ nodes
1	1	3
2	2	5

Figure 2.21   Output summary table for the connected components.

Sometimes, the links will reveal more useful insights about the social or corporate networks than the nodes themselves. Sometimes it is not about the entities in the network, but how these entities relate to each other.

2.3 Biconnected Components

A biconnected component of a graph $G = (N, L)$ is a subgraph that cannot be broken into disconnected pieces by deleting any single node and its incidents links. If a component can be broken into

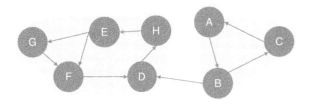

Figure 2.22 Input graph with directed links and the identified connected components.

other connected components, there is a node that if removed from the graph, will increase the number of connected components. This type of node is commonly called an articulation point.

Biconnected components can be seen as a variation of the connected components concept, meaning the graph may have some important nodes that can break down a component into smaller components. Sometimes the articulation points are also referred to as bridges, as these nodes create bridges between different groups or components within the network.

Articulation points often play an important role when analyzing graphs that represent any sort of communications network. For example, suppose a graph $G = (N, L)$ and an articulation point $i \in N$. If the articulation point i is removed from the graph G, it will break down the graph into two components C^1 and C^2. All paths in G between nodes in C^1 and nodes C^2 must pass through node i. The articulation point i is crucial to keep the communication flow within the original component or graph.

Depending on the context, particularly in scenarios associated with communication networks, the articulation points can be viewed as points of vulnerability in a network. Roads and rails represented by graphs, supply chain processes, social and political networks, among other types, are graphs where the articulation points represent vulnerabilities within the network, either breaking down the components into more connected components, or stopping the communication flow within the network. Imagine a bridge connection two locations. These two locations can represent articulation points and can stop the network flow not just between these two locations, but among the locations that are adjacent to them. Terrorist networks are also a fitting example of the importance of the articulation points. Commonly, targeting and eliminating individuals that represent articulation points within the network can strategically break down the power of a cohesive component into smaller and disorganized groups.

In proc network, the BICONNECTEDCOMPONENTS statement invokes the algorithm that solves the biconnected component problem, or the algorithm that identifies all biconnected components within a given graph. Different from the connected components, which can be identified in both directed and undirected graphs, the biconnected components can be identified only for the undirected graphs. In order to compute and identify the biconnected components, proc network uses a variant of the depth-first search algorithm. This algorithm can scale large graphs.

The final results for the biconnected component algorithm in proc network are reported in the output table defined by the options OUT =, OUTNODES =, and OUTLINKS =. The option OUT = within the BICONNECTEDCOMPONENTS statement produces an output containing all the biconnected components and the number of links associated with each of them.

- biconcomp: the biconnected component identifier.
- links: the number of links contained in the biconnected component.

The biconnected components are more associated with the links within the graph than to the nodes. Therefore, the option OUTNODES = within the PROC NETWORK statement produces an output containing all the nodes identification, and a flag to determine whether or not each node is an articulation point.

- node: the node identifier.
- artpoint: identify whether or not the node is an articulation point. The value of 1 determines that the node is an articulation point. The value of 0 determines that the node is not an articulation point.

The option OUTLINKS = within the PROC NETWORK statement produces an output containing all the links and the biconnected component each link belongs to.

- from: the origin node identifier.
- to: the destination node identifier.
- biconcomp: the biconnected component identifier for the link.

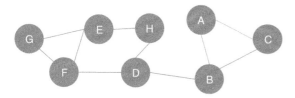

Figure 2.23 Input graph with undirected links for the biconnected components identification.

Let's see an example of connected components for an undirected graph.

2.3.1 Finding the Biconnected Components

Let's consider the same graph we have used for the connected components example. That graph is shown in Figure 2.23 and has 8 nodes and 10 links. Similar to the connected components, links weights are not required when searching for biconnected components.

The following code describes the links dataset creation. No weights are required, just the origin and destination nodes for the undirected graph.

```
data mycas.links;
   input from $ to $ @@;
datalines;
A B   B C   C A   B D   D H   H E   E F   E G   G F   F D
;
run;
```

Based on the links dataset, we can invoke the biconnected component algorithm to reach for the biconnected components and the articulation point nodes within the graph. The following code describes how to do that.

```
proc network
   direction = undirected
   links = mycas.links
   outlinks = mycas.outlinks
   outnodes = mycas.outnodes
   ;
   linksvar
      from = from
      to = to
   ;
   biconnectedcomponents
      out = mycas.biconcompout
   ;
run;
```

The following picture shows the outputs for the biconnected components algorithm. Figure 2.24 reports a summary of the execution process, showing the network composed of 8 nodes and 10 links, as an undirected graph. A solution was found with three biconnected components being identified and with two nodes being determined as articulation points. The last part of the output describes the three output tables created, the links output table, the nodes output table, and the biconnected components output table.

Figure 2.25 shows the OUTLINKS table, which describes the links and the biconnected components each link belongs to. The output table describes the original links variables from and to, and the biconnected component identifier assigned to the link.

Notice that there are three biconnected components. The biconnected component 1 comprises three links, C→A, A→B, and B→C. The biconnected component 2 comprises one single link, B→D. Finally, the biconnected component 3 comprises six links, E→F, D→H, G→F, H→E, F→D, and E→G.

Figure 2.26 shows the OUTNODES table, which describes the nodes and the articulation point flag.

Two nodes were identified as articulation points, which means they can break down the connected component into smaller components within the input graph. The nodes B and D can represent bridges in the original network splitting the original graph into multiple subgraphs.

The NETWORK Procedure

Problem Summary

Number of Nodes	8
Number of Links	10
Graph Direction	Undirected

The NETWORK Procedure

Solution Summary

Problem Type	Biconnected Components
Solution Status	OK
Number of Components	3
Number of Articulation Points	2
CPU Time	0.00
Real Time	0.00

Output CAS Tables

CAS Library	Name	Number of Rows	Number of Columns
CASUSER(Carlos.Pinheiro@sas.com)	OUTLINKS	10	3
CASUSER(Carlos.Pinheiro@sas.com)	OUTNODES	8	2
CASUSER(Carlos.Pinheiro@sas.com)	BICONCOMPOUT	3	2

Figure 2.24 Output results by proc network for the biconnected components algorithm.

	from	to	biconcomp
1	C	A	1
2	E	F	3
3	A	B	1
4	D	H	3
5	G	F	3
6	B	C	1
7	H	E	3
8	F	D	3
9	B	D	2
10	E	G	3

Figure 2.25 Output links table for the biconnected components.

Finally, the output table BICONCOMPOUT shows the summary of the biconnected component algorithm. It is presented in Figure 2.27. This table presents the biconnected components identifiers and the number of links in each biconnected component. As described above, biconnected component 3 has six links, biconnected component 2 has one single link, and biconnected component 1 has three links.

« OUTNODES	Table Rows: 8	Columns: 2 of 2	Rows 1 to 8		
	⚠ node			⊞ artpoint	
1	A			0	
2	B			1	
3	D			1	
4	H			0	
5	G			0	
6	F			0	
7	C			0	
8	E			0	

Figure 2.26 Output nodes table for the biconnected components.

« BICONCOMPOUT	Table Rows: 3	Columns: 2 of 2	Rows 1 to 3		
		⊞ biconcomp		⊞ links	
1		3		6	
2		2		1	
3		1		3	

Figure 2.27 Output summary table for the biconnected components.

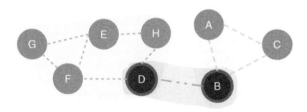

Figure 2.28 Input graph with the identified biconnected components and the articulation point nodes.

Figure 2.28 shows the biconnected components highlighted within the original input graph, as well as the articulation point nodes.

Notice on the left that six links create one biconnected component. On the right, three links create another biconnected component. In the middle of the graph, a single link creates the last biconnected component. On the left biconnected component, if we remove node E, nodes G and F still can reach out to node H through node D. If we remove node H, node D still can reach out to node E through node F. Therefore, nodes G, F, E, H, and D form a unique biconnected component. By removing any of these nodes, we still cannot break apart the original component into two or more smaller components. The same thing happens for nodes A, B, and C. Removing any of these nodes does not break apart the original component into smaller components. The remaining nodes will still be able to connect to each other. Only nodes D and B have the ability to break apart the original component into smaller components. If we remove node D, the nodes G, F, E, and H will form a component, and node A, B, and C will for another one. If we remove node B, nodes A and C will form a component, and nodes G, F, E, H, and D will form another one. For this reason, nodes D and B are the articulation points within this input graph.

2.4 Community

Community detection is a crucial step in network analysis. It breaks down the entire network into smaller groups of nodes. These groups of nodes are more connected among them than between that group and the others. The idea of community detection is to join nodes based on their average link weights. The average link weights inside the communities should be greater than the average link weight between the communities. Nodes are gathered based on the strength of their relationships in comparison to the rest of their connections.

Communities are densely connected subgraphs within the network where nodes belonging to the same community are more likely to be connected to nodes within their own communities than to nodes outside their communities. A network can contain as many communities as the number of nodes, or one single community representing the entire network. Often, something in between that captures the overall characteristics of different groups of nodes while keeping the network topology structure. Similar to clustering analysis, the community detection algorithm assigned to each node to a unique community.

The idea of communities within networks is very intuitive and quite useful when analyzing large graphs. As groups of nodes densely connected, the concept of communities is commonly found in networks representing social interactions, chemistry and biological structures, and grids of computers, among other fields. Community detection is important to identify customers consuming the same service, products being purchased together (like in association rules or market basket analysis), companies most frequently making transactions together, or just groups of people who are densely related to each other. Notice that community detection is not clustering analysis. Both techniques combine observations into groups. Clustering works on the similarity of the observations' attributes. For example, customers' demographics or consuming habits. Customers with no relation can fit into the same group just based on their similar attributes. Community detection combines observations (nodes) based on their relations, or the strength of their connections to each other. In fact, this is the fundamental difference between common machine learning or statistical models and network analysis. The first always looks into observations attributes, and the second always looks into observations connections.

There are two primary methods to detect communities within networks, the agglomerative methods, and the divisive methods. Agglomerative methods add links one by one to subgraphs containing only nodes. The links are added from the higher link weight to the lower link weight. Divisive methods perform a completely opposite approach. In divisive methods, links are removed one by one from the subgraphs. Both methods can find multiple communities upon the same network, with different numbers of members in them. It is hard to say which method is the best, as well as what is the best number of communities within a network and the average number of members in them. Mathematically, we can use metrics of resolution and modularity to narrow this search, but in most cases, a domain knowledge is important to understand what makes sense for each particular network. For example, let's say that in a telecommunications network, the average number of relevant distinct connections for each subscriber is around 30. When detecting communities within this network, we should try to find a particular number of communities that represent a similar number of mem'ers in each community on average. Suppose that carrier has around 120 million subscribers. Finding 2 or 3 million communities wouldn't be absurd. That number would give us around 40-60 as an average number of members in each community. Community detection is important in understanding and evaluating the structure of large networks, like the ones associated with communications, social media, banking, and insurance, among others.

The community detection algorithm was initially created to be used on undirected graphs. Currently, there are some algorithms that provide alternative methods to compute community detection on directed graphs. If we are working on directed graphs and still want to perform a community detection using algorithms for undirected graphs, we need to aggregate the links between the pair of nodes, considering both directions, and sum up their weights. Once the links are aggregated and the links weighted are summed up, we can invoke a community detection algorithm designed to detect communities on undirected graphs. Figure 2.29 describes that scenario.

The method of partitioning the network into communities is an iterative process, as described before. Two types of algorithms to partition a graph into subgraphs can be used upon undirected links. Proc network implements both heuristics methods, called Label Propagation and Louvain. Proc network also implements a particular algorithm to partition a graph into communities based on directed links, called Parallel Label Propagation. All three methods for finding communities in directed and undirected graphs are heuristic algorithms, which means they search for an optimal solution, speeding up the searching process by employing practical methods that reach satisfactory solutions. The solution found doesn't mean that the solution is perfect.

The Louvain algorithm aims to optimize modularity, which is one of the most popular merit functions for community detection. Modularity is a measure of the quality of a division of a graph into communities. The modularity of a division is defined to be the fraction of the links that fall within the communities, minus the expected fraction if the links were distributed at random, assuming that the degree of each node does not change.

Figure 2.29 Aggregating directed links into undirected links.

The modularity can be expressed by the following equation:

$$Q = \frac{1}{2W_L} \sum_{i,j \in N} \left(w_{ij} - \frac{w_i w_j}{2W_L} \right) \Delta(C_i, C_j)$$

where Q is the modularity, W_{ij} is the sum of all link weights in the graph $G = (N, L)$, w_{ij} is the sum of the link weights between nodes i and j, C_i is the community to which node i belongs, C_j is the community to which node j belongs, and $\Delta(C_i, C_j)$ is the Kronecker delta defined as:

$$\Delta(C_i, C_j) = \begin{cases} 1, if \ C_u = C_v \\ 0, otherwise \end{cases}$$

The following is a brief description of the Louvain algorithm:

1) Initialize each node as its own community.
2) Move each node from its current community to the neighboring community that increases modularity the most. Repeat this step until modularity cannot be improved.
3) Group the nodes in each community into a super node. Construct a new graph based on super nodes. Repeat these steps until modularity cannot be further improved or the maximum number of iterations has been reached.

The label propagation algorithm moves a node to a community to which most of its neighbors belong. Extensive testing has empirically demonstrated that in most cases, the label propagation algorithm performs as well as the Louvain method. The following is a brief description of the label propagation algorithm:

1) Initialize each node as its own community.
2) Move each node from its current community to the neighboring community that has the maximum number of nodes. Break ties randomly, if necessary. Repeat this step until there are no more movements.
3) Group the nodes in each community into a super node. Construct a new graph based on super nodes. Repeat these steps until there are no more movements or the maximum number of iterations has been reached.

The parallel label propagation algorithm is an extension of the basic label propagation algorithm. During each iteration, rather than updating node labels sequentially, nodes update their labels simultaneously by using the node label information from the previous iteration. In this approach, node labels can be updated in parallel. However, simultaneous updating of this nature often leads to oscillating labels because of the bipartite subgraph structure often present in large graphs. To address this issue, at each iteration the parallel algorithm skips the labeling step at some randomly chosen nodes in order to break the bipartite structure.

For directed graphs, the modularity equation changes slightly.

$$Q = \frac{1}{2W_L} \sum_{i,j \in N} \left(w_{ij}^{out} - \frac{w_i^{out} w_j^{in}}{W_L} \right) \Delta(C_i, C_j)$$

where w_{ij}^{out} is the sum of the weights of the links departing from node i and arriving at node j, w_i^{out} is the sum of the weights of the links outgoing from node i, and w_j^{in} is the sum of the weights of the links incoming to node j.

All three algorithms adopt a heuristic local optimization approach. The final result often depends on the sequence of nodes that are presented in the links input data set. Therefore, if the sequence of nodes in the links data set changes, the result can be different.

The graph direction is relevant in the community detection process. For undirected graphs, both algorithms; Louvain and label propagation; find communities based on the density of the subgraphs. For directed graphs, the algorithm parallel label propagation finds communities based on the information flow along the directed links. It propagates the community identification along the outgoing links of each node. Then, when using directed graphs, if the graph does not have a circle structure where nodes are connected to each other along outgoing links, the nodes are likely to switch between communities during the computation. As a result, the algorithm might not converge well and then a good community structure within the graph might not be established.

We always need to think about the graph direction depending on the problem we are analyzing. For example, in a social network, the definition of friendship is represented by undirected links. For a telecommunications or banking network, the call, text, or money transfer are represented by directed links. However, the community detection approach can still be performed by both directed and undirected graphs. For instance, in the telecommunications network, even though the relationships between subscribers are represented by directed links, it makes sense to detect communities based on an undirected graph, as both directions for calls and texts ultimately create the overall relation among people. Perhaps for the banking network, detect communities based on directed links makes more sense as the groups of accounts would be based on the information flow, or the sequence of transactions.

A crucial step in community detection is always to analyze the outcomes and verify if the groups found make sense in the context of the business or the problem to be solved. For example, If the parallel label propagation algorithm does not converge for the directed graphs and poor communities are detected, a reasonable approach is to perform the community detection based on undirected graphs. If the label propagation does not work well in finding proper communities upon undirected graphs, we can always try the Louvain algorithm, and vice versa. Changing parameters for the recursive option, or the resolution used are also options to better detect reasonable communities. The community detection is ultimately a trial and error process. We need to spend time executing multiple runs and analyzing the outcomes until we decide the best structure and partitioning.

Another point of attention is the size of the network. For large networks, a power law distribution is commonly observed in terms of the average number of nodes within the communities, which results in few communities with a large number of members, and most of the communities with a small number of nodes. There are some approaches to reduce this effect, by using some of the parameters mentioned before, like the resolution list and the recursive methods. A combination of approaches can also be deployed, usually providing reliable results. For example, we can first perform community detection based on the resolution list, and then adjust the partitioning by using the recursive method considering solely the best resolution in the last step.

In proc network, the COMMUNITY statement invokes the algorithm to detect communities within an input graph. There are many options when running community detection in proc network. The first and probably the most important one is the decision about which algorithm to use. The option ALGORITHM = specifies the algorithm to be used during the community detection process, one of the three algorithms previously mentioned. ALGORITHM = LOUVAIN specifies that the community detection will be based on the algorithm Louvain. It runs only for undirected graphs. ALGORITHM = LABELPROP specifies that the community detection will be performed based on the label propagation algorithm. It can be used only for undirected graphs. Finally, the option ALGORITHM = PARALLELLABELPROP specifies that the community detection will be based on the parallel label propagation algorithm. Although this algorithm is designed to directed graphs, it can also run based on undirected graphs.

The FIX = option specifies which variable in the nodes input table will define the group of nodes that need to be together in a community. This option is only used for undirected graphs, which means it is available when the ALGORITHM = option is LOUVAIN or LABELPROP. The idea behind this option is to allow users to define a group of nodes to be together, in the same community, no matter the result of the community detection based on the partitioning algorithm. Usually, community detection algorithms start by assigning each node as one community and then progressively changing nodes between communities based on the type of the algorithm and the set of parameters defined. The variable specified in the FIX = option should be in the nodes input table defined in the NODES = option. This variable must contain a nonnegative integer value. Nodes with the same value are forced to be in the same community. Regardless of the partitioning process, they cannot be split into different groups along the community detection process. These nodes that are forced into the same community still can be merged or split into larger or smaller communities, but they must be together in the same community. For example, a nodes dataset *NodeSetIn* has a variable called *comm* to determine the nodes that must be together. The option NODES = NodeSetIn defines the nodes input table. The option FIX = comm specifies the variable to be used during the partitioning process. Let's say nodes A, B, and C have *comm* equal to 1, and nodes X, Y, and Z have *comm* equal to 2. All other nodes have missing values for *comm*. During the community detection process, proc network will force that nodes A, B, and C fall into the same community, and nodes X, Y, and Z fall into the same community as well. They can be in the same community (all of them together), or in two distinct communities (A/B/C together in one and X/Y/Z together in another), but they cannot be split into multiple communities. The community detection algorithm basically treats the group of nodes that are required to be in the same community as a single unit, like a "supernode". A straightforward consequence of fixing nodes to be in the same community by using the FIX = option is that proc network will have fewer options to move these nodes around into different communities. The overall number of node movements will be drastically reduced according to the number of nodes set to be together. The other consequence is that less communities with more members may be found.

The WARMSTART = option works similarly to the FIX = option, at least at the beginning of the process. This option specifies which variable in the nodes input table will define the group of nodes that need to be together when the partitioning process starts. This option defines the initial partition for warm starting community detection. This initialization can be based on prior executions of community detection, or according to existing knowledge about the data, the problem or the business scenario involved. Similar to the FIX = option, the nodes that have the same value in the variable defined in the option WARMSTART are placed in the same community. All other nodes with missing values in that variable or not specified in the nodes input table, are placed in their own community. The warm start just places the nodes together at the beginning of the partitioning process. During the community detection execution, those nodes can move around to different communities and can end up falling into distinct groups. For example, a nodes dataset *NodeSetIn* has a variable called *startcomm* to determine the nodes that must be placed together when the graph partitioning process starts. The option NODES = NodeSetIn defines the nodes input table. The option FIX = startcomm specifies the variable to be used during the community detection process. Let's say nodes A, B, and C have *startcomm* equal to 1, and nodes X, Y, and Z have *startcomm* equal to 2. All other nodes have missing values for *startcomm*. At the beginning of the partitioning, nodes A, B, and C will form a single community. Nodes X, Y, and Z will form another community. All other nodes will be their own community. Nodes A, B, and C, as well as X, Y, and Z, will be treated as a "supernode" ABC and XYZ, respectively. They can change communities along the partitioning, but they need to be placed together at the beginning.

Notice that both FIX = and WARMSTART = options can be used at the same time. When both options are specified simultaneously, the values of the variables defined for fix and warm start are assumed to be related. For example, if nodes A and B have a value of 1 for the fix, and node C has a value of 1 for the warm start, then they are initialized in the same community. However, during the partitioning, the algorithm can move node C to a different community than nodes A and B, but nodes A and B are forced to move together to another community. At the end of the process, A and B are forced to be in the same community, while C can end up in a different one.

The RECURSIVE option recursively breaks down large communities into smaller groups until some certain conditions are satisfied. The RECURSIVE option provides a combination of three possible sub options, MAXCOMMSIZE, MAXDIAMETER, and RELATION. These options can be used individually or simultaneously. For example, we can define RECURSIVE (MAXCOMMSIZE = 100) to define a tentative maximum of 100 members for community, or RECURSIVE (MAXCOMMSIZE = 500 MAXDIAMETER = 3 RELATION = AND) to define the maximum number of members for each community as 500 and the maximum distance between any pair of nodes in the community as 3. Both conditions must be satisfied during the graph partitioning as the sub option RELATION is defined as AND. The sub option MAXCOMMSIZE defines the maximum number of nodes for each community. The sub option MAXDIAMETER defines the maximum number of links within the shortest paths between any pair of nodes within the community. The diameter represents the largest number of links to be traversed in order to travel from one node to another. It is the longest shortest path of a graph. The sub option RELATION defines the relationship between the other two sub options, MAXCOMMSIZE and MAXDIAMETER. If RELATION = AND, the recursive partitioning continues until both constrains defined in MAXCOMMSIZE and MAXDIAMETER sub options are satisfied. If RELATION = OR, the partitioning continues until either the maximum number of members in any community is satisfied or the maximum diameter is reached.

At the first step, the community detection algorithm processes the network data with no recursive option. At the end of each step, the algorithm checks if the community outcomes satisfy the criteria defined in the recursive option. If the criteria are satisfied, the recursive partitioning stops. If the criteria are not satisfied yet, the recursive partition continues on each large community (the ones do not satisfy the criteria) as an independent subgraph and the recursive partitioning is applied to each one of them. The criteria define the maximum number of members for each community and the maximum distance between any pair of nodes within each community. However, the recursive method only attempts to limit the size of the community and the maximum diameter. The criteria are not guaranteed to be satisfied as in certain cases; some communities can no longer be split. Because of that, at the end of the process, some communities can have more members than defined in MAXCOMMSIZE or more links between a pair of nodes than defined in MAXDIAMETER.

The RESOLUTIONLIST option specifies a list of values that are used as a resolution measure during the recursive partitioning. The list of resolutions is interpreted differently depending on the algorithm used for the community detection. The algorithm Louvain uses all values defined in the resolution list and computes multiple levels of recursive partitioning. At the end of the process, the algorithm produces different numbers of communities, with different numbers of members in them.

For example, higher resolution tends to produce more communities with less members, and low resolution tends to produce less communities with more members. This algorithm detects communities at the highest level first, and then merges communities at lower levels, repeating the process until the lowest level is reached. All levels of resolution are computed simultaneously, in one single execution. This allows the process to be computed extremely fast, even for large graphs. On the other hand, the recursive option cannot be used. If the recursive option is used in conjunction with the resolution list, only the default resolution value of 1 is used and the list of multiple resolutions is ignored. Also, because of the nature of this optimization algorithm, two different executions most likely will produce two different results. For example, a first execution with a resolution list of 1.5, 1.0, and 0.5, and a second execution with a resolution list of 2.0 and 0.5 can produce different outcomes at the same resolution level of 0.5. This happens mostly because of the recursive merges happening throughout the process from higher resolutions to lower ones.

The algorithm parallel label propagation executes the recursive partitioning multiple times, one for each resolution specified. Because of that, the resolution list in the parallel label propagation algorithm is fully compatible with the recursive option. Every execution of the community detection algorithm will use one resolution value and the recursive criteria. In the parallel label propagation, the resolution value is used when the algorithm decides to which neighboring communities a node should be fit. A node cannot fit to a community if the density of the community, after adding the node, is less than the resolution value r. The density of a community is defined as the number of existing links in the community divided by the total number of possible links, considering the current set of nodes within it. As this algorithm runs in parallel when deciding about which community to fit a node, the final communities are not guaranteed to have their densities larger than the resolution value at all times. The parallel label propagation algorithm may not converge properly if the resolution value specified is too large. The algorithm will probably create many small communities, and the nodes will tend to change between communities along the iterations. On the other hand, if the resolution value is too small, the algorithm will probably find a small number of communities large. The use of the recursive option could be an approach to break down those large communities into smaller ones. Notice that, the larger the network, the longer it will take to properly run this process.

The algorithm label propagation is not compatible with the resolution list option. If a resolution list is specified when the label propagation algorithm is defined, the default resolution value of 1 is used instead.

The resolution can be interpreted as a measure to merge communities. Two communities are merged if the sum of the weights of intercommunity links is at least r times the expected value of the same sum if the graph is reconfigured randomly. Large resolutions produce more communities with a smaller number of nodes. On the contrary, small resolutions produce fewer communities with a greater number of nodes. Particularly for the algorithm Louvain, the resolution list enables running multiple community detections simultaneously. At the end of the recursive partitioning process multiple communities can be found at the same execution. This approach is interesting when searching for the best topology, particularly in large networks. However, this approach requires further analysis on the outcomes in order to identify the best community distribution. There are two main approaches to analyze the community distribution. The first one is by looking at the modularity value for each resolution level. Mathematically, the higher the modularity, the better. The resolution that gives the modularity closest to 1 provides the best topology in terms of community distribution. The second approach relies on the domain knowledge and the context of the problem. The number of communities and the average number of members in each community can be driven based on previous executions or upon a business understanding of the problem. The best approach is commonly a combination of the mathematical and business methods looking for a good modularity but restricting the number of communities and the average number of members in each community based on the domain knowledge.

The LINKREMOVALRATIO = option specifies the percentage of links with small weights will be removed from each node neighborhood. This option can dramatically reduce the execution time when running community detection on large networks by removing unimportant links between nodes. A link is removed if its weight is relatively smaller than the weights of the neighboring links. For example, suppose there are links A→B, A→C, and A→D with link weights 100, 80, and 1, respectively. When nodes are merged into communities, links A→B and A→C are more important than link A→D, as they contribute more to the overall modularity value. Because of that, link A→D can be removed from the network. By removing link A→D, node A will be connected only to nodes B and C. Node D will be disconnected to A when the algorithm merge nodes into communities. It doesn't mean that node D is removed from the network. Perhaps node D has strong links to other nodes in the graph and will be connected to them in a different community. If the weight of any link is less than the number specified in the LINKREMOVALRATIO = option divided by 100 and multiplied by the maximum link weight among all links incident to the node, that link will be removed from the network. The default value for the link removal ratio is 10.

The MAXITERS = option specifies the maximum number of iterations the community detection algorithm runs during the recursive partitioning process. For the Louvain algorithm, the default number of times is 20. For the parallel label propagation and the label propagations algorithms, the default number of times is 100.

The RANDOMFACTOR = option specifies the random factor for the parallel label propagation algorithm in order to avoid nodes oscillating in parallel between communities. At each iteration of the recursive partitioning, the number specified in the option is the percentage of the nodes that are randomly skipped to the step. The number varies from 0 to 1. For example, 0.1 means that 10% of the nodes are not considered during each iteration of the algorithm.

The RANDOMSEED = option works together with the random factor in the parallel label propagation algorithm. The random factor defines the percentage of the nodes that will be not considered at each iteration of the partitioning. In order to consider a different set of random nodes at each iteration, a number must be specified to create the seed. The default seed is 1234.

The TOLERANCE = option or the MODULARITY = option specifies the tolerance value to stop the iterations when the algorithm searches for the optimal partitioning. The Louvain algorithm stops iterating when the fraction of modularity gain between two consecutive iterations is less than the number specified in the option. The label propagation and parallel label propagation stop iterating when the fraction of community changes for all nodes in the graph is less than the number specified in the option. The value for this option varies from 0 to 1. The default value for the Louvain and the label propagation algorithms is 0.001, and for the parallel label propagation algorithm is 0.05.

The community detection process can produce up to six outputs as a result. These datasets describe the network structure in terms of the communities found based on the set of nodes, the set of links, the different levels of executions when the community detection uses the algorithm Louvain with multiple resolutions, the properties of the communities, the intensity of the nodes within the communities, and how the communities are connected. To produce all these outcomes, the community detection algorithm may consume considerable time and storage, particularly for large networks. In that case, we may consider suppressing some of these outcomes that are mostly intended for post community detection analysis.

The option OUTNODES = within the COMMUNITY statement produces an output containing all nodes and their respective community. In the case of multiple resolutions, the output reports the community for each node at each resolution level. After the topology is selected, all nodes are uniquely identified to a community, at each resolution level. Post analysis of the communities can be performed based on the nodes and their communities upon each level executed.

- node: identifies the node.
- community_i: identifies the community the node belongs to, where i identifies the resolution level defined in the resolution list, and i is ordered from the greatest resolution to the lowest in the list.

The option OUTLINKS = contains all links and their respective communities. Again, if multiple resolutions are specified during the community detection process, the output reports the community for each link at each resolution level. If the link contains the *from* node assigned to one community and the *to* node assigned to a different community, the community identifier receives a missing value to describe that link belongs to inter-communities.

- from: identifies the originating node.
- to: identifies the destination node.
- community_i: identifies the community the link belongs to, where i identifies the resolution level defined in the resolution list and is ordered from the greatest to the lowest resolution in the list.

The option OUTLEVEL = contains the number of communities and the modularity values associated with the different resolution levels. This outcome is probably one of the first outputs to be analyzed in order to identify the possible network topology in terms of communities. The balance between the modularity value, the number of communities found, and the average number of members for all communities is evaluated at this point.

- level: identifies the resolution level.
- resolution: identifies the value in the resolution list.
- communities: identifies the number of communities detected on that resolution list.
- modularity: identifies the modularity achieved for the resolution level when the community detection process is executed.

The option OUTCOMMUNITY = contains several attributes to describe the communities found during the community detection process.

- level: identifies the resolution level.
- resolution: identifies the value in resolution list.
- community: identifies the community.
- nodes: identifies the number of nodes in the community.
- intra_links: identifies the number of links that connect two nodes in the community.
- inter_links: identifies the number of links that connect two nodes within different communities. For example, *from* node in one community and *to* node in a different community than *from* node.
- density: the intra_links divided by the total number of possible links within the community.
- cut_ratio: the inter_links divided by the total number of possible links from the community to nodes outside the community.
- conductance: the fraction of links from a node in the community that connects to nodes in a different community. For undirected graphs, it is the inter_links divided by the sum of inter_links, plus two times the intra_links. For directed graph, it is the inter_links divided by the sum of inter_links, plus intra_links.

These measures are especially useful in analyzing the communities, particularly in large graphs. The density and the intra links show how dense and well connected the community is, no matter its size or the number of members. The inter links, the cut ratio, and the conductance describe how isolated the community is from the rest of the network, or how interconnected it is to other communities within the network. Communities more interconnected to the rest of the network tend to spread the information flow faster and stronger. This characteristic can be used to diffuse marketing and sales information. Spare or highly isolated communities can be unexpected and can indicate possible suspicious behaviors. All these analyzes certainly depend on the context of the problem.

The option OUTOVERLAP = contains information about the intensity of each node. Even though a node belongs to a single community, it may have multiple links to other communities. The intensity of the node is calculated as the sum of the link weights it has with nodes within its community, divided by the sum of all link weights it has (considering link to nodes to the same community and to nodes within different communities). For directed graphs, only the outgoing links are taken into consideration. This output can be expensive to compute and store, particularly in large networks. For that reason, this output contains only the results achieved by the smallest value of the resolution list. This output table is one of the first options to be suppressed when working on large networks.

- node: identifies the node.
- community: identifies the community.
- intensity: identifies the intensity of the node that belongs to the community.

Finally, the option OUTCOMMLINKS contains information about how the communities are interconnected. This information is important to understand how cohesive the network is in a community's perspective.

- level: identifies the resolution level.
- resolution: identifies the value in the resolution_list.
- from_community: identifies the community of the *from* community.
- to_community: identifies the community of the *to* community.
- link_weight: identifies the sum of the link weights of all links between *from* and *to* communities.

Let's look into some different examples of community detection for both directed and undirected networks.

2.4.1 Finding Communities

Let's consider the following graph presented in Figure 2.30 to demonstrate the community detection algorithm. This graph has 13 nodes and 15 links. Link weights in community detection is crucial and defines the recursive partitioning of the network into subgraphs. The links in this network are first considered undirected. Multiple algorithms will be used and therefore, different communities will be detected.

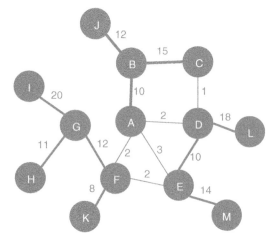

Figure 2.30 Input graph with undirected weighted links.

The following code creates the input dataset for the community detection algorithm. The links dataset contains origin and destination nodes plus the weight of the links. Both algorithms Louvain and label propagation are used. Later, the same network is be used as a directed graph to demonstrate the parallel label propagation algorithm.

```
data mycas.links;
    Input from $ to $ weight @@;
datalines;
A B 10 B C 15 B J 12
A F 2 A E 3 A D 2 C D 1 F E 2
G F 12 G I 20 G H 11 K F 8
D L 18 D E 10 M E 14
;
run;
```

As the input dataset is defined, we can now invoke the community detection algorithm using proc network. The following code describes how to define the algorithm and other parameters when searching for communities. In the LINKSVAR statement the variables from and to receive the origin and the destination nodes, even though the direction is not used at this first run. The variable weight receives the weight of the link. In the COMMUNITY statement, the ALGORITHM option defines the Louvain algorithm. The RESOLUTIONLIST option defines three resolutions to be used during the recursive partitioning. Notice that all three resolutions will be used in parallel, identifying distinct communities' topologies at a single run. Resolutions 2.5, 1.0, and −0.15 are defined for this search.

```
proc network
    direction = undirected
    links = mycas.links
    outlinks = mycas.outlinks
    outnodes = mycas.outnodes
    ;
    linksvar
        from = from
        to = to
    ;
    community
        algorithm=louvain
        resolutionlist = 2.5 1.0 0.15
        outlevel = mycas.outcommlevel
        outcommunity = mycas.outcomm
        outoverlap = mycas.outoverlap
        outcommLinks = mycas.outcommlinks
    ;
run;
```

The following pictures show the several outputs for the community detection algorithm. Figure 2.31 reports the summary for the execution process. It provides the size of the network, with 13 nodes and 15 links, and the direction of the graph. The solution summary states that a solution was found and provides the name of all output tables generated during the process.

Particularly for the Louvain algorithm, when multiple resolutions are used, the first output table analyzed is the OUTCOMMLEVEL. This output shows the result of each resolution used, the number of communities found, and the modularity achieved. It is presented in Figure 2.32.

As described before, the higher the modularity closest to one, the better the distribution of the nodes throughout the communities identified. The highest modularity for this execution is 0.5915, assigned to the level 2, using resolution 1.0, where 3 communities were found.

The NETWORK Procedure

Problem Summary

Number of Nodes	13
Number of Links	15
Graph Direction	Undirected

The NETWORK Procedure

Solution Summary

Problem Type	Community Detection
Solution Status	OK
CPU Time	0.00
Real Time	0.00

Output CAS Tables

CAS Library	Name	Number of Rows	Number of Columns
CASUSER(Carlos.Pinheiro@sas.com)	OUTLINKS	15	6
CASUSER(Carlos.Pinheiro@sas.com)	OUTNODES	13	4
CASUSER(Carlos.Pinheiro@sas.com)	OUTCOMMLINKS	10	5
CASUSER(Carlos.Pinheiro@sas.com)	OUTCOMM	10	9
CASUSER(Carlos.Pinheiro@sas.com)	OUTCOMMLEVEL	3	4
CASUSER(Carlos.Pinheiro@sas.com)	OUTOVERLAP	16	3

Figure 2.31 Output summary results for the community detection on an undirected graph.

| « OUTCOMMLEVEL | Table rows: 3 | Columns: 4 of 4 | Rows 1 to 3 | ↑ ↑ ↓ ↓ | ⟳ ▾ | ⋮ |

⟩ Enter expression

	level	resolution	communities	modularity
1	1	2.5	5	0.5485459184
2	2	1	3	0.5915306122
3	3	0.15	2	0.4419387755

Figure 2.32 Output communities level table for the community detection.

The OUTNODES table presented in Figure 2.33 shows the nodes and the communities they belong to for each resolution used. For example, for the best resolution (level 2), the variable community_2 specifies which node is assigned to each community.

For the level 2, nodes A, B, C, and J fall into community 1, nodes D, E, L, and M fall into community 2 and nodes F, G, H, I, and K fall into community 3.

The OUTLINKS table presented in Figure 2.34 shows all links and the community they belong to for each resolution used. Notice that some links do not belong to any community and do not have a community identifier assigned. These links connect nodes that fall into different communities. For example, nodes A and D are connected in the original network but belong to different communities, 1 and 2. Then, these particular links do not belong to any community and receive a missing value.

The OVERLAP table presented in Figure 2.35 shows the nodes, which community or communities they belong to, and their intensity on each community. For example, node A belongs to community 1 with intensity 0.88 and to community 2 with intensity 0.11. Remember that the intensity of a node measures the ratio of the sum of the link weights with nodes

	▲ node	⊞ community_1	⊞ community_2	⊞ community_3
1	A	1	1	1
2	B	1	1	1
3	C	1	1	1
4	D	2	2	1
5	E	3	2	1
6	F	4	3	2
7	G	5	3	2
8	H	5	3	2
9	I	5	3	2
10	J	1	1	1
11	K	4	3	2
12	L	2	2	1
13	M	3	2	1

Figure 2.33 Output nodes table for the community detection.

	▲ from	▲ to	⊞ weight	⊞ community_1	⊞ community_2	⊞ community_3
1	A	B	10	1	1	1
2	A	D	2	.	.	1
3	A	E	3	.	.	1
4	A	F	2	.	.	.
5	B	C	15	1	1	1
6	B	J	12	1	1	1
7	C	D	1	.	.	1
8	D	E	10	.	2	1
9	D	L	18	2	2	1
10	F	E	2	.	.	.
11	G	F	12	.	3	2
12	G	H	11	5	3	2
13	G	I	20	5	3	2
14	K	F	8	4	3	2
15	M	E	14	3	2	1

Figure 2.34 Output links table for the community detection.

within the same community by the sum of the link weights with all nodes. The intensity of node A in community 1 is much higher than the intensity in community 2, and in this case, it makes more sense that node A belongs to community 1.

The OUTCOMM table presented in Figure 2.36 shows the communities' attributes, for each level, or the resolution specified, the community identifier, the number of nodes, the intra links, the inter links, the density, the cut ration, and the conductance:

« OUTOVERLAP   Table rows: 16 | Columns: 3 of 3 | Rows 1 to 16   ↑ ↑ ↓ ↓ | ↺ ▾ | ⋮

	△ node	⊞ community	⊞ intensity
1	A	1	0.8823529412
2	A	2	0.1176470588
3	B	1	1
4	C	1	1
5	D	1	1
6	E	1	0.9310344828
7	E	2	0.0689655172
8	F	1	0.1666666667
9	F	2	0.8333333333
10	G	2	1
11	H	2	1
12	I	2	1
13	J	1	1
14	K	2	1
15	L	1	1
16	M	1	1

Figure 2.35   Output overlap table for the community detection.

« OUTCOMM   Table rows: 10   Columns: 9 of 9   Rows 1 to 10   ↑ ↑ ↓ ↓ | ↺ ▾ | ⋮

	⊞ level	⊞ resolution	⊞ community	⊞ nodes	⊞ intra_links	⊞ inter_links	⊞ density	⊞ cut_ratio	⊞ conductance
1	1	2.5	1	4	3	4	0.5	0.1111111111	0.4
2	1	2.5	2	2	1	3	1	0.1363636364	0.6
3	1	2.5	3	2	1	3	1	0.1363636364	0.6
4	1	2.5	4	2	1	3	1	0.1363636364	0.6
5	1	2.5	5	3	2	1	0.6666666667	0.0333333333	0.2
6	2	1	1	4	3	4	0.5	0.1111111111	0.4
7	2	1	2	4	3	4	0.5	0.1111111111	0.4
8	2	1	3	5	4	2	0.4	0.05	0.2
9	3	0.15	1	8	9	2	0.3214285714	0.05	0.1
10	3	0.15	2	5	4	2	0.4	0.05	0.2

Figure 2.36   Output communities' attributes table for the community detection.

The OUTCOMMLINKS table presented in Figure 2.37 shows how the communities are connected, providing the links between the communities and the sum of the link weights.

Figure 2.38 shows the best topology for the community detection, highlighting the three communities identified and their nodes.

The Louvain algorithm found three different community topologies based on three distinct resolutions. The best resolution, the one which achieved the highest modularity, and identified three communities. Let's see how the other algorithms work and what happens when other options like the recursive approach are used.

	⊞ level	⊞ resolution	⊞ from_community	⊞ to_community	⊞ link_weight
1	1	2.5	1	2	3
2	1	2.5	1	3	3
3	1	2.5	1	4	2
4	1	2.5	2	3	10
5	1	2.5	3	4	2
6	1	2.5	4	5	12
7	2	1	1	2	6
8	2	1	1	3	2
9	2	1	2	3	2
10	3	0.15	1	2	4

Table rows: 10 | Columns: 5 of 5 | Rows 1 to 10

« OUTCOMMLINKS

Figure 2.37 Output communities' links table for the community detection.

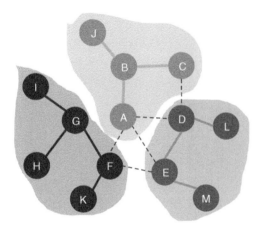

Figure 2.38 Communities identified by the Louvain algorithm.

The following code describes how to invoke the label propagation algorithm. A list of resolutions can be defined, but only the first one is used. If the resolution list is not specified, the default value of 1 is used. Any other value can be used but the resolution list must be specified with the value wanted.

```
proc network
    direction = undirected
    links = mycas.links
    outlinks = mycas.outlinks
    outnodes = mycas.outnodes
    ;
    linksvar
        from = from
        to = to
    ;
    community
        algorithm=labelprop
        resolutionlist = 1.0
        outlevel = mycas.outcommlevel
        outcommunity = mycas.outcomm
        outoverlap = mycas.outoverlap
        outcommLinks = mycas.outcommlinks
    ;
run;
```

The OUTCOMMLEVEL table presented in Figure 2.39 shows the final result for the community detection based on the label propagation algorithm. Resolution 1.0 represents level 1, where 4 communities were found. The modularity achieved was 0.5738, lower than the 0.5915 achieved by the Louvain algorithm based on the same resolution.

The OUTNODES table presented in Figure 2.40 shows the distribution of the nodes throughout the communities found. Now, nodes A, B, C, and J form the community 1, nodes D and L form the community 2, nodes F, G, H, I, and K form community 3, and finally nodes E and M form community 4. Notice that in the previous community detection, based on the Louvain algorithm, nodes D, L, E, and M formed one single community, and now they are split into two distinct communities.

	⊞ level	⊞ resolution	⊞ communities	⊞ modularity
1	1	1	4	0.5738520408

Figure 2.39 Output communities' level table for the community detection.

« OUTNODES   Table rows: 13   Columns: 2 of 2   Rows 1 to 13 ⋮

	⚠ node	⊞ community_1
1	A	1
2	B	1
3	C	1
4	D	2
5	E	4
6	F	3
7	G	3
8	H	3
9	I	3
10	J	1
11	K	3
12	L	2
13	M	4

Figure 2.40 Output nodes table for the community detection.

Figure 2.41 shows the final community topology for the label propagation algorithm, highlighting the 4 communities detected.

Now, let's use the recursive option and see the results for the community detection. The algorithm used is still the label propagation. The recursive option specifies a maximum of 4 members for each community and the maximal diameter of 3 (the distance between any pair of nodes within the same community).

```
Proc network
   direction = undirected
   links = mycas.links
   outlinks = mycas.outlinks
   outnodes = mycas.outnodes
   ;
   linksvar
      from = from
      to = to
   ;
```

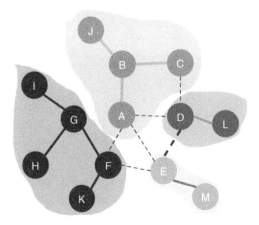

Figure 2.41 Communities identified by the label propagation algorithm.

```
community
    algorithm=labelprop
    resolutionlist = 1.0
    recursive(maxcommsize=4 maxdiameter=3 relation=and)
    outlevel = mycas.outcommlevel
    outcommunity = mycas.outcomm
    outoverlap = mycas.outoverlap
    ;
run;
```

The OUTCOMMLEVEL table presented in Figure 2.42 shows the final result for this particular community detection. Based on the label propagation algorithm and the recursive option, 5 communities were found.

The OUTNODES table presented in Figure 2.43 shows the distribution of the nodes throughout the communities. Nodes A, B, C, and J form the community 1, nodes D and L form the community 2, nodes E and M form community 3, nodes G, H, and I form community 4, and nodes F and K form community 5. Finally, notice that in the previous community detection, based on the Louvain algorithm, nodes D, L, E, and M formed one single community, and now they are split into two distinct communities. Also, notice that the previous community 3 was split into two distinct communities, 4 and 5.

Figure 2.44 shows the final community topology for the label propagation algorithm with the recursive option, highlighting the 5 communities detected.

The parallel label propagation algorithm is designed specifically for directed graphs. In order to demonstrate the process, let's use the same input graph, presented in Figure 2.45, but now considering the order of the links definition in the dataset as the direction of the connection.

« OUTCOMMLEVEL Table rows: 1 | Columns: 4 of 4 | Rows 1 to 1

	level	resolution	communities	modularity
1	1	1	5	0.5485459184

Figure 2.42 Output communities' level table for the community detection.

« OUTNODES Table rows: 13 | Columns: 2 of 2 | Rows 1 to 13

	node	community_1
1	A	1
2	B	1
3	C	1
4	D	2
5	E	3
6	F	5
7	G	4
8	H	4
9	I	4
10	J	1
11	K	5
12	L	2
13	M	3

Figure 2.43 Output nodes table for the community detection.

The following code shows how to invoke the parallel label propagation algorithm for the directed graph. Differently than the label propagation algorithm, the parallel label propagation can use a list of resolutions when detecting the communities. The difference between the Louvain algorithm and the parallel label propagation is that the first one executes all resolutions at the same time, and the second executes each resolution one at a time. Here we are using one single resolution, 1.0.

```
proc network
   direction = directed
   links = mycas.links
   outlinks = mycas.outlinks
   outnodes = mycas.outnodes
   ;
   linksvar
      from = from
      to = to
   ;
   community
      algorithm=parallellabelprop
      resolutionlist = 1.0
      outlevel = mycas.outcommlevel
      outcommunity = mycas.outcomm
      outoverlap = mycas.outoverlap
      outcommLinks = mycas.outcommlinks
   ;
run;
```

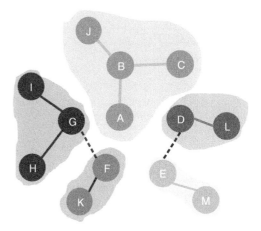

Figure 2.44 Communities identified by the label propagation algorithm with the recursive option.

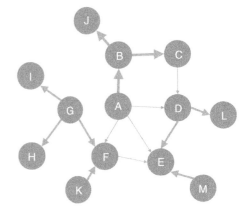

Figure 2.45 Input graph with directed weighted links.

The OUTCOMMLEVEL table presented in Figure 2.46 shows the final result for the community detection based on the parallel label algorithm upon a directed graph. This execution identified 3 communities. The modularity of 0.3 is the lowest considering all algorithms used before.

The OUTNODES table presented in Figure 2.47 shows the distribution of the nodes throughout the communities. Based on a directed graph, nodes A, C, B, and J form the community 1, nodes E, I, M, D, L, G, K, and F form community 2, and node H is isolated in community 3.

Figure 2.48 shows the final community topology for the parallel label propagation algorithm based on a directed graph, highlighting the 3 communities detected.

As we can see, different algorithms can detect distinct communities. The direction of the graph also plays a critical role, causing the algorithms to identify different topologies. As an optimization algorithm, different executions may end up with different results, particularly on large networks.

Finally, let's see two additional methods that can be used during the community detection process. The first method forces nodes to be in the same community at the beginning of the recursive partitioning. These nodes defined to be in the same

	level	resolution	communities	modularity
1	1	1	3	0.3005102041

Figure 2.46 Output communities' level table for the community detection.

<< OUTNODES   Table rows: 13   Columns: 2 of 2   Rows 1 to 13   ↑ ↑ ↓ ↓   ⋮

▽ | Enter expression   ⌕

	△ node	⊞ community_1
1	A	1
2	E	2
3	I	2
4	M	2
5	D	2
6	H	3
7	L	2
8	C	1
9	G	2
10	K	2
11	B	1
12	F	2
13	J	1

Figure 2.47 Output nodes table for the community detection.

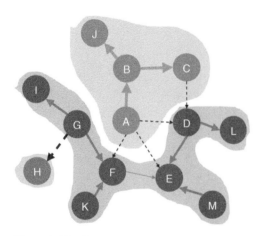

Figure 2.48 Communities identified by the parallel label propagation algorithm based on a directed input graph.

community at the beginning of the recursive partitioning can definitely interchange communities throughout the execution. The second method forces nodes to be in the same community at the end of the partitioning process.

The first part of the following code defines the set of nodes that needs to be in the same community at the beginning of the recursive partitioning process. Nodes A, L, E, and M need to be in the same community at the beginning of the community detection. The variable warm defines the community identifier for the nodes at the beginning of the process. In this example, all nodes have the value 1, indicating that they will start all together at the same community. We can define multiple sets of nodes by specifying different values for the variable warm. For example, nodes A and B with warm value 1, nodes C and D with warm values equal to 2, and nodes E and F with warm value 3. In this case, the first partitioning step considers nodes A and B in the same community, as well as nodes C and D and nodes E and F. They can change communities throughout the partitioning process and can end up in different communities at the final topology.

```
data mycas.nodeswarm;
   input node $ warm @@;
datalines;
D 1 L 1 E 1 M 1
;
run;
```

The following code invokes the community detection algorithm considering an initial set for nodes D, L, E, and M. The set of nodes used at the beginning of the recursive partitioning are specified by using the option NODES = in the NETWORK statement. The initial communities for the warm nodes are specified by using the option WARMSTART = in the COMMUNITY statement. The warm variable created in the nodes' dataset definition is referenced in that option.

```
proc network
   direction = undirected
   nodes = mycas.nodeswarm
   links = mycas.links
   outlinks = mycas.outlinks
   outnodes = mycas.outnodes
   ;
   linksvar
      from = from
      to = to
   ;
   nodesvar
      node = node
   ;
   community
      warmstart=warm
      algorithm=labelprop
      resolutionlist = 1.0
      outlevel = mycas.outcommlevel
      outcommunity = mycas.outcomm
      outoverlap = mycas.outoverlap
      outcommLinks = mycas.outcommlinks
   ;
run;
```

The OUTCOMMLEVEL table presented in Figure 2.49 shows the final result for the community detection based on the label algorithm using a warm set of nodes to initiate the recursive partitioning. This execution identified 3 communities. The final modularity is around 0.5915.

The OUTNODES table presented in Figure 2.50 shows the distribution of the nodes in different communities based on a warm start. Nodes A, B, C, and J form the community 2, nodes D, E, L, and M form community 3 and nodes F, G, H, I, and K form community 1.

Notice that the warm process started with nodes D, L, E, and M. A previous execution of the label propagation algorithm with resolution 1.0 ended up with four communities, where nodes D and L formed one community and nodes E and M formed another one. Now, because of the warm start, only three communities were found, and nodes D, L, E, and M ended up in the same community.

Figure 2.51 shows the final result for the label propagation algorithm using the warm start for the community detection.

They could have ended up in different communities as the warm start only forces nodes to initiate the recursive partitioning, not finish at the same community.

The method to force a set of nodes to finish at the same community is applied by using the option FIX =. The following code creates the nodes dataset with the community identifier for each one of them. For example, nodes D and L are set to end up at the same community, and nodes F and E are set to end up at another one. The first set is defined by the value 1 for the variable fixcomm, and the second set is defined by the value 2. Many different sets can be defined in the nodes' dataset definition.

Figure 2.49 Output communities' level table for the community detection.

	⚠ node	⊕ community_1
« OUTNODES	Table rows: 13	Columns: 2 of 2	Rows 1 to 13						
1	A	2
2	B	2
3	C	2
4	D	3
5	E	3
6	F	1
7	G	1
8	H	1
9	I	1
10	J	2
11	K	1
12	L	3
13	M	3

Figure 2.50   Output nodes table for the community detection.

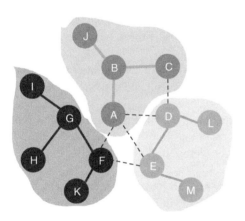

Figure 2.51   Communities identified by the label propagation algorithm with the warm start.

```
data mycas.nodesfix;
    input node $ fixcomm @@;
datalines;
D 1 L 1
F 2 E 2
;
run;
```

The following code invokes the community detection algorithm considering the nodes that are fixed to end up in the same communities. The set of nodes are set by using the option NODES = in the NETWORK statement. The final communities for the fixed nodes are specified by using the option FIX = in the COMMUNITY statement. The variable fixcomm specified in the nodes' dataset definition is referenced in the FIX = option.

```
proc network
    direction = undirected
    nodes = mycas.nodesfix
    links = mycas.links
    outlinks = mycas.outlinks
    outnodes = mycas.outnodes
    ;
    linksvar
        from = from
        to = to
    ;
```

```
nodesvar
    node = node
;
community
    fix=fixcomm
    algorithm=labelprop
    resolutionlist = 1.0
    recursive(maxcommsize=4 maxdiameter=3 relation=and)
    outlevel = mycas.outcommlevel
    outcommunity = mycas.outcomm
    outoverlap = mycas.outoverlap
;
run;
```

The OUTCOMMLEVEL table presented in Figure 2.52 shows the final result for the community detection based on the label algorithm using a fixed set of nodes ending up in the same communities. This execution identified 4 communities. The final modularity is around 0.5277.

The OUTNODES table presented in Figure 2.53 shows the distribution of the nodes throughout the communities. Based on the sets of fixed nodes, nodes A, C, B, and J form the community 1; nodes D and L form community 3; as defined in the

	level	resolution	communities	modularity
1	1	1	4	0.5277295918

Figure 2.52 Output communities' level table for the community detection.

	node	community_1
1	A	1
2	B	1
3	C	1
4	D	3
5	E	4
6	F	4
7	G	2
8	H	2
9	I	2
10	J	1
11	K	4
12	L	3
13	M	4

Figure 2.53 Output nodes table for the community detection.

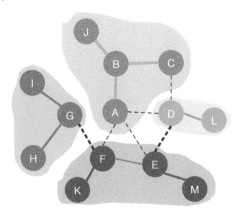

Figure 2.54 Communities identified by the label propagation algorithm with the fixed nodes.

fixed nodes option; nodes E, F, K, and M form community 4; having nodes F and E in the same community as defined in the fixed nodes option; and nodes G, H, and I form the final community 2.

Figure 2.54 shows the final result for the label propagation algorithm using the fixed nodes for the community detection process.

Notice that the regular execution with the label propagation algorithm ended up with nodes D and L forming a unique community, as defined for one set of fixed nodes. However, the node F ended up in a community with nodes I, G, H, and K, and node E ended up in a community with node M. As the nodes F and E were fixed to be in the same community and based on the sum of the link weights between the nodes, nodes F, E, K, and M ended up in one community, and nodes I, G, and H were set in a different one.

There are many methods and approaches to detecting communities, especially in large networks. Most of the time, the process is iterative, based on executions and post analyzes. As stated before, modularity is commonly used to select the best community topology among several executions and upon different resolutions. Recursive methods to limit the size of each community and the average distance between the nodes within the same community are also used in order to mitigate large partitions. A combination of algorithms like Louvain and label propagation are also possible to narrow down the selection process of the best community topology. Often, the Louvain algorithm can be executed on multiple resolutions, and the label propagation can be later executed using the best resolution and the recursive options. Also, even though the network is represented by a directed graph, in some cases the execution of Louvain or label propagation on an undirected graph can make sense and produce better results. The network metrics can still be computed based on the directed links even though the communities are identified based on undirected connections. Finally, business context and the domain knowledge can be used in the process to identify the best community topology according to the particular goal and the problem definition. There is no right or wrong in community detection. The best community distribution may vary according to multiple variables and different scenarios.

2.5 Core

A k-core decomposition is an undirected graph where every subgraph has nodes with degree at most k. In other words, some nodes in the subgraphs are linked to k or less other nodes throughout the subgraphs' links. A k-core of an undirected graph can also be seen as the maximal connected component of the graph G where all nodes have degree at least k.

K-core decomposition has many applications in multiple industries, like pharmaceutical companies, information technology, transportation, telecommunications, public health, etc., along various disciplines such as biology, chemistry, epidemiology, computer science, social studies, etc. It is commonly applied to visualize network structures, to interpret cooperative connections in social networks, to capture structural variety in social contagion, to analyze network hierarchies, among many others.

The core algorithm finds cohesive groups within a network by extracting k-cores. It can be seen as an alternative process for community detection. The core decomposition is not as powerful as the community detection algorithms in searching groups of nodes based on the strength of their relationships, but it can find a rough approximation of cohesive structures within the network with a low computational cost.

The core of a graph $G = (N, L)$ is defined as an induced graph on nodes S, where $G_S = (S, L_S)$.

The subgraph G_S is a k-core only if every node in S has a degree centrality greater than or equal to k, and G_S is the maximum subgraph with this property.

An important concept assigned to the k-core decomposition is that all the cores found in a network are nested. For example, if G_{S_k} is a k-core of size k, then $G_{S_{k+1}}$ is contained in G_{S_k}.

In proc network, the CORE statement invokes the k-core decomposition algorithm. This algorithm determines the core hierarchy considering a straightforward property. From a given graph $G = (N, L)$ it recursively deletes all nodes and links incident with them, of degree less than k. The remaining graph is the k-core.

The final results for the core decomposition algorithm in proc network are reported in the output nodes data table specified by the option OUTNODES =. The k-core decomposition algorithm can be executed based on undirected and directed graphs. For an undirected graph, each node in the output nodes data table is assigned to a core through the variable *core_out*, which identifies the core number, and the highest-order core that contains this node. For a directed graph, each node in the output nodes data table is assigned to a core-in and core-out through the variables *core_in* and *core_out*, respectively.

The only option for the CORE statement is the MAXTIME = option. This option specifies the maximum amount of time to spend calculating the core decomposition. The type of time (either CPU time or real time) is determined by the value of the TIMETYPE = option in the PROC NETWORK statement. The default is the largest number that can be represented by a double.

Let's see an example of k-core decomposition in both directed and undirected graphs.

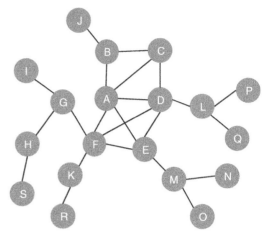

Figure 2.55 Input graph with undirected links.

2.5.1 Finding k-Cores

Let's consider the following graph to demonstrate the k-core decomposition algorithm. This graph is an extension of the graph used previously to demonstrate the community detection algorithm. It has 19 nodes and 23 links, slightly larger than the previous one. Also, the link weights in k-core decomposition are not necessary and then they were suppressed from the links dataset. Figure 2.55 shows the input undirected graph.

The following code creates the input dataset for the k-core decomposition algorithm. The links dataset contains only the origin and destination nodes.

```
data mycas.links;
    input from $ to $ @@;
datalines;
A B B C B J A F A E A D A C C D
F D F E G F G I G H K F D L D E
M E M N M O L P L Q K R H S
run;
```

Once the input dataset is created, we can now invoke the k-core decomposition algorithm using proc network. The following code describes how to do that. In the LINKSVAR statement, the variables *from* and *to* receive the origin and the destination nodes, even though the direction is not used at this first run. The final result for the k-core decomposition is reported in the output table defined by the OUTNODES = option.

```
proc network
    direction = undirected
    links = mycas.links
    outnodes = mycas.outnodes
    ;
    linksvar
        from = from
        to = to
    ;
    core
    ;
run;
```

The following pictures show the outputs for the k-core decomposition algorithm. Figure 2.56 reports the summary for the execution process. It provides the size of the network, with 19 nodes and 23 links, and the direction of the graph, the algorithm executed (the problem type) and the output tables.

The NETWORK Procedure

Problem Summary

Number of Nodes	19
Number of Links	23
Graph Direction	Undirected

The NETWORK Procedure

Solution Summary

Problem Type	Core Decomposition
Solution Status	OK
CPU Time	0.00
Real Time	0.00

Output CAS Tables

CAS Library	Name	Number of Rows	Number of Columns
CASUSER(Carlos.Pinheiro@sas.com)	OUTNODES	19	2

Figure 2.56 Output summary results for the k-core decomposition on an undirected graph.

The OUTNODES table presented in Figure 2.57 shows each node within the network and which core it belongs to.

Notice that the nodes in the same core don't necessarily need to have the same degree. As nested or hierarchical cores, nodes in the same k-core need to have a degree greater than or equal k. For example, based on the previous undirected input graph, it starts from the less cohesive cores to the most cohesive ones. Starting off with core 1, all nodes must have a degree greater than or equal 1. This is the case for nodes J, I, S, R, O, N, Q, and P with degree of 1, node K with degree of 2, and nodes G, M, and L with degree of 3. They all fall into core 1. Core 2 comprises nodes B and C with degree of 3. The question here is why nodes G, M, and L, which also have degree of 3 are not in the same core 2? The answer is because they are not connected. Core is a matter of interconnectivity, which accounts for the degree of connections but also the existing connections among nodes. Finally, core 3, the most cohesive one, has nodes A, D, and F with degree of 5 plus node E with degree of 4.

Figure 2.58 shows the nested cores highlighted.

Let's see how the k-core decomposition works on directed graphs. In addition to the interconnectivity, the k-core decomposition in directed graphs also takes into consideration the information flow created by the directed links.

Let's use the same graph defined previously. The direction of the links will follow the definition of the links data set. For example, the definition of the link A B means will create a directed link A→B.

The code to execute the k-core decomposition is exactly the same as used before, except by the DIRECTION = option, which will be specified as DIRECTED. The following code line shows the definition.

```
direction = directed
```

The method to find cores within directed graphs is basically a search for circuits, as the directed links create unique paths within the network.

Based on the current links set, the original input graph has one single core, as shown by the output table OUTNODES presented in Figure 2.59.

Notice that as a directed graph, two core identifiers are provided, the *core_out* and the *core_in* variables. Both identifiers in these examples are zero, representing that the entire network is a single core.

Let's change one single direction in the original input graph, switching the link A→D to the link D→A. Now, two cores are identified. There are two sets of nodes identified by the core out, and a different two sets of nodes identified by the core in. Figure 2.60 shows the output table OUTNODES.

>> OUTNODES   Table rows: 19   Columns: 2 of 2   Rows 1 to 19   ↑ ↑ ↓ ↓   ⋮

▽ Enter expression   ⌕

	△ node	⊞ core_out
1	A	3
2	B	2
3	E	3
4	F	3
5	D	3
6	G	1
7	H	1
8	M	1
9	L	1
10	Q	1
11	C	2
12	K	1
13	N	1
14	R	1
15	J	1
16	O	1
17	S	1
18	I	1
19	P	1

Figure 2.57   Output for the k-cores and the assigned nodes.

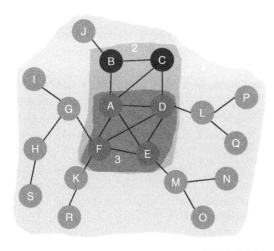

Figure 2.58   The undirected input graph highlighting the cores and the nodes for each core.

Based on the core out, nodes A, B, C, D, F, G, and K fall into core 1. Based on the direction of the links between them, a circuit comprising nodes A, B, C, and D is found. Nodes K, G, and F can reach out to this circuit. They all form the core out 1.

Figure 2.61 shows the two cores based on the core out identifier. The direction of the links is shown, and the link that has the directed switched is highlighted.

Based on the core in, there are also two cores, but the set of nodes in each one of them is different than the sets based on core out. Now, nodes A, B, C, D, E, F, J, L, P, and Q fall into the same core in. Similarly, there is a circuit between nodes A, B, C, and D, but now nodes F, E, J, L, P, and Q can be reached from that circuit. They all form the core in 1. Figure 2.62 highlights this core within the directed graph.

Notice that many algorithms in network analysis, particularly in terms of subgraph detection, are intended, or at least designed to be undirected graphs. That is the case for connected components, biconnected components, community detection, and k-core decomposition. However, there are algorithms available to compute the same type of subgraphs considering directed graphs. The main distinction between them is that a directed graph has unique paths among nodes, which changes the information flow throughout the network.

OUTNODES   Table rows: 19 | Columns: 3 of 3 | Rows 1 to 19 | ⬆ ↑ ↓ ⬇ | ⟳ ▾ | ⋮

▽ [Enter expression   🔍]

#	△ node	⊕ core_out	⊕ core_in
1	A	0	0
2	F	0	0
3	C	0	0
4	D	0	0
5	G	0	0
6	I	0	0
7	E	0	0
8	L	0	0
9	P	0	0
10	B	0	0
11	H	0	0
12	M	0	0
13	Q	0	0
14	K	0	0
15	N	0	0
16	R	0	0
17	J	0	0
18	O	0	0
19	S	0	0

Figure 2.59 Cores identified by the k-core decomposition algorithm.

2.6 Reach Network

Reach networks, also known as ego or egocentric networks, focus on the analysis of groups of nodes based on a particular set of nodes, or a single node. This method considers the individual node, or a set of nodes, rather than the entire structure of the network. The node starting the analysis is referred to as the ego, or in the case of a set of nodes, each node in the initial set will create an ego network. All connections of the ego nodes are captured and analyzed in order to identify the related nodes, and therefore, create its neighborhood, or its particular subnetwork. The analysis of the connections incident to the ego nodes is important to highlight how the subnetwork can affect the ego nodes, but mostly how the ego nodes can affect the rest of the subnetwork. Often, the ego nodes are selected based on their network centralities. The more connected, or more central nodes, are targeted as ego nodes and the reach network is computed. This approach normally highlights possible viral effects or cascade events throughout the network over time, which is especially useful in marketing, sales, and relationships campaigns, but also in fraud detection processes. The analysis concentrates on how many nodes the ego nodes can reach out, considering one single connection or degree of separation, or eventually more connections (second degree, third degree, and so forth). In reach networks, the focal or initial nodes are called ego, and the links incident to the egos are called alters.

Although the structure created by the ego network does not represent the entire network, egocentric data can be particularly important in network analysis. Particularly in large networks, this method is useful in understanding some subgraphs features or some specific social structures within the entire network. Very often, the analysis of the ego networks reveals possible influencers and how wide they can affect their social structures. The influencers, or the ego nodes, varies depending on the context or problem. They can be customers, companies, organizations, countries, etc. Therefore, their particular ego networks can represent a company's market or customer base, business relations, or international trading between countries or organizations, and financial transactions, among many others.

OUTNODES Table rows: 19 | Columns: 3 of 3 | Rows 1 to 19   ↑ ↑ ↓ ↓   ⟳ ▾   ⋮

▽ Enter expression   🔍

	⚠ node	⊞ core_out	⊞ core_in
1	A	1	1
2	B	1	1
3	E	0	1
4	F	1	1
5	D	1	1
6	G	1	0
7	H	0	0
8	M	0	0
9	L	0	1
10	Q	0	1
11	C	1	1
12	K	1	0
13	N	0	0
14	R	0	0
15	J	0	1
16	O	0	0
17	S	0	0
18	I	0	0
19	P	0	1

Figure 2.60 Cores out identified by the k-core decomposition algorithm.

In some specific applications such as financial risk, fraud events, etc., the ego networks can expose a possible viral effect of an illegal transaction or operation. Once an entity (company, group, individual) is identified as committing fraud, the ego network associated with that entity indicates how wide the illegal actions can be. Additional suspicious actors, especially in fraud, can be highlighted by the analysis of the ego networks. Particularly in fraud, in most cases, the business scenario consists of an entire structure behind the scenes to make it work. Beginning by looking at a small set of individuals or entities can be beneficial in narrowing down the fraud detection process exposing other actors that might be involved.

Formally, the reach network of a graph $G = (N, L)$ is defined as a subgraph $G_H^R = (N_H^R, L_H^R)$, induced by the set of nodes N_H^R that are reachable in H steps or hops from a set of nodes S called source nodes. Reach network can be executed in both undirected and directed graphs. Particularly for directed graphs, the set of nodes N_H^R that can be reached by the set of source nodes S is identified by using only the outgoing links. The induced subgraph created by the outgoing links is often called the out-reach network.

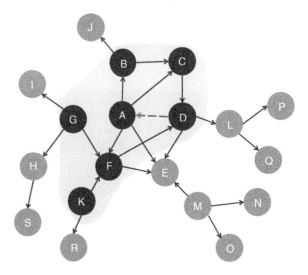

Figure 2.61 Cores out highlighted in the directed network.

In proc network, the REACH statement invokes the algorithm to identify the ego networks within the network starting from an initial set of ego nodes. There are two methods to define this initial set of ego nodes. Commonly, the set of ego nodes used to

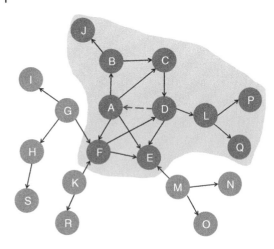

Figure 2.62 Cores in highlighted in the directed network.

compute the reach networks are defined in the data table consisting of a subset of nodes. The NODESSUBSET = option specifies that particular subset of ego nodes. This option to define a subset of nodes to begin an algorithm can be used to find the reach network, but also to the algorithms to find paths, shortest paths, and nodes similarity. The data table used to specify the initial set of ego nodes can define distinct sets of sources nodes. Each set of source nodes is used to find one specific reach network. For example, suppose we are interested in three distinct groups of nodes, and we want to evaluate the reach network of each one of these three sets. Then, in the data table used to define the ego nodes (the nodes subset) we can specify the nodes and the subset they belong to. The alternative method to compute the reach network is to specify all nodes. The option EACHSOURCE determines that proc network will search the reach network for all nodes within the network, or the nodes presented in the links dataset.

The number of steps from the ego node, or the source node, to the related node, or the sink node, can also be determined when proc network computes the reach network for each subset of ego nodes. For example, we can define that the reach network for a particular ego node will consider the straight connected nodes, or the straight connected nodes plus the nodes connected to them. The first case represents a reach of 1, and the second case represents a reach of 2. As we increase the maximum reach, the ego network produced by the algorithm can be quite large. The outcome from the reach network algorithm, especially considering a large network and a great amount of ego nodes can be computationally intensive.

The final results for the reach network algorithm in proc network are reported in the output tables defined by the options OUTREACHNODES =, OUTREACHLINKS =, and OUTCOUNTS =. These output tables may contain several columns, may vary according to the direction of the graph, and whether the nodes have weights defined in the dataset representing the nodes.

The OUTREACHNODES = option defines the output table containing the nodes of each reach network associated with all subsets of ego nodes specified in the NODESSUBSET = option. This output contains the following columns:

- Reach: the identifier of the reach network associated with each subset of source nodes.
- node: the label of the node for each node in each reach network.

The OUTREACHLINKS = option defines the output table containing the links of each reach network associated with all subsets of ego nodes specified in the NODESSUBSET = option. This output contains the following columns:

- reach: the identifier of the reach network associated with each subset of source nodes.
- from: the label of the *from* node for each link in each reach network.
- to: the label of the *to* node for each link in each reach network.

The OUTCOUNTS = option defines the output table containing the number of nodes of each reach network associated with all subsets of ego nodes specified in the NODESSUBSET = option. This output contains the following columns:

- reach: the identifier of the reach network associated with each subset of source nodes.
- node: the label of the node for each node in each reach network.
- count: the number of nodes in the (out-)reach network from the subset of source nodes.
- count_not: the number of nodes not in the (out-)reach network from the subset of source nodes.

As we noticed before, nodes and links can be weighted in order to represent their importance within the network. For the reach algorithm, if the nodes have weights, the output table specified by the option OUTCOUNTS = contains some additional columns:

- count_wt: the sum of the weights of the nodes counted in the (out-)reach network.
- count_not_wt: the sum of the weights of the nodes not counted in the (out-)reach network.

The reach network can be computed for both directed and undirected graphs. For a directed graph, proc network has an option to compute the direct counts when the number of steps from the source node and the sink node is limited to 1. In that case, the output table specified by the option OUTCOUNTS = contains some additional columns:

- count_in: the number of nodes that reach some node in the subset of source nodes using at most one step.
- count_out: the number of nodes in the out-reach network from the subset of source nodes.
- count_in_or_out: the number of nodes in the union of the subset of nodes counted in count_in and count_out.
- count_in_and_out: the number of nodes in the intersection of the subset of nodes counted in count_in and count_out.

In the case of the defined weight of the nodes, and the directed graph, the output table defined by the option OUTCOUNTS = contains more additional columns:

- count_in_wt: the sum of the weights of the nodes counted in count_in.
- count_out_wt: the sum of the weights of the nodes counted in count_out.
- count_in_or_out_wt: the sum of the weights of the nodes counted in count_in_or_out.
- count_in_and_out_wt: the sum of the weights of the nodes counted in count_in_and_out.

As mentioned before, if the network has directed links, it is possible to compute the number of directed counts for each reach network. The option DIGRAPH forces proc network to calculate the directed reach counts when computing the reach networks. The directed counts are reported in the output table specified in the option OUTCOUNTS =. The directed counts are calculated when the graph is directed, and the maximum reach is 1.

The option EACHSOURCE treats each node in the network as a source node (ego node), and then calculates the reach network from each node in the input graph. For a large network, this option can be computationally intensive. The output results when using this option in large networks can also be extremely large. This is the second method to define the source nodes when computing the reach networks. Often, a subset of nodes is defined, including the possibility of specifying distinct sets of source nodes, and then the option NODESSUBSET = is used.

The option MAXREACH = specifies the maximum number of steps from each source node in a reach network. For example, MAXREACH = 1 specifies that all nodes directed connected to a source node are counted in the reach network. MAXREACH = 2 add to the reach network all nodes linked to the nodes connected to that source node. The default for the option MAXREACH = is 1.

Let's see some examples of reach networks for both directed and undirected graphs.

2.6.1 Finding the Reach Network

Let's consider a graph similar to the one used to demonstrate the k-core decomposition, presented in Figure 2.63. This graph has 12 nodes and 12 links. Initially this graph will be considered an undirected graph, with no link weights.

The following code describes the creation of the links dataset and the subset of ego or source nodes. The reach algorithm will use each subset of source nodes to compute the different reach networks. The links dataset has only the nodes identification, specified by the from and to variables. There are two subsets of ego nodes. The first one has just node A, identified by the variable reach equals 1, and the second subset contains nodes D and G, identified by the variable reach equals 2.

```
data mycas.links;
    input from $ to $ @@;
datalines;
A B B J A G G I G H A D D C D L A F F K A E E J
;
run;

data mycas.egonodes;
    input node $ reach @@;
datalines;
A 1 G 2 D 2
;
run;
```

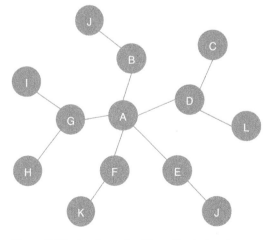

Figure 2.63 Input graph with undirected and unweighted links.

Once the input links dataset is defined, we can invoke the reach network algorithm in proc network. The following code describes the process. Notice the links definition using the LINKSVAR statement. Also notice the NODESSUBSET option to define the subsets of ego nodes to start off the reach network computation. All three possible output tables are specified, for links, nodes, and counts.

```
proc network
   direction = undirected
   links = mycas.links
   nodessubset = mycas.egonodes
   ;
   linksvar
      from = from
      to = to
   ;
   reach
      outreachlinks = mycas.outreachlinks
      outreachnodes = mycas.outreachnodes
      outcounts = mycas.outreachcounts
   ;
run;
```

Figure 2.64 shows the summary output for the reach network algorithm. It shows the size of the network, (12 nodes and 12 links), the direction of the graph, the algorithm used, the time to process, and the final output tables.

The OUTREACHNODES table presented in Figure 2.65 shows all reach networks and the nodes on them. The variable reach identifies the reach network, and the variable node identifies which node belongs to each reach network.

The NETWORK Procedure

Problem Summary

Number of Nodes	12
Number of Links	12
Graph Direction	Undirected

The NETWORK Procedure

Solution Summary

Problem Type	Reach
Solution Status	OK
CPU Time	0.00
Real Time	0.01

Output CAS Tables

CAS Library	Name	Number of Rows	Number of Columns
CASUSER(Carlos.Pinheiro@sas.com)	OUTREACHLINKS	11	3
CASUSER(Carlos.Pinheiro@sas.com)	OUTREACHNODES	13	2
CASUSER(Carlos.Pinheiro@sas.com)	OUTREACHCOUNTS	3	4

Figure 2.64 Output results by proc network running the reach network algorithm on an undirected graph.

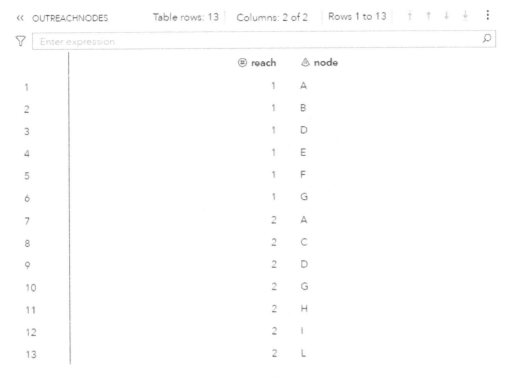

Figure 2.65 Output nodes table for the reach network.

	reach	node
1	1	A
2	1	B
3	1	D
4	1	E
5	1	F
6	1	G
7	2	A
8	2	C
9	2	D
10	2	G
11	2	H
12	2	I
13	2	L

Figure 2.66 Output links table for the reach network.

	reach	from	to
1	1	A	D
2	1	A	B
3	1	A	F
4	1	A	G
5	1	A	E
6	2	A	G
7	2	A	D
8	2	D	L
9	2	D	C
10	2	G	I
11	2	G	H

Notice that there are two subsets of nodes. The first one comprises only the node A. Its reach network consists of node A, of course, plus nodes B, D, E, F, and G. The second subset of ego nodes comprises nodes D and G. The second reach network consists of nodes D and G, plus nodes A, I, H, C, and L.

The OUTREACHLINKS table presented in Figure 2.66 shows the links within each reach network identified.

Finally, the OUTCOUNTS table presented in Figure 2.67 shows the number of nodes in each reach network by each ego node. It shows the number of nodes within the reach network and also the number of nodes outside the ego network.

	⊞ reach	⚠ node	⊞ count	⊞ count_not
1	1	A	6	6
2	2	D	7	5
3	2	G	7	5

« OUTREACHCOUNTS   Table rows: 3 | Columns: 4 of 4 | Rows 1 to 3

Figure 2.67 Output links table for the reach network.

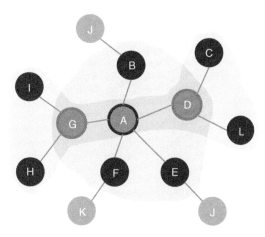

Figure 2.68 Input graph with the reach networks highlighted.

For example, the reach network 1, based on the ego node A, has 6 nodes within it and 7 nodes outside of it. The reach network 2, based on the ego nodes D and G, has 7 nodes within it and 5 nodes outside of it.

Figure 2.68 shows a graphical representation of the two reach networks.

The MAXREACH = option plays an important role when computing the reach network. Depending on the level of connectivity of the network, a high maximum reach can represent a large ego network. For example, using this small network, let's increase the maximum number of steps of hops from 1 to 2. Let's use just node A as the source node. The subset node would contain only node A.

```
data mycas.egonodes;
   input node $ reach @@;
datalines;
A 1
;
run;
```

Also, the option MAXREACH = needs to be set within the proc network.

```
proc network
...
   reach
      maxreach = 2
...
run;
```

Because of the small size and of the high connectivity network, the ego network of node A considering 2 hops is the entire network. The OUTREACHCOUNTS table presented in Figure 2.69 shows the count for the reach network assigned to node A. All 12 nodes within the network were counted.

The use of the EACHSOURCE option forces proc network to compute the reach network for all nodes in the network. For large networks, the time to compute all reach networks as well as the size of the output can be extremely large. This option must be used with caution. The following code shows only the part of the REACH statement including the EACHSOURCE option.

```
proc network
...
   reach
      eachsource
...
run;
```

The OUTREACHCOUNTS table presented in Figure 2.70 shows the counts for all reach networks. Notice that all 12 nodes in the original network are represented there.

« OUTREACHCOUNTS		Table rows: 1	Columns: 4 of 4	Rows 1 to 1	↑ ↑ ↓ ↓ ⋮

	⊞ reach	△ node	⊞ count	⊞ count_not
1	1	A	12	0

Figure 2.69 Counts table for the node A using maximum reach equals 2.

« OUTREACHCOUNTS		Table rows: 12	Columns: 4 of 4	Rows 1 to 12	↑ ↑ ↓ ↓ ⋮

	⊞ reach	△ node	⊞ count	⊞ count_not
1	1	A	6	6
2	2	B	3	9
3	3	C	2	10
4	4	D	4	8
5	5	E	3	9
6	6	F	3	9
7	7	G	4	8
8	8	H	2	10
9	9	I	2	10
10	10	J	3	9
11	11	K	2	10
12	12	L	2	10

Figure 2.70 Counts for all reach networks considering all nodes as source.

Finally, the direction of the link plays a crucial role when computing reach networks. The paths created by the directed links are considered to assess when a sink node can be reached out by a source node. For example, using the same network, let's just invert the definition of two links, from the original D→C and D→L, to C→D and L→D. Let's also use only node D as the ego node. See the following parts of the code to proceed with these changes.

```
data mycas.links;
   input from $ to $ @@;
datalines;
... C D L D ...
;
run;

data mycas.egonodes;
   input node $ reach @@;
datalines;
D 1
;
run;

proc network
   direction = directed
...
run;
```

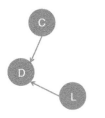

Figure 2.71 Directed links for node D.

Figure 2.71 shows the portion of the network with the two directed links.

The OUTREACHCOUNTS table presented in Figure 2.72 shows the reach network for node D. Notice that there is only 1 count for it, which is node D itself. Nodes C and L, previously counted in its ego network on the undirected graph are not counted now because node D cannot reach out to nodes C and L.

The additional columns in the outputs when the links and nodes are weighted are suppressed here because they actually don't change the search for reach network. They do change the outcomes by adding all weight information in the output tables.

2.7 Network Projection

Bipartite graphs are graphs with two distinct partitions of nodes. The nodes assignment is mutually exclusive, which means nodes must be assigned to one partition or the other. In unipartite graph, all nodes and links within the network belong to the same set or partition. This single set or partition contains unique nodes and links. Neither nodes nor links can be duplicated in the network. In bipartite graphs, links can only occur between nodes of different partitions. That means, nodes within the same partition are not connected to each other. They can be connected only to nodes in the other partition. There are multiple applications based on bipartite graphs. For example, every time a set of agents or workers must be assigned to tasks, a bipartite graph can be used to represent that problem or scenario. Workers are not connected to each other. Tasks are not connected to each other. The only connections that can occur are between workers and tasks. Connections are from two distinct sets of nodes (agents and tasks), or between two distinct partitions of the graph. This scenario can be extended to many other business situations, like customers consuming services, users watching movies, criminals committing crimes, students enrolling to courses, and so forth. When two different types of nodes coexist in a network, where the same type of nodes do not connect to each other, a bipartite graph can be used to represent the problem.

The main idea of the network projection is then to reduce a bipartite graph with two distinct sets of nodes into a single network with one projected set of nodes. The projected set of nodes are connected by links when they share a common neighbor in the original bipartite network. Once the network is projected, all algorithms designed to be used in unipartite networks can be normally applied.

As an example, suppose there is a set of students and a list of courses. Imagine that students are not connected to each other, nor are the courses connected to each other. Students are connected to courses. This is a common example of a bipartite network, with two distinct sets of nodes, each set of one particular type, and the nodes within each set are not connected, only nodes between the two distinct sets. Let's also assume that the students will represent the projected nodes. Notice that the set of courses can also be used as the projected node set. Considering the students as the projected node set, students will be connected in the network projection only if they share the same course. That means every time any pair of students is connected to the same course in the bipartite network, they will be connected in the unipartite network, or in the projected network. If the same pair of students is enrolled in multiple courses together, the number of courses can be used to define the weight of the link in the projected network. Thinking on the course side is also possible. Courses will be connected in the projected network if they share the same students in the bipartite network. The number of same students enrolled in a pair of courses will determine the link weight between the courses in the projected network. Courses sharing multiple students will be strongly connected in the projected network.

‹‹ OUTREACHCOUNTS	Table rows: 1	Columns: 4 of 4	Rows 1 to 1	↑ ↑ ↓ ↓ ⋮
	⊞ reach	⚠ node	⊞ count	⊞ count_not
1	1	D	1	11

Figure 2.72 Counts for the reach network assigned to node D.

The network projection algorithm can also be seen as a process to reduce multiple partition graphs to a single partition graph. If a network is projected based on a subset of nodes U through another subset of nodes V, the unipartite network produced by this projection will contain only the nodes from the subset U. The links connecting the nodes in the projected network, the one based on the subset of nodes U, are created if these nodes share a common neighbor in the bipartite network, or they are connected to the same nodes in the subset V. The network projection algorithm assumes that the graph that is induced by the union of the two subsets of nodes $U \cup V$ in the original network is bipartite. Therefore, any possible existing link between the nodes in U or any possible links between the nodes in V, are suppressed.

In proc network, the PROJECTION statement invokes the algorithm to compute the network projection. The algorithm computes the network projection based on nodes specified in the nodes input table. The nodes dataset is defined in proc network by the option NODES =. The nodes dataset contains the subsets of nodes for each partition by specifying the nodes identifiers and the partition flag. In the PROJECTION statement, the option PARTITION = is used to define the partition of each node. The value assigned to that option must be 0, 1, or missing. In order to create a network projection on the nodes set U through the nodes set V, in the nodes dataset the partition flag must be 1 to all nodes from the set U and 0 to all nodes from the set V. Nodes with missing values for the partition flag in the nodes dataset will be ignored, as well as the nodes from the original bipartite network that are not specified in the dataset.

The strength of the connections between the nodes in the projected network can be determined by the number of neighbors they have in common in the bipartite network. Proc network uses different similarity measures to quantify the link weights in the projected network. The current similarity measures are Adamic-Adar, common neighbors, cosine, and Jaccard.

Network projection can be computed in both undirected and directed graphs. When the graph is based on undirected links, the evaluation of neighbors is straightforward. If nodes a and b from set U are connected to node i from subset V, then nodes a and b will be connected in the projected network. However, if the graph is based on directed links, there are different methods to manage the direction of the links and the concept of neighborhood.

The final results for the network projection algorithm in proc network are reported in the output tables defined by the options OUTPROJECTIONLINKS =, OUTPROJECTIONNODES =, and OUTNEIGHBORSLIST =.

The OUTPROJECTIONLINKS = option defines the output data table containing the links of the projected graph. Additionally, this table can also report the similarity measures computed between the pairs of nodes that are represented by a link in the projected graph. These similarity measures quantify how the pairs of nodes in the subset with partition flag 1 are connected through the nodes in the subset with partition flag 0. This output table contains the following variables:

- from: the from node label of the link.
- to: the to node label of the link.
- adamicAdar: the Adamic-Adar similarity score between the from and to nodes.
- commonNeighbors: the common neighbors similarity score between the from and to nodes.
- cosine: the cosine similarity score between the from and to nodes.
- jaccard: the Jaccard similarity score between the from and to nodes.

The OUTPROJECTIONNODES = option defines the output table containing the nodes of the projected graph. This output contains the following column:

- node: the node label.

The OUTNEIGHBORSLIST = option defines the output table containing the list of neighbors that are common to each pair of nodes that are joined by a link in the projected graph. Each link listed in the output table defined by the option OUTPROJECTIONLINKS = creates one or multiple records in that neighbor list table. Each neighbor that is common to both *from* and *to* in the links table creates a record neighbor list table. This output contains the following columns:

- from: the from node label.
- to: the to node label.
- neighbor_id: the sequential index of the neighbor.
- neighbor: the node label of the common neighbor of the from and to nodes.

The network projection algorithm in proc network contains some additional options to determine what is to be considered when computing the projected network from a bipartite graph.

The option ADAMICADAR = specifies whether to calculate the Adamic-Adar node similarity measure when quantifying the number of neighbors a pair of nodes have in common. The values for this option are TRUE or FALSE. If true, this node

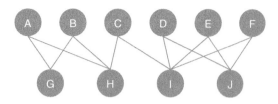

Figure 2.73 Input bipartite graph.

similarity measure is reported in the output table specified by the OUTPROJECTIONLINKS = option. If false, proc network does not calculate Adamic-Adar node similarity. The default value is false.

The option COMMONNEIGHBORS = specifies whether to calculate common neighbors node similarity when quantifying the number of neighbors nodes have in common. The options are TRUE or FALSE. If true, proc network calculates the common neighbors node similarity and saves the results in the output table specified by the option OUTPROJECTIONLINKS =. If false, proc network does not calculate common neighbors node similarity. The default value is false.

The option COSINE = specifies whether to calculate cosine node similarity to quantify the common neighbors for the pairs of nodes. The options are TRUE or FALSE. If true, proc network calculates the cosine node similarity and reports the results in the output table specified by the option OUTPROJECTIONLINKS =. If false, proc network does not calculate cosine node similarity, and the default value is false.

Network projection can be computed based on both undirected and directed graphs. On directed graphs, due to the direction of the links and the possible paths of connections throughout the network, different methods may be used to handle the existing relations. The option DIRECTEDMETHOD = specifies the method to be used to compute the network projection when the graph is based on directed links. This option is used only for directed graphs when the option DIRECTION = is specified as DIRECTED. For directed graphs, proc network provides three different methods to handle the directed links. The method CONVERGING considers only the out-neighbors of nodes in U, which means only the nodes in V that are connected by outgoing links. The method DIVERGING considers only in-neighbors of nodes in U, which means only nodes in V that are connected by ingoing links. Finally, the method TRANSITIVE considers only paths of length two, which means paths that start at some node in U, traverse through some node in U, and end at some node in U. The default method used in proc network is CONVERGING.

The option JACCARD = specifies whether to calculate the Jaccard node similarity. The possible values are TRUE or FALSE. If true, proc network calculates the Jaccard node similarity and reports the result in the output table specified in the OUTPROJECTIONLINKS = option. if false, node similarity is not calculated, and the default value is false.

The option PARTITION = specifies the name of the variable for the partition flag. The partition variable is defined in the input nodes dataset. The value can be 1 or 0 (or missing). All nodes with the partition variable equal to 1 define the set of nodes that the input graph will be projected on. The links in the projected graph are created based on the nodes that are connected through common neighbors among the set of nodes that have the partition flag of 0. Nodes with the partition flag missing are excluded from the calculation.

Let's see some examples of network projection for both directed and undirected graphs.

2.7.1 Finding the Network Projection

Figure 2.73 shows a bipartite graph containing 10 nodes and 12 links. Nodes A, B, C, D, E, and F do not connect to each other. They represent one partition of the bipartite graph. Similarly, nodes G, H, I, and J do not connect to each other and therefore represent another partition in this bipartite graph. Nodes from one partition (A,B,C,D,E,F) do connect to nodes to the other partition (G,H,I,J).

The following code describes how to create the links dataset. For the network projection, the nodes partitions must also be provided. The nodes dataset with the partition flag is also created. The projected network is based on the subset of nodes A, B, C, D, E, and F. This subset of nodes is set in the nodes dataset with the partition id 1. The nodes projection will be through the common neighbors in the subset of nodes G, H, I, and J. This subset is flagged with partition id 0.

```
data mycas.links;
   input from $ to $ @@;
datalines;
A G A H B G B H C H C I D I D J E I E J F I F J
;
run;
```

```
data mycas.partitionnodes;
   input node $ partitionid @@;
datalines;
A 1 B 1 C 1 D 1 E 1 F 1 G 0 H 0 I 0 J 0
;
run;
```

Once the input links and nodes are defined, proc network can be invoked to compute the network projection on this bipartite input graph. The following code describes how to do that. For this particular execution, all methods to compute the nodes similarities are used, Adamic-Adar, common neighbors, Cosine, and Jaccard.

```
proc network
   direction = undirected
   links = mycas.links
   nodes = mycas.partitionnodes
   ;
   linksvar
      from = from
      to = to
   ;
   projection
      partition = partitionid
      outprojectionlinks = mycas.outprojectionlinks
      outprojectionnodes = mycas.outprojectionnodes
      outneighborslist = mycas.outneighborslist
      adamicadar = true
      commonneighbors = true
      cosine = true
      jaccard = true
   ;
run;
```

Figure 2.74 shows the summary output for the network projection algorithm. It shows the size of the network, 10 nodes and 12 links, the direction of the graph, the algorithm used, the time to process, and the final output tables.

The OUTPROJECTIONNODES table presented in Figure 2.75 shows all nodes projected in the network projection. These nodes belong to one partition and are projected based on their common neighbors within the other partition in the bipartite input graph.

The OUTPROJECTIONLINKS table presented in Figure 2.76 shows the links connecting the projected nodes. The nodes similarity scores for all four methods are reported in this output table. Notice the common neighbors measure. It shows how many common neighbors a particular link has. For example, nodes A and B are connected in the projected network because they have 2 common neighbors, nodes G and H in the other partition (A↔G/A↔H and B↔G/B↔H).

The OUTNEIGHBORSLIST table presented in Figure 2.77 shows the common neighbors between the projected nodes based on each link. For example, the link between nodes A and B are created based on the common neighbors G and H (lines 1 and 2 in the output table).

Figure 2.78 shows the final projected network considering only the projected nodes and their links based on the common neighbors.

Let's see how the network projection works on a directed graph, and also change the set of nodes to be projected. Based on the links input data, let's invert the directions of some links to evaluate the final results of the network projection.

The links G→B, H→B and I→E had their original directions inverted (instead of B–G, B–H, and E–I). The subset of nodes to be projected now considers only the nodes A, B, E, and F (nodes C and D were suppressed). The following code describes these changes.

The NETWORK Procedure

Problem Summary

Number of Nodes	10
Number of Links	12
Graph Direction	Directed

The NETWORK Procedure

Solution Summary

Problem Type	Projection
Solution Status	OK
CPU Time	0.03
Real Time	0.01

Output CAS Tables

CAS Library	Name	Number of Rows	Number of Columns
CASUSER(Carlos.Pinheiro@sas.com)	OUTPROJECTIONLINKS	9	6
CASUSER(Carlos.Pinheiro@sas.com)	OUTPROJECTIONNODES	6	1
CASUSER(Carlos.Pinheiro@sas.com)	OUTNEIGHBORSLIST	13	4

Figure 2.74 Output results by proc network running the network projection algorithm on an undirected graph.

« OUTPROJECTIONNODES	Table rows: 6	Columns: 1 of 1	Rows 1 to 6	⊤ ↑ ↓ ⊥ ⋮

∇	Enter expression	🔍

	⚠ node
1	A
2	E
3	B
4	C
5	D
6	F

Figure 2.75 Output nodes for the projected network.

```
data mycas.links;
   input from $ to $ @@;
datalines;
A G A H G B H B C H C I D I D J I E E J F I F J
;
run;

data mycas.partitionnodes;
   input node $ partitionid @@;
datalines;
A 1 B 1 E 1 F 1 G 0 H 0 I 0 J 0;
run;
```

<< OUTPROJECTIONLINKS Table rows: 9 Columns: 6 of 6 Rows 1 to 9

Enter expression

	from	to	adamicAdar	commonNeighbors	cosine	jaccard
1	A	B	5.4178313692	2	1	1
2	A	C	2.0959032743	1	0.5	0.3333333333
3	E	F	3.7568673217	2	1	1
4	B	C	2.0959032743	1	0.5	0.3333333333
5	C	D	1.6609640474	1	0.5	0.3333333333
6	C	E	1.6609640474	1	0.5	0.3333333333
7	C	F	1.6609640474	1	0.5	0.3333333333
8	D	E	3.7568673217	2	1	1
9	D	F	3.7568673217	2	1	1

Figure 2.76 Output links for the projected network.

<< OUTNEIGHBORSLIST Table rows: 13 Columns: 4 of 4 Rows 1 to 13

Enter expression

	from	to	neighbor_id	neighbor
1	A	B	1	G
2	A	B	2	H
3	A	C	1	H
4	E	F	1	I
5	E	F	2	J
6	B	C	1	H
7	C	D	1	I
8	C	E	1	I
9	C	F	1	I
10	D	E	1	I
11	D	E	2	J
12	D	F	1	I
13	D	F	2	J

Figure 2.77 Output common neighbors for the projected network.

The new bipartite input graph is presented in Figure 2.79. Nodes C and D, and their respective links are faded out.

One single but important change is made in the code to invoke the network projection algorithm. The direction now is set to directed, and the common neighbors node similarity is calculated. The following code shows these changes.

```
proc network
   direction = directed
...
   projection
      partition = partitioned
      directedmethod = converging
...
      commonneighbors = true
   ;
run;
```

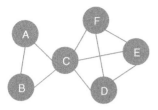

Figure 2.78 Projected network.

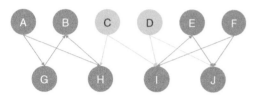

Figure 2.79 Input bipartite graph.

The OUTPROJECTIONNODES table presented in Figure 2.80 shows only the four nodes projected in partition 1, nodes A, B, E, and F.

The OUTPROJECTIONLINKS table presented in Figure 2.81 shows the single link connecting the projected nodes E and F. They have one single common node (node J in partition 0).

The default method for directed graphs is converging, which considers only the outgoing links from the projected nodes to the common neighbors. Notice that node A points to node G, but node G points to node B. Then, node G is not a common neighbor between nodes A and B based on a directed graph. The same happens for nodes E and F with the former common neighbor node I. Now, node I points to node E and node F points to node I. Node I is no longer a common neighbor between nodes E and F in that configuration. The only projection that remains valid is between nodes E and F through node J. Nodes E and F point to node J, and then, node J is still a common neighbor between those two projected nodes.

The OUTNEIGHBORSLIST table presented in Figure 2.82 shows the only common neighbor (node J) between the projected nodes E and F.

Figure 2.83 shows the final projected network considering the projected nodes and their links based on the common neighbors.

This project network based on the directed graph has only one pair of nodes with a single link between them, as opposed to the previous projected network on the undirected graph with six nodes and nine links.

Figure 2.80 Output nodes for the projected network.

Figure 2.81 Output links for the projected network.

« OUTNEIGHBORSLIST   Table rows: 1 | Columns: 4 of 4 | Rows 1 to 1

	from	to	neighbor_id	neighbor
1	E	F	1	J

Figure 2.82 Output common neighbors for the projected network.

2.8 Node Similarity

Network analysis is a fundamental tool to evaluate relationships and uncover hidden information about entities. Particularly in social networks, the analysis of the nodes' rela-

Figure 2.83 Projected network.

tionships can reveal useful insights about the nodes' characteristics or attributes. That means in network analysis, analyzing the links can disclosure relevant information about the nodes. In other words, analyzing the relations between nodes can describe the roles or functions of the nodes. One important concept about the nodes based on their links is the node similarity. Node similarity has many applications in business, science, and social studies, among others. Companies can use node similarity to make better recommendations upon customers' relationships with products and services, or even between them. Government agencies can use node similarity to predict connections between similar nodes and possibly identify terrorists' relationships. Functions and roles within networks or subnetworks can be better explained by nodes similarities. In medical research, node similarity can be used to study diseases and treatments. In the social domain, node similarity can be used to identify unusual behavior or behavioral changes within communities or throughout different groups. Most of the techniques in node similarity consider the neighborhood of the nodes to identify or quantify a measure of similarity. For example, nodes with the same neighbors would be considered similar. These nodes would have similar links to other nodes. The labels or roles within the network are usually propagated through the links, and then those nodes would be similar because they would have similar labels, roles, or functions with the network.

In graph theory, node similarity measures the similarity of nodes neighborhoods. There are several methods to measure this similarity between nodes. Different algorithms evaluate and compute the similarity between two nodes in particular ways. The most common method is based on the immediate neighborhood of two nodes. That means two nodes are similar if they have the same neighbors. Other methods are based on the structural role proximity of the network around two particular nodes. Instead of searching for the same neighbors between two nodes, these methods search for the similar network structure assigned to these two nodes. The network structure is commonly represented by the first and second order proximity (higher orders of proximity can also be accounted for). Based on these methods, two nodes located far away from each other within the network can still be similar if they have the same structural role proximity. For example, both nodes are center of a star structure, connecting the same number of nodes in first and second order proximity. Even though they are far away apart, they are similar in roles within the network, and then they are considered similar. These methods open up a wide range of possibilities when searching for similar nodes in multiple industries or disciplines in science. For instance, consider a network representing global trading between countries. Two countries may be similar if they trade with the same countries (share the same neighbors). However, two countries far away from each other with no shared neighbors, can also be similar if they trade with other countries and have the same structural role proximity (same number of first and second order proximity creating a similar network structure).

Figure 2.84 shows two similar nodes A and B based on their same set of neighbors.

Figure 2.85 shows two similar nodes A and T based on their same structure role proximity, even though they might be far away from each other in the network.

In proc network, the NODESIMILARITY statement invokes the algorithm to quantify the node similarity within the network. There are six algorithms to quantify the node similarity. Four of them use the concept of same neighbors to evaluate how similar two nodes are. Two others use the concept of the structural role proximity to quantify how similar two nodes are. The four algorithms to quantify the node similarity based on the nodes' neighbors are Common Neighbors similarity, Jaccard similarity, Cosine similarity, and Adamic-Adar similarity.

A straightforward method to quantify node similarity within a network is to assess the number of neighbors that two nodes have in common. In other words, two nodes are similar based on a proportion of the size of their neighborhoods that overlapped.

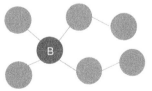

Figure 2.84 Similar nodes based on their neighborhood.

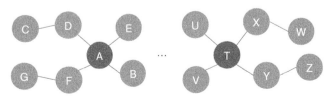

Figure 2.85 Similar nodes based on their structural role proximity.

For example, let's assume an undirected graph $\hat{N}_i$ that represents the set of neighbors of node i, excluding the node i itself. That means unique nodes that are connected by links to node i. In a directed graph, there are two sets of neighbors of node i, considering the outgoing links and the incoming links incident to node i. The first set $\hat{N}_i^{out}$ represents the set of out-neighbors of node i, excluding node i itself. In that case, it considers all unique nodes that are connected to node i by outgoing links. The second set $\hat{N}_i^{in}$ represents the set of in-neighbors, excluding node i itself. That set considers all unique nodes that are connected to node i by incoming links. The common neighbors measure can be calculated for both undirected and directed graphs. For undirected graphs, the measure to quantify the similarity between nodes i and j based on their neighborhood is defined by the following equation:

$$S_{ij}^{CN} = |\hat{N}_i \cap \hat{N}_j|$$

For directed graphs, as the similarity considers the outgoing and incoming links to nodes i and j differently, the measure is calculated by the following equation:

$$S_{ij}^{CN} = |\hat{N}_i \cap \hat{N}_j|$$

The measure based on common neighbors favors nodes with multiple connections. Then, nodes with high degree centrality are most likely to have large similarity scores, even when they have a small number of common neighbors connected to them.

The Jaccard similarity aims to correct the bias created by the high degree centrality. This measure is similar to the common neighbors similarity, but it divides the common neighbors measure by the number of nodes within the two nodes i and j neighbors sets, which means the union of the two sets of neighborhoods. For undirected graphs, the Jaccard similarity measure is defined by the following equation:

$$S_{ij}^{J} = \frac{|\hat{N}_i \cap \hat{N}_j|}{|\hat{N}_i \cup \hat{N}_j|}$$

For directed graphs, the Jaccard similarity considers the out-neighbors and the in-neighbors sets created by the outgoing and incoming links incident to nodes i and j. The measure is calculated by the following equation:

$$S_{ij}^{J} = \frac{|\hat{N}_i^{out} \cap \hat{N}_j^{out}|}{|\hat{N}_i^{out} \cup \hat{N}_j^{out}|}$$

If nodes i and j are the same, then the Jaccard similarity measure equals to 1. If the nodes i and j are different but the union of their neighborhoods is empty, the Jaccard similarity measure equals to 0.

The Cosine similarity is another method used to reduce the bias created by the high degree centrality on the common neighbors. The Cosine similarity uses the cosine of the angle between the two vectors that represent the sets of neighbors for nodes i and j. For undirected graphs, the Cosine similarity measure is defined by the following equation:

$$S_{ij}^{C} = \frac{\sum_{k \in \hat{N}_i \cap \hat{N}_j} \overline{w}_{ik} \overline{w}_{jk}}{\sqrt{\sum_{k \in \hat{N}_i} \overline{w}_{ik}^2} \sqrt{\sum_{k \in \hat{N}_j} \overline{w}_{jk}^2}}$$

In the Cosine similarity measure, $\overline{w}_{ik}$ and $\overline{w}_{jk}$ represent the parallel link weights between the nodes i and k, and j and k, respectively. This parallel link weight can be formulated by the following equation:

$$\overline{w}_{ij} = \sum_{ij \in \delta_i \cap \delta_j} w_{ij}$$

where $\overline{w}_{ij}$ is the sum of the parallel link weights between the nodes i and j.

For directed graphs, the Cosine similarity considers the out-neighbors and in-neighbors separately. The measure is calculated by the following equation:

$$S_{ij}^{C} = \frac{\sum_{k \in \hat{N}_i^{out} \cap \hat{N}_j^{out}} \overline{w}_{ik} \overline{w}_{jk}}{\sqrt{\sum_{k \in \hat{N}_i^{out}} \overline{w}_{ik}^2} \sqrt{\sum_{k \in \hat{N}_j^{out}} \overline{w}_{jk}^2}}$$

The parallel link weights in directed graphs consider the outgoing and incoming links between the nodes. The sum of parallel link weights in directed graphs is defined by the following equation:

$$\overline{w}_{ij} = \sum_{ij \in \delta_i^{out} \cap \delta_j^{in}} w_{ij}$$

If a graph is defined with no link weights, w_{ij} is set by convention to 1. If nodes i and j are the same, then the Cosine similarity measure equals to 1. If the nodes i and j are different but either of the two neighborhoods sets is empty, the Cosine similarity equals to 0.

The Adamic-Adar similarity measure is a third way to correct the bias created by the high degree centrality in the common neighbors similarity measure. The Adamic-Adar similarity aims to discount the contribution of common neighbors that have high degree centrality. The Adamic-Adar is quite different than the similarity measures Jaccard and Cosine. Both Jaccard and Cosine similarities are relative measures assigning a score to a pair of nodes i and j. This score is relative to the proportion of the neighborhoods sets of i and j that overlap. Adamic-Adar similarity on the other hand is an absolute measure. It assigns to each common neighbor of nodes i and j an inverse log weight. By doing this, Adamic-Adar similarity measure discounts connections that are very weak. For example, two nodes i and j have mutual connections to a particular node k, which, for example, is a hub in the network with a very high degree centrality. The Adamic-Adar similarity measure will discount the importance of this common neighbor for nodes i and j because of its high degree centrality. For undirected graphs, the Adamic-Adar similarity measure is defined by the following equation:

$$S_{ij}^{AA} = \sum_{k \in \hat{N}_i \cap \hat{N}_j} \frac{1}{\log |\hat{N}_k|}$$

For directed graphs, the Adamic-Adar similarity considers the out-neighbors for each pair of nodes and the in-neighbors for the common neighbor nodes. The measure is calculated by the following equation:

$$S_{ij}^{AA} = \sum_{k \in \hat{N}_i^{out} \cap \hat{N}_j^{out}} \frac{1}{\log |\hat{N}_k^{in}|}$$

The previous four similarity measures consider the common neighbors that a pair of nodes have. Another method to compare nodes within the network is by evaluating their structural roles in the graph, particularly how they are connected to other nodes by first and second orders of proximity. This method commonly embeds the nodes within a graph by creating a multidimensional vector space that preserves the original network structure. These multidimensional vectors aim to capture the first and second order proximity between nodes. The proximity between nodes corresponds to the proximity between the vector space embeddings. The first order proximity is the local pairwise proximity between two connected nodes. It can be represented w_{ij}, which is the sum of the link weights between nodes i and j. The first order proximity captures the direct neighbor relationships between nodes. The second order proximity captures the two-step relations between each pair of nodes. For each pair of nodes, the second order proximity is determined by the number of common neighbors shared by the two nodes, which can be measured by the two-step transition probability from node i to node j equivalently. The second order proximity between a pair of nodes i and i is the similarity between their neighborhood structures. Ultimately, the second order proximity, or even higher order, captures the similarity of indirectly connected nodes with similar structural context.

The first order proximity of node i to all other nodes in the graph can be represented by the embedded vector $p_i = (w_{i1}, ..., w_{i|n|})$. The second order proximity between nodes i and j is determined by the similarity between their embedded vectors p_i and p_j. The second-order proximity assumes that the nodes that share many connections to other nodes are similar to each other in their structural roles. This method calculates a multidimensional vector u_i for each node i by minimizing a corresponding objective function. The objective function for the first order proximity is defined by the following equation:

$$O_1 = -\sum_{ij \in L} w_{ij} \log p_1(from(ij), to(ij))$$

where *from(ij)* is the from (source) node of the link *ij*, *to(ij)* is the to (sink) node of link *ij*, and p_1 is the joint probability of a link between nodes *i* and *j*. The join probability between two nodes can be calculated by using the following formula:

$$p_1(i,j) = \frac{1}{1 + \exp(-u_i.u_j)}$$

For the second order proximity, the multidimensional vector u_i and the multidimensional context embeddings v_i are calculated by minimizing the objective function defined by the following equation:

$$O_2 = -\sum_{ij \in E} w_{ij} \log p_2(to(ij) \mid from(ij))$$

Where the multidimensional context $p_2(j \mid i)$ can be calculated by the following formula:

$$p_2(j \mid i) = \frac{\exp(v_j.u_i)}{\sum_{k \in N} \exp(v_k.u_i)}$$

The optimization method used to minimize the objective function is based on the stochastic gradient descent algorithm. This algorithm uses a random sampling of the links from the input link dataset defined to describe the graph. The stochastic gradient descent algorithm runs multiple iterations updating the embedding vector space. Once the objective function stops improving, the algorithm converges, and the optimization process terminates before running all samples specified. The tolerance for this convergency can also be specified, defining the rate of changes in the embedding vectors and how the objective functions are minimized throughout the iterations. Optimizing the embedding vectors without any change in the formulation can be expensive. A more efficient way of deriving the embedding vectors is by using negative sampling. The negative sampling is in fact a different objective function and is used to approximate the loss. The embedding vector algorithm uses a random distribution over all nodes in the input graph. Instead of normalizing with respect to all nodes, the negative samples just normalize against *k* random nodes. In this way, to compute the loss function, only *k* negative samples proportional to the degree can be used.

The embedding vector space is computed for each node within the graph. Then, the vector similarity between each pair of nodes *i* and *j* is calculated as a function of the angle between the embedded vectors for *i* and *j*. The angle between the embedded vectors between a pair of nodes is calculated by using the following equation:

$$S_{ij}^V = \frac{\cos(\theta_{u_i u_j}) + 1}{2}$$

where $\theta_{u_i u_j}$ is the angle between u_i and u_j.

Computing node similarity, particularly for large graphs, can be expensive. By default, proc network calculates the node similarity for every pair of nodes within the network. To diminish the computation cost for this calculation, a subset of nodes can be defined when calculating the node similarity. Two different subsets of nodes can be defined to drive the node similarity calculation. The first set defines the source (origin) nodes. The second set defines the sink (destination) nodes. For undirected graphs, source and sink nodes do not represent the beginning nor the end of the links. For directed graphs, the source nodes do represent the beginning of the links, its origin, and the sink nodes represent the end of the links, its destination. The NODESSUBSET = option in proc network define the subset of nodes to be used when computing the node similarity, considering a subset of source nodes and distinctly a subset of sink nodes. Finally, a specific source node and/or sink node can be defined. By doing this, the node similarity will be computed only for this specific pair of nodes. The options SOURCE = and SINK = define a single pair of nodes to be used when computing the node similarity.

The final results for the node similarity algorithm in proc network are reported in the output tables defined by the options OUTSIMILARITY =, OUTNODES =, and OUTCONVERGENCE =. These output tables contain several columns, providing detailed information about the outcomes of the node similarity algorithm.

The OUTSIMILARITY = option defines the output table containing the similarity score for each pair of nodes within the network, or specified in the options NODESSUBSET =, SOURCE =, or SINK =. This output table can be large depending on the size of the network and the number of pairs of nodes to have the node similarity calculated. Depending on the size of the network, the time to produce this output table can be higher than to compute the node similarity. To minimize the cost to produce this output table when computing the node similarity for large networks, proc network allows a filter for sparse similarity. The option SPARSE = defines when to output the similarity score for sparse nodes, or for nodes with no common neighbors, or when the pair of nodes has the same source and sink node. If TRUE, these pairs of nodes are excluded from the output. If FALSE, these pairs of nodes are included to the output table. Proc network can also define when to compute the embed vector for the nodes by specifying the option VECTOR =. When calculating the embed vectors, it is not possible to exclude from the output the spare pairs of nodes. This output contains the following columns:

- source: the label of the source node of this pair.
- sink: the label of the sink node of this pair.
- link: a flag indicating if this pair of nodes exists in the links dataset. If the pair of nodes exists as a link in the input dataset, the flag is 1, or 0 otherwise.
- order: the ranking of each pair of nodes ordered by decreasing similarity score. The score is reported when the options TOPK = or BOTTOMK = are specified.
- jaccard: the Jaccard similarity score when the JACCARD = option is specified.
- commonNeighbors: the common neighbors similarity score when the COMMONNEIGHBORS = is specified.
- cosine: the cosine similarity score when the COSINE = option is specified.
- adamicAdar: the Adamic-Adar similarity score when the ADAMICADAR = option is specified.
- vector: the vector similarity score when the VECTOR = option is specified.

The OUTNODES = option defines the output table containing the vector representations for each node within the input graph. This output table is produced when the option VECTOR = is TRUE. The columns are reported based on the specification of other options. The vector embedding columns are defined as vec_1 through vec_d, where d is the number of dimensions computed in the embed vector specified in the option NDIMENSIONS =. The embed vector can use the first and the second order of proximity. If the option PROXIMITYORDER = is specified as SECOND, the output table also contains the context embedding columns, defined as ctx_1 through ctx_d, where d is the number of dimensions computed in the embed vector.

The OUTCONVERGENCE = option defines the output table containing the convergence curves for the vector embeddings for all nodes during the training process. The convergence value is periodically reported to the output table during the training process. Similar to the similarity scores, this output table can be quite large depending on the size of the network. To reduce the cost of reporting this outcome, the algorithm reports only at certain sample intervals. This output contains the following columns:

- worker: the identifier for the machine that reports the convergence value.
- sample: the number of samples that have been completed when the convergence value is reported.
- convergence: the reported convergence value.

Proc network produces six similarity measures. The different measures can be computed in parallel, in the same execution of proc network. Each similarity measure has an option to specify and invoke the correct algorithm to compute the different scores.

The option COMMONNEIGHBORS = specifies whether to calculate common neighbors node similarity. If the option is TRUE, proc network calculates the common neighbors node similarity and reports the results in the output table that specified in the OUTSIMILARITY = option. If the option is FALSE, the common neighbors node similarity is not calculated.

The option JACCARD = specifies whether to calculate the Jaccard node similarity. If the option is TRUE, proc network calculates the Jaccard node similarity and reports the results in the output table specified in the OUTSIMILARITY = option. If the option is FALSE, the Jaccard node similarity is not calculated.

The option COSINE = specifies whether to calculate the cosine node similarity. If the option is TRUE, proc network calculates the cosine node similarity and reports the results in the output table specified in the OUTSIMILARITY = option. If the option is FALSE, the cosine node similarity is not calculated.

The option ADAMICADAR = specifies whether to calculate the Adamic-Adar node similarity. If the option is TRUE, proc network calculates the Adamic-Adar node similarity and reports the results in the output table specified in the OUTSIMILARITY = option. If the option is FALSE, the Adamic-Adar node similarity is not calculated.

Those are the four methods to compute the node similarity based on the neighborhood of each node. The other two methods are based on the structural roles' proximity within the network.

The option VECTOR = specifies whether to calculate vector node similarity. If the option is TRUE, proc network calculates the vector node similarity and reports the results in the output table specified in the OUTSIMILARITY = option. If the option is FALSE, the vector node similarity is not calculated. When calculating the vector node similarity (VECTOR = TRUE), other options are available to define the type of the calculation.

The option PROXIMITYORDER = specifies the type of the proximity to use in the vector algorithm. If the option is FIRST, proc network uses the first-order proximity when computing the embed vector. If the option is SECOND, proc network uses the second-order proximity when computing the embed vector.

The option EMBED = specifies whether to calculate vector embeddings. If the option is TRUE, proc network calculates the vector embeddings and reports the results in the output table specified in the OUTNODES = option. If the option is FALSE, the vector embeddings is not calculated.

The option EMBEDDINGS = specifies the variables to be used as precalculated vector embeddings. Multiple columns can be specified in the embeddings. The values assigned to the columns used in the embeddings must be numeric (no missing) and must exist for each node in the input graph. The variables used in the embeddings must be specified in the input dataset defined by the option NODES =. In order to compute the embeddings using a list of variables, the options VECTOR = and EMBED = must be TRUE.

The option NDIMENSIONS = specifies the number of dimensions for node-embedding vectors.

The option NSAMPLES = specifies the number of training samples for the vector algorithm. The value assigned to the number of samples by default is 1000 times the number of links in the input graphs.

The option NEGATIVESAMPLEFACTOR = specifies a multiplier for the number of negative training samples per positive training sample for the vector algorithm. The vector algorithm uses the number specified in the option NSMAPLES = as positive samples and that number times the number assigned to the option NEGATIVESAMPLEFACTOR = as negative samples. For example, if NSAMPLES = 1000 and NEGATIVESAMPLEFACTOR = 5, the vector algorithm uses 1000 positive samples and 5,000 negative samples. Recall that the negative sample is smaller and proportional to the degree, instead of a random sample against to all nodes in the input graph. In practice, the k used as a factor for the negative sample is between 5 and 20. The default in proc network is 5. The default number of samples is 1000.

The option CONVERGENCETHRESHOLD = specifies the convergence threshold when computing the vector similarity. If the convergence value drops below this threshold before proc network completes the number of samples specified in the option NSAMPLES =, the vector embeddings training terminates early.

The option ORDERBY = specifies the similarity measure to use when ranking the pairs of nodes reported in the output table specified in the option OUTSIMILARITY =. The options are the different similarity measures computed by proc network, COMMONNEIGHBORS, JACCARD, COSINE, ADAMICADAR, and VECTOR. The ORDERBY = option is used in conjunction with the options TOPK = and BOTTOMK =.

The option TOPK = specifies the maximum number of highest-ranked similarity pairs to be reported in the output table specified in the option OUTSIMILARITY =. If multiple similarity measures are computed, the option ORDERBY = must be used to define which similarity measure is used to rank the pairs in the output table.

Similarly, the option BOTTOMK = specifies the maximum number of lowest-ranked similarity pairs to be reported in the output table defined by the option OUTSIMILARITY =. Again, when multiple similarity measures are calculated, the option ORDERBY = defines the similarity measure used to rank the pairs in the output table.

The option MAXSCORE = specifies the maximum similarity score to report in the output table specified in the option OUTSIMILARITY =. The similarity score ranges from 0 to 1. This option is supported only for the Jaccard and vector similarity measures. By default, the possible maximum score 1 is used.

The option MINSCORE = specifies the minimum similarity score to report in the output table specified in the option OUTSIMILARITY =. Similarly, this option is supported only for the Jaccard and vector similarity measures. By default, the minimum possible score 0 is used.

Let's see some examples of node similarity for both directed and undirected graphs.

2.8.1 Computing Node Similarity

In order to demonstrate the node similarity algorithm, let's consider a graph similar to the one presented at the beginning of this section, shown in Figure 2.86, with some additional nodes and links. This graph has 18 nodes and 18 links. Initially this graph will be considered an undirected graph, with no link weights.

The following code describes how to create the input links dataset.

```
data mycas.links;
   input from $ to $ @@;
datalines;
A E A B A F A D A H A I D C F G
T U T V T X T Y T R T S X W Y Z
E V B U
;
run;
```

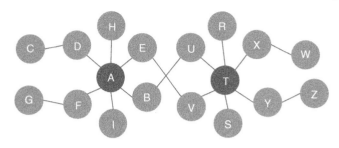

Figure 2.86 Input undirected graph.

Once the input links dataset is defined, we can invoke the node similarity algorithm in proc network. The following code describes the computation of all possible node similarity measures:

```
proc network
   direction = undirected
   links = mycas.links
   outnodes = mycas.outnodesvectors
   ;
   linksvar
      from = from
      to = to
   ;
   nodesimilarity
      commonneighbors = true
      jaccard = true
      cosine = true
      adamicadar = true
      vector = true
      ndimensions = 100
      proximityorder = second
      outsimilarity = mycas.outsimilarity
   ;
run;
```

Figure 2.87 shows the summary output for the node similarity algorithm. It shows the size of the network, the direction of the graph and the output tables created.

The OUTNODESVECTORS table presented in Figure 2.88 shows all nodes from the original input graph and the values for each vector dimension. In the proc network code, the option NDIMENSIONS = 100 requests the algorithm to compute a vector with 100 dimensions. Each dimension is represented by a column name *vec_x*, where *x* varies from 1 to 100. This vector can be used to warm start the node similarity calculation when the EMBEDDINGS = option is defined.

The OUTSIMILARITY table presented in Figure 2.89 shows the pairs of nodes from the input graph where the node similarity measures were computed. In our case, five measures were calculated, Common Neighbors, Jaccard, Cosine, Adamic-Ada, and the vector node similarity with second order proximity. This output table also shows whether or not the pair of nodes have a link between them. Notice that this outcome can be extensive, particularly when large networks are used as input for the node similarity algorithm. Also, an auto-relationship is assumed making the algorithm to compute the node similarity for pairs of nodes where source and sink are the same. For example, the node similarity for E as a source node and E as a sink node is 1. The number of common neighbors in that case is the number of connections this particular node has.

The NETWORK Procedure

Problem Summary

Number of Nodes	18
Number of Links	18
Graph Direction	Undirected

The NETWORK Procedure

Solution Summary

Problem Type	Node Similarity
Solution Status	OK
CPU Time	0.66
Real Time	0.26

Output CAS Tables

CAS Library	Name	Number of Rows	Number of Columns
CASUSER(Carlos.Pinheiro@sas.com)	OUTNODESVECTORS	18	11
CASUSER(Carlos.Pinheiro@sas.com)	OUTSIMILARITY	171	8

Figure 2.87 Output results by proc network running the node similarity algorithm.

	OUTNODESVECTORS		Table rows: 18	Columns: 11 of 11	Rows 1 to 18		

	node	vec_1	vec_2	vec_3	vec_4	vec_5	
1	A	0.013156...	-0.05183...	-0.43828...	0.368573...	0.054925...	0.3268
2	B	-0.40545...	-0.08810...	-0.28750...	0.245128...	0.453025...	0.3605
3	C	-0.11839...	-0.41486...	-0.21784...	-0.09967...	-0.35853...	0.3463
4	D	0.005759...	-0.50264...	-0.48555...	0.043557...	-0.05895...	0.3454
5	E	0.161131...	0.377051...	0.204269...	0.192728...	0.182426...	0.3345
6	F	0.143017...	0.294501...	-0.63102...	0.393563...	-0.04562...	-0.293
7	G	0.293329...	0.334857...	-0.52718...	0.306861...	-0.15708...	-0.562
8	H	0.436879...	0.181756...	-0.52302...	0.146850...	0.017679...	0.3330
9	I	-0.22744...	-0.31676...	-0.02664...	0.448887...	-0.15393...	0.3209
10	R	0.244156...	-0.04087...	-0.31455...	0.381326...	0.412382...	-0.147
11	S	-0.43463...	0.101361...	-0.28192...	0.160164...	0.275573...	0.4324
12	T	-0.04321...	-0.19582...	-0.01146...	0.101599...	0.537869...	0.2837
13	U	-0.30633...	-0.32991...	-0.25730...	-0.16820...	0.491364...	0.2837
14	V	0.138079...	0.251575...	0.534885...	0.081696...	0.319000...	0.2355
15	W	0.163024...	-0.52063...	0.185039...	0.131412...	0.063092...	0.3560
16	X	0.223555...	-0.5369894	0.040119...	-0.03056...	0.312982...	0.5033
17	Y	-0.20250...	-0.08381...	0.352974...	0.349115...	0.487794...	0.0854
18	Z	-0.26397...	0.249090...	0.179974...	0.482646...	0.474559...	0.3089

Figure 2.88 Output vectors for the node similarity.

OUTSIMILARITY — Table rows: 171 | Columns: 8 of 8 | Rows 1 to 171

Enter expression

	source	sink	link	jaccard	commonNeighbors	cosine	
1	E	E	0	1	2	1	4.
2	E	F	0	0.33333...	1	0.5	1.
3	E	G	0	0	0	0	
4	E	H	0	0.5	1	0.7071...	1.
5	E	I	0	0.5	1	0.7071...	1.
6	E	R	0	0	0	0	
7	E	S	0	0	0	0	
8	E	T	0	0.14285...	1	0.2886...	3.
9	E	U	0	0	0	0	
10	E	V	1	0	0	0	
11	E	W	0	0	0	0	
12	E	X	0	0	0	0	
13	E	Y	0	0	0	0	
14	E	Z	0	0	0	0	
15	F	F	0	1	2	1	1.79
16	F	G	1	0	0	0	
17	F	H	0	0.5	1	0.7071...	1.
18	F	I	0	0.5	1	0.7071...	1.

Figure 2.89 Output node similarity measures.

We are not actually interested in the similarity of the same node, or the similarity measure between the node and itself. The following code creates a similar output without the pairs of nodes where source and sink are the same:

```
data mycas.outsimilarityunique;
   set mycas.outsimilarity(where=(source ne sink));
run;
```

The OUTSIMILARITYUNIQUE table shows the node similarity measures for the real pair of nodes (not considering the assumed auto-relationships).

Figure 2.90 shows only the top 2 pairs of nodes based on the Jaccard node similarity.

OUTSIMILARITYUNIQUE — Table rows: 153 | Columns: 8 of 8 | Rows 1 to 153

Enter expression

	source	sink	link	jaccard ↓	commonNeighbors	cosi
1	H	I	0	1	1	
2	R	S	0	1	1	

Figure 2.90 Output node similarity for the top Jaccard measure.

Nodes H and I, as well as nodes R and S have the Jaccard node similarity equal 1, the maximum value possible. Node H has only one connection, to node A, and node I has only one connection, also to node A. These nodes are similar because they have the same neighbors, in this case, only one.

We can also compute the node similarity based on a specific pair of source and sink nodes. A minor change in the code invokes proc network to compute the node similarity only between the nodes A and T, as shown in the following excerpt of code:

```
proc network
...
    nodesimilarity
        source = 'A'
        sink = 'T'
...
run;
```

The OUTSIMILARITY table presented in Figure 2.91 shows the node similarity measures for one single pair of nodes, considering node A as the source node and node T as the sink node.

Notice that even though nodes A and T have node connection between them, and no common neighbor, they are sort of similar because of their structural role proximity.

In addition to the SOURCE = and SINK = options to define a single pair of nodes, proc network can use a definition of a subset of nodes to consider a set of nodes as source nodes and a different set as sink nodes. The following code describes how to define a subset of source and sink nodes to be used in proc network to compute the node similarity to a set of pairs of nodes instead of the entire network:

```
data mycas.nodessim;
    input node $ source sink @@;
datalines;
A 1 0 D 1 0 F 1 0
T 0 1 X 0 1 Y 0 1
;
run;
```

Once the subset of nodes is defined, it must be referenced in proc network. The following code excerpt describes how to do that.

```
proc network
    direction = undirected
    links = mycas.links
    nodessubset = mycas.nodessim
...
    nodesimilarity
...
run;
```

The NODESSIM table defines 3 nodes as sources by using the variable source equals 1 and the variable sink equals 0. Nodes A, D, and F are defined as source nodes. Nodes T, X, and Y are defined as sink nodes by setting the variable source equals 1 and the variable sink equals 0.

	source	sink	link	jaccard	commonNeighbors	cosine	adamicAdar	vector
1	A	T	0	0	0	0	0	0.5139166303

Figure 2.91 Output node similarity measures for the pair of nodes A and T.

	⚠ source	⚠ sink	⊞ link	⊞ jaccard	⊞ commonNeighbors
1	A	T	0	0	0
2	A	X	0	0	0
3	A	Y	0	0	0
4	D	T	0	0	0
5	D	X	0	0	0
6	D	Y	0	0	0
7	F	T	0	0	0
8	F	X	0	0	0
9	F	Y	0	0	0

Figure 2.92 Output nodes similarity.

Figure 2.93 Node similarity outcomes.

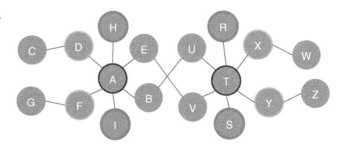

The OUTSIMILARITY output table presented in Figure 2.92 now shows only the node similarity measures for the pairs of nodes defined in the subset of nodes presented before.

Figure 2.93 shows some pairs of nodes that were considered similar based on some node similarity measures. For example, nodes H and I, as well as nodes R and S have the highest Jaccard similarity measure. Nodes A and T have a high vector similarity measure. Nodes D and F, as well as nodes X and Y have a high common neighbors similarity measure.

Finally, node similarity can also be calculated for directed graphs. The only change needed is the definition of the graph direction in proc network, as shown by the following code excerpt:

```
proc network
   direction = directed
...
run;
```

The directed graph considers the information flow created by the direction of the graphs, which affects the concept of neighbors, and then the calculation of the node similarity. Considering for example exactly the same network, by using the links creation to define the direction of the connections, the node similarity algorithm presents a quite different result. Based on an undirected graph, several pairs of nodes got the maximum node similarity measure, for example, Jaccard equals 1. However, based on a directed graph, considering the same network, only 2 pairs of nodes got a Jaccard measure as 0.16 and all other pairs got 0 as node similarity.

The output table OUTSIMILARITYUNIQUE presented in Figure 2.94 selects all rows from the original OUTSIMILARITY table excluding the observations where the source node equals to the sink node.

Node similarity can be particularly useful to find nodes with common neighbors or nodes with similar structural role proximity. This information can be used to search similar behaviors or connections within large networks. Suppose that an illegal transaction is detected in the transactional system. The entity committing this transaction probably have

		source	sink	⊞ link	⊞ jaccard ↓	⊞ commonNeighbors	⊞ cosine	⊞ ad
1		B	T	0	0.1666666667	1	0.408248...	3.32
2		E	T	0	0.1666666667	1	0.408248...	3.32
3		G	X	0	0	0	0	
4		C	Y	0	0	0	0	
5		I	U	0	0	0	0	

OUTSIMILARITYUNIQUE — Table rows: 153 | Columns: 8 of 8 | Rows 1 to 153

Figure 2.94 Output node similarity measures for a directed graph.

connections to other entities. By computing the node similarity, similar types of connections or similar relationship behaviors can be found in a large network and highlight possible illegal transactions that were not captured yet by the transaction systems.

2.9 Pattern Matching

Node similarity is a method to search similar actors or entities within network structures that somehow present comparable or alike connections, common neighbors, or structural roles proximity. Another method to find similarities within networks is to search for whole graph structures. Graph pattern matching has many applications, from plagiarism detection, to biology, social networks, image recognition, intelligence analysis, and fraud investigation, among many others.

Pattern matching or graph matching is the problem of finding a homomorphic or isomorphic subgraph within an input graph. This subgraph is called the patterns in the given input graph, called the target. This problem is also referred to as subgraph a homomorphism or subgraph isomorphism problem. The pattern matching is commonly defined in terms of the subgraph isomorphism, or the search for subgraphs of a given input graph that matches to a query graph. For example, suppose a social network structure composed by friends, where they live, and what soccer team they support. The nodes in this network are then persons, locations, and teams. The links are "is friends of," "lives in," and "support." For example, a node *Carlos* (person) is connected to a node *Cary* (location) by a link "lives in" and is connected to a node *Fluminense* (team) by a link "support." A common pattern matching search in this social network would be to find friends of Carlos who also support Fluminense and live in the same city. The query graph would be any person node connected to team node Fluminense by the link "support" and connected to the city node Cary by the link "lives in." The main drawback in pattern matching is that searching for subgraphs isomorphism can be complex and extremely time consuming.

Formally, pattern matching can be defined considering a main graph and a query graph. Subgraph isomorphism given a main graph G and a query graph Q is the problem of finding all subgraphs Q' of G that are isomorphic to Q. The subgraph Q' must have the same network structure or topology of the subgraph Q. The pattern matching problem considers the attribute of nodes and links in the network structure, like in the example previously described. Nodes may be of multiple types, like person, location, and team, as well as the links, as friends of, lives in, and support. The problem is the process of finding all subgraphs Q' of G isomorphic to subgraph Q where all attributes for the nodes and links defined for the subgraphs Q are preserved in the subgraphs Q'.

In proc network, the PATTERNMATCH statement invokes the algorithm to search for the isomorphic subgraphs of a query graph given an input graph. The query graph is defined by using the options LINKSQUERY = and NODESQUERY =. These options can be used independently to define the query graphs. For example, just the links attributes can be defined as a query graph, just the nodes attributes, or both. The attributes definition for the query graph follows a similar criterion to define attributes for the main input graph. The graphs are created as links and nodes input tables. The main graph is defined by using the options LINKS = and NODES = in proc network. The query graph is defined by using the options LINKSQUERY = and NODESQUERY =. The attributes of the links and nodes in the main graph are defined by using the options LINKSVAR = and NODESVAR in proc network. The attributes of the nodes and links in the query graph are defined by using the options LINKSQUERYVAR = and NODESQUERYVAR.

The final results for the pattern matching algorithms in proc network are reported in the output tables defined by the options OUTMATCHGRAPHLINKS=, OUTMATCHGRAPHNODES=, OUTMATCHLINKS=, OUTMATCHNODES=, OUTQUERYLINKS=, OUTQUERYNODES=, and OUTSUMMARY=. These output tables may contain several columns and vary based on the definition of the query graph and the pattern matching searching, which are defined by the multiple possible parameters in proc network.

The OUTMATCHGRAPHLINKS = option defines the output table containing the links in the induced subgraph of matches. The induced subgraph is created based on the union of the sets of nodes in all matches for each query key or across all query keys. The first approach to create the induced subgraph (each query key) is defined by the option QUERYKEYAGGREGATE = equals FALSE. The second approach is defined by the option QUERYAGGREGATE = equals TRUE. This output table contains the following columns:

- queryKey: the query key when QUERYKEY = option is defined and the option QUERYKEYAGGREGATE = is FALSE.
- from: the label of the from node for each link in the main graph that is in the induced subgraph of matches.
- to: the label of the to node for each link in the main graph that is in the induced subgraph of matches.

The OUTMATCHGRAPHNODES = option defines the output table containing the nodes in the induced subgraph of matches. The induced subgraph is created based on the union of the sets of nodes in all matches for each query key or across all query keys. Similar to the graph links, the first approach is defined by the option QUERYKEYAGGREGATE = equals FALSE. The second approach is defined by the option QUERYAGGREGATE = equals TRUE. This output table contains the following columns:

- queryKey: the query key when the QUERYKEY = option is defined and the option QUERYKEYAGGREGATE = is FALSE.
- node: the label of the node for each node in the main graph that is in the induced subgraph of matches.

The OUTMATCHLINKS = option defines the output table containing the subgraph in the main graph for each pattern match. This output table contains the following columns:

- queryKey: the query key when the QUERYKEY = option is defined.
- match: the identifier of the match.
- from: the label of the from node for each link in the main graph.
- to: the label of the to node for each link in the main graph.

The OUTMATCHNODES = option defines the output table containing the mappings from the nodes in the query graph to the nodes in the main graph for each pattern match. This output table contains the following columns:

- queryKey: the query key when the QUERYKEY = option is defined.
- match: the identifier of the match.
- nodeQ: the label of the node for each node in the query graph.
- node: the label of the node for each node in the main graph.

The OUTQUERYLINKS = option defines the output table containing the links for the generated queries. This output table contains the following columns:

- queryKey: the query key.
- from: the label of the from node for each link in the query graph.
- to: the label of the to node for each link in the query graph.

The OUTQUERYNODES = option defines the output table containing the nodes for the generated queries. This output table contains the following columns:

- queryKey: the query key.
- node: the label of the node for each node in the query graph.

Finally, the OUTSUMMARY = option defines the output table containing the summary information about the executed queries, including the number of matches that were found, and the time consumed during the searching. This output table contains the following columns:

- queryKey: the query key when the QUERYKEY = option is defined.
- nodes: the number of nodes in the query graph.
- links: the number of links in the query graph.
- matches: the number of matches that were found.
- realTime: the elapsed time in seconds for the pattern match search.

The pattern matching algorithm in proc network has an extensive number of options to define the query graph and the pattern matching methods within the network. Two of these options were mentioned before and directly affect the output tables created to report the results of the pattern matching algorithm.

The option QUERYKEY = specifies the name of the data variable in the nodes and links datasets that will be used as the query key. The pattern matching algorithm in proc network runs parallel searches in a single execution. By doing this, it is possible to search for more than one pattern in the main graph in a single pass. This option specifies the variable in the nodes and links dataset that indicates multiple patterns to be searched in one single pass. This parallel running improves the overall performance as the pattern matching algorithm can be executed across multiple machines and threads. The option QUERYKEY = is useful when searching for sequential patterns. For example, a pattern to be searched involves a path of sequential transactions that starts at a node X and ends at a node Y. Suppose that the length of the path associated with the query pattern can have between two and five links. A straightforward approach is to define four different query graphs containing a path of sequential transactions with, two, three, four, and five links. Based on that, it is possible to execute the pattern matching algorithm in proc network four times, one for each query graph defined. Another way to search for these different patterns, and much more efficient, is to use the option QUERYKEY = and process these four queries in one single pass of the pattern matching algorithm.

Figure 2.95 shows an example of these four possible patterns.

The option QUERYKEYAGGREGATE = specifies whether to aggregate the nodes across the query keys in order to create the induced subgraphs. If the option is TRUE, proc network aggregates all the nodes across the query keys when creating the induced subgraphs. If the option is FALSE, the algorithm does not aggregate the nodes.

There are two ways to create the query graphs when executing the pattern matching algorithm. The first method is to define the query graph as input data tables, specifying the graph topology by a set of nodes and its attributes, and a set of links and its attributes. The second method is to use a concept called Function Compiler. Function Compiler enables end users programming in SAS to create, test and store specific functions, call routines, and subroutines, before these functions are used in other procedures or data steps. The Function Compiler builds these functions and stores them in a dataset, using data step syntax. This feature enables programmers to read, write, and maintain complex code with independent and reusable subroutines. In network optimization, particular for proc network, the Function Compiler provides a substantial flexibility to define the query graphs. In proc network, the Function Compiler provides a set of Boolean functions that, when associated with a specific pattern match query, defines additional conditions that a subgraph from the main graph must satisfy in order to be considered a match. The Function Compiler applies to the attributes of the query graph, or the main graph, when defining the conditions of the pattern match. The Function Compiler permits exact and inexact attribute matching for individual nodes and links. These different types of attribute matching can be designed to specific nodes and links in the query graph. The ability to use functions that apply to pairs of nodes and links enables the definition of conditions for which the scope is more global than the definition of conditions or attributes of individual nodes and links. Refinement in the query graph can be applied by using pair filters on the set of candidate matching nodes and links.

The option CODE = specifies the code that defines the Function Compiler (FCMP) functions that will be applied to define the query graph.

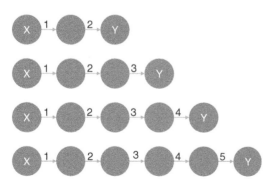

Figure 2.95 Four patterns in a path of sequential transactions.

The option INDUCED = specifies whether to filter matches by using the induced subgraph of the match. If the option is TRUE, proc network keeps only matches whose induced subgraph is topologically equivalent to the query graph. If the option is FALSE, proc network does not filter by using the induced subgraph of the match.

For the nodes, there are two types of filters. The option LINKFILTER = specifies the name of the FCMP function that will be applied for link filters. The option LINKPAIRFILTER = specifies the name of the FCMP function that will be applied for link-pair filters.

For the nodes, there are also two types of filters. The option NODEFILTER = specifies the name of the FCMP function that will be applied for node filters. The option NODEPAIRFILTER = specifies the name of the FCMP function that will be applied for node-pair filters.

The option MATCHFILTER = specifies the name of the FCMP function for a filter that is based on a potential match. That means the match is based on any subset of nodes or links or both.

The option MAXMATCHES = specifies the maximum number of matches for the pattern matching algorithm to report in the output tables. This option can be any number greater than or equal to one. If the option is ALL, which is the maximum value that can be represented by a 64-bit integer, all matches (limited to that maximum number) are reported.

The option MAXTIME = specifies the maximum amount of time to spend in the pattern matching algorithm. The type of time can be either CPU time or real-time. The type of the time is determined by the value of the TIMETYPE = option in the PROC NETWORK statement. The default value is the largest number that can be represented by a double.

The pattern matching problem involves finding and extracting subnetworks from a larger network that perfectly match certain criteria. In other words, the pattern matching creates subnetwork by querying a large network and selecting all the subnetworks that satisfy a set of predefined rules. Conceptually, the pattern matching method resembles the process of querying a traditional database, selecting only the rows where a criterion is satisfied. In that way, instead of **selecting** all observations **from** a table **where** some criteria, we can think of this process as **selecting** all subgraphs **from** the main graph **where** links and nodes match to a specific pattern. The table in SQL procedures is the main graph, and the where clause in the SQL statement is the filter defined by the pattern on the links and nodes. The observations returned as the outcome of the SQL procedure are the subgraphs returned as the outcome of the pattern matching.

2.9.1 Searching for Subgraphs Matches

Let's consider that the following graph presented in Figure 2.96 exemplifies the searching for subgraph matches. This graph has 19 nodes and 26 links. Initially this graph will be considered as an undirected graph. All the links within this input graph are weighted. The link weight is important when the query graph is defined and searched in the main graph.

The following code describes how to create the links dataset The links dataset has only the nodes identification, specified by the from and to variables, plus the weight of each connection.

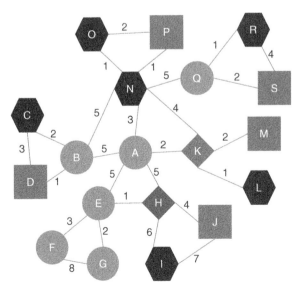

```
data mycas.links;
    input from $ to $ weight @@;
datalines;
A B 5 B C 2 C D 3 D B 1 N A 3 A K 2 N K 4 M K 2 L K 1
A E 5 E H 1 H A 5 H I 6 H J 4 J I 7 E F 3 E G 2 F G 8
B N 5 N O 1 O P 2 P N 1 N Q 5 Q R 1 R S 4 Q S 2
;
run;
```

The following code describes the nodes dataset. In the pattern matching, the nodes are important when the query graph considers some particular nodes attributes. The nodes dataset defines the node identification and the attribute that will be used in the query graph, the shape.

Figure 2.96 The main graph with undirected links.

```
data mycas.nodes;
   input node $ shape $ @@;
datalines;
A circle B circle C hexagon D square E circle F circle G circle
H diamond I hexagon J square K diamond L hexagon M square
N hexagon O hexagon P square Q circle R hexagon S square
;
run;
```

The following code defines the nodes used in the query. The query graph specifies a particular set of nodes and their attributes. For example, this definition specifies that the query graph must have six nodes (1 to 6), where two nodes with square shapes (2 and 6) and another two nodes with hexagon shapes (1 and 5). The other two nodes (3 and 4) could have any shape.

```
data mycas.nodesquery;
   input node shape $ @@;
datalines;
1 hexagon
2 square
5 hexagon
6 square
;
run;
```

Finally, the following code defines the links used in the query graph when the pattern matching algorithm searches for the subgraphs. In addition to the set of nodes, the query graph also specifies a particular set of links and their attributes. For example, this definition specifies that the query graph consists of seven links, connecting nodes 1 and 2, 2 and 3, and then 3 and 1. The missing values for the link weights specify that any weight between these nodes is accepted in the pattern matching search. The connection of these three nodes creates a triangle cycle (1→2→3→1). The nodes identification as 1, 2, and 3 means that any triangle cycle within the input graph satisfy the criteria for the query graph, as long as the first node is a hexagon, and the second node is a square. The third node could be of any shape. The links dataset for the query graph defines another triangle cycle, specified by the links between nodes 4 and 5, 5 and 6, and then 6 and 4 (cycle 4→5→6→4). Again, the missing values for the link weights define that any weight between these nodes is accepted by the query graph. Finally, the query graph defines that these two triangle cycles must be connected by a link with a weight of 5. The link between nodes 3 and 4 defines the connection between the triangles with a link weight of 5.

```
data mycas.linksquery;
   input from to weight @@;
datalines;
1 2 .
2 3 .
3 1 .
3 4 5
4 5 .
5 6 .
6 4 .
;run;
```

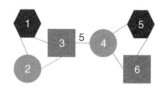

Figure 2.97 The query graph.

Figure 2.97 shows the query graph defined by the nodes and links datasets associated with the pattern matching. This query graph means that the pattern searched in the main graph consists of two triangles connected by a link with weight of 5. Each triangle must contain a hexagon, a circle, and a square.

As the main graph is defined by the links and nodes datasets, and the query graph is defined by the nodes query and links query datasets, we can invoke the pattern matching algorithm in proc network to search for all subgraphs within the input graph that matches to the pattern specified. The following code describes the process. Notice that, in addition to the nodes and links definitions by using the options NODES = and LINKS =, we also have the definition of the query graph by using the options NODESQUERY = and LINKSQUERY. We specify the variables used to define the main graph by using the statements NODESVAR and LINKSVAR. The option VARS = in the NODESVAR statement specifies the attribute of the nodes that will be used during the pattern matching, which is the shape of the node, defined by the variable shape in the nodes' dataset. Similarly, we need to specify the variables used to define the query graph. Now we use the statements NODESQUERYVAR and LINKSQUERYVAR. The VARS = option in the NODESQUERYVAR statement defines the attributes of the nodes that need to be matched during the search. The variable shape in the nodes query dataset defines the shapes of the nodes comprised in the triangles, hexagon, circle, and square. Finally, the VARS = option in the LINKSQUERYVAR statement defines the attribute of the links that must be matched during the search. The variable weight in the links query dataset defines the link weight of the connection between the two specified triangles as a weight of 5.

```
proc network
   direction = undirected
   nodes = mycas.nodes
   links = mycas.links
   nodesquery = mycas.nodesquery
   linksquery = mycas.linksquery
   ;
   nodesvar
      node = node
      vars = (shape)
   ;
   linksvar
      from = from
      to = to
      weight = weight
   ;
   nodesqueryvar
      node = node
      vars = (shape)
   ;
   linksqueryvar
      from = from
      to = to
      vars = (weight)
   ;
   patternmatch
      outmatchnodes = mycas.outmatchnodes
      outmatchlinks = mycas.outmatchlinks;
run;
```

Figure 2.98 shows the summary output for the pattern matching algorithm. It shows the size of the network, 19 nodes and 26 links, the direction of the graph, and the number of matches found along the output tables reported.

The OUTMATCHNODES table presented in Figure 2.99 shows all the matches found, with the association of the query graph node (1 through 6) to the main graph node (A through S). That determines, for this particular match, what role the node in the main graph role plays in the query graph. Finally, this output shows the node attribute used when searching for the matches. The shape of the node plays a key role when the algorithm matches the query graph to the main graph.

As an undirected graph, the number of matches is doubled, as the links in both main and query graphs have both directions. For example, matches 1 and 2 are the same. The only difference is the direction of the links when the pattern matching algorithm searches for the query graph in the main graph. The same happens for the matches 3 and 4. They are exactly the same match with different links directions. Let's consider the first match for example. The nodes C, D, and B represent the first triangle cycle, where one node must be a hexagon (node C), one node must be a square (node D), and the third node can

The NETWORK Procedure

Problem Summary

Number of Nodes	19
Number of Links	26
Graph Direction	Undirected

The NETWORK Procedure

Solution Summary

Problem Type	Pattern Match
Solution Status	OK
Number of Matches	4
CPU Time	0.00
Real Time	0.01

Output CAS Tables

CAS Library	Name	Number of Rows	Number of Columns
CASUSER(Carlos.Pinheiro@sas.com)	OUTMATCHLINKS	28	4
CASUSER(Carlos.Pinheiro@sas.com)	OUTMATCHNODES	24	4

Figure 2.98 Output results by proc network running the pattern matching algorithm.

be of any shape. For this particular triangle, node B is a circle. The second triangle cycle in this match is represented by the nodes N, O, and P. Similarly, the triangle must have a hexagon node, which is node N, and a square node, which here could be either node N or node O, as both are hexagons nodes. The third node in the triangle can be of any shape, which means it can be either node N or node O. The last criterion for the pattern match is the link weight between these two triangles cycles. It must be a link with a weight of 5. This link is the connection between nodes B and N. This link will be reported in the output table for the links within the matches. The same concept is applied to the second unique match, the match 3. It is represented by one triangle cycle formed by the nodes O, P, and N (with either node O or N as the hexagon and node P as the square), another triangle cycle formed by the nodes Q, R, and S (with node R as the hexagon and node S as the square) and the connection between these two triangles cycles represented by the link N-Q with weight of 5.

The OUTMATCHLINKS table presented in Figure 2.100 shows the links between the nodes that represent the matches found.

The matches in the nodes output are associated with the matches in the links output. That means the match 1 in the OUTMATCHLINKS shows the links between all nodes in the match 1 in the OUTMATCHNODES output. The important link though here is the link that represents the connection between the two triangles specified in the query graph. This link in the match 1 is the link between nodes B and N, which must have weight of 5. That is described by line 3 in the output table. The same for the second unique match 3, which is represented by the link between nodes N and Q with link weight of 5. This link is in the output table, but it is suppressed in the picture.

Figure 2.101 shows a graphical representation of the two matches found by the pattern matching algorithm.

Notice that this picture shows only the unique two matches found, matches 1 and 3. Matches 2 and 4 are the same as 1 and 3, only with different links directions. Match 1 is represented by nodes C, D, and B, connected to the nodes O, P, and N. The connecting link between the two triangle cycles is represented by the link (B,N,5). Match 3 is represented by nodes O, P, and N, connected to the nodes R, S, and Q. The connecting link between the two triangle cycles is represented by the link (N,Q,5). This picture also highlights a possible match. The triangle E-F-G is connected to the triangle H-I-J. The first triangle cycle however doesn't match the criterion of having one square and one hexagon nodes. The second triangle cycle matches that criterion. However, this subgraph also doesn't match the criterion of having a link between the two triangle cycles with weight of 5. The link between them in this subgraph has link weight of 1.

« OUTMATCHNODES   Table rows: 24 | Columns: 4 of 4 | Rows 1 to 24 ↑ ↑ ↓ ↓ ⟳ · ⋮

⚲ Enter expression

	⊞ match	⊞ nodeQ	△ node	△ shape
1	1	1	C	hexagon
2	1	2	D	square
3	1	3	B	circle
4	1	4	N	hexagon
5	1	5	O	hexagon
6	1	6	P	square
7	2	1	O	hexagon
8	2	2	P	square
9	2	3	N	hexagon
10	2	4	B	circle
11	2	5	C	hexagon
12	2	6	D	square
13	3	1	O	hexagon
14	3	2	P	square
15	3	3	N	hexagon
16	3	4	Q	circle
17	3	5	R	hexagon
18	3	6	S	square
19	4	1	R	hexagon
20	4	2	S	square
21	4	3	Q	circle
22	4	4	N	hexagon

Figure 2.99 Output nodes table for the pattern matching algorithm.

Let's see how the pattern matching works when the graph is defined by directed links. Let's consider the exact same graph definition we used before. The only difference here is that the order of the nodes in the links dataset defines the direction of the link by the *from* node and the *to* node. The same is applied for the query links dataset. The links definition in the query links table also defines the direction of the link represented by the variable *from* and *to*.

Figure 2.102 shows the main graph but now considering the direction of the links.

As stated before, the direction of the links also affects the query graph. Figure 2.103 shows the query graph considering directed links.

We can invoke proc network to execute the pattern matching algorithm within the defined links and nodes sets, as well as the query links and query nodes sets with a single change to consider a directed graph instead of an undirected one. The following code shows the change in the DIRECTION = option.

```
proc network
    direction = directed
...
run;
```

The OUTMATCHNODES table presented in Figure 2.104 shows the single match found. This match is the same match 1 found on the undirected graph. This match comprises the first triangle cycle formed by nodes C, D, and B, and the second triangle cycle formed by nodes N, O, and P.

      ▽   Enter expression   ⌕

	⊞ match	⟁ from	⟁ to	⊞ weight
2	1	D	B	1
3	1	B	N	5
4	1	N	O	1
5	1	O	P	2
6	1	B	C	2
7	1	P	N	1
8	2	O	P	2
9	2	P	N	1
10	2	B	N	5
11	2	B	C	2
12	2	C	D	3
13	2	N	O	1
14	2	D	B	1
15	3	O	P	2
16	3	P	N	1
17	3	N	Q	5
18	3	Q	R	1
19	3	R	S	4
20	3	N	O	1
21	3	Q	S	2
22	4	R	S	4

Figure 2.100 Output links table for the pattern matching algorithm.

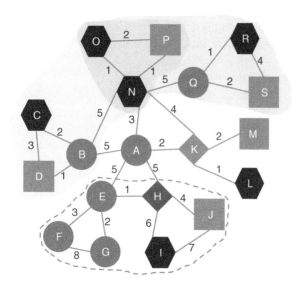

Figure 2.101 Pattern matching outcome.

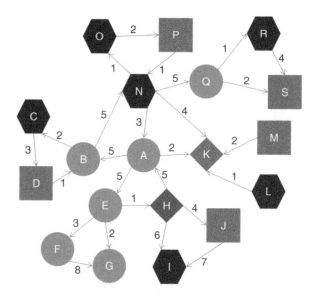

Figure 2.102 The main graph with directed links.

Figure 2.103 The query graph with directed links.

<< OUTMATCHNODES   Table rows: 6   Columns: 4 of 4   Rows 1 to 6   ↑ ↑ ↓ ↓   ↺ ▾   ⋮

▽ Enter expression   ⌕

	⊞ match	⊞ nodeQ	△ node	△ shape
1	1	1	C	hexagon
2	1	2	D	square
3	1	3	B	circle
4	1	4	N	hexagon
5	1	5	O	hexagon
6	1	6	P	square

Figure 2.104 Output nodes table for the pattern matching algorithm for a directed graph.

<< OUTMATCHLINKS   Table rows: 7   Columns: 4 of 4   Rows 1 to 7   ↑ ↑ ↓ ↓   ↺ ▾   ⋮

▽ Enter expression   ⌕

	⊞ match	△ from	△ to	⊞ weight
1	1	C	D	3
2	1	D	B	1
3	1	B	N	5
4	1	N	O	1
5	1	O	P	2
6	1	B	C	2
7	1	P	N	1

Figure 2.105 Output links table for the pattern matching algorithm for a directed graph.

The OUTMATCHLINKS table presented in Figure 2.105 shows the links between the nodes that represent the single match found. It shows the links that create the two triangle cycles (C→D→B→C and O→P→N→O) along the link that connect them (B→N,5).

Figure 2.106 shows a graphical representation of the single match found by the pattern matching algorithm when running on the directed graph.

Pattern matching is getting more and more attention in network science and analytical modeling, particularly in business applications related to fraud detection and suspicious transactions. This algorithm can be applied to multiple business scenarios, considering different industries.

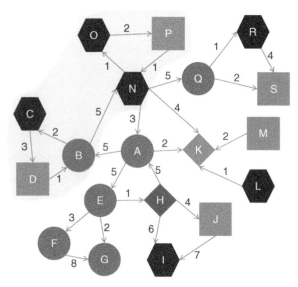

Figure 2.106 Pattern matching outcome on the directed graph.

2.10 Summary

The concept of subgraphs is important in network analysis and network optimization. Many network problems can be divided into smaller problems looking into subnetworks instead of the entire original graph. Also, particularly for large networks, there are multiple distinct patterns hidden throughout smaller groups of actors and relations in the original graph. Analyzing subgraphs instead of the entire graph can, therefore speed up the overall data analysis and make the network outcomes easier to interpret.

Concepts on connected components and how each member of the connected component can reach out to any other member makes this type of subgraph relevant within the network. Some networks don't show many connected components, and if we find some, there should be worth to deeply analyzing them.

The concept of biconnected components is similar to the connected component except for the fact that there is an important actor on it. The articulation point can play a significant role in the network, splitting the network into two or more groups of members. The articulation point plays as a bridge in connecting different groups within the original network and by removing them from the graph, we can completely change the structure of the original network. This role is, therefore, very important, and any articulation point in the network should be further analyzed or closely monitored.

The core, or the k-core decomposition, is an extremely important concept that divides the original network into groups of nodes based on their interconnectivity. Nodes with a high level of interconnectivity will be gathered into the same k-core, and nodes with low level of connectivity will be fitted into different cores. This approach allows us to identify groups of individuals or entities within the network that play similarly in terms of relationships. The most cohesive core represents the nodes with the same higher interconnectivity, and the nodes in the less cohesive core will be the nodes with the lower level of interconnectivity or relations. K-core decomposition has been used in many applications in life sciences, healthcare, and pharmaceutical industries, identifying relevant patterns in diseases spread and contagious viruses.

Reach or ego networks are instrumental when we want to highlight the behavior of a small group of individuals, or even a single one in particular. The reach network represents how many individuals can be reached out from a single person, in one or more waves or connections, and can be important in marketing campaigns when thinking about business, or in terms of the relevance of a node in spreading out something, when thinking about healthcare and life sciences. For example, targeting individuals with a substantial reach network can be effective a marketing campaigns, as well as identifying cells with a great reach network can be effective in terms of possible treatments.

Community detection and analysis is one of the most common and important concepts in subgraphs. Data scientists and data analysts use communities to better understand the network they are analyzing and efficiently identify the nuances of the entire graph in terms of structure and topology. Specially in large networks, the concept of communities can help us in computing network metrics, dividing the overall computation from one single huge graph to multiple smaller subgraphs. Some of the concepts we are going to see in the next chapter, like closeness and betweenness centralities, can be impractical or even infeasible in large networks, but they can make much more sense when computed upon communities. Multiple communities can be detected within large networks, and often they have very different shapes, structures, and patterns of members relationships. Analyzing different communities by different perspectives allows us to define more precise and specific approaches when applying network analysis outcomes. Some communities can be large and dense, some can be large and sparse, some can be small and dense, some small and sparse. Different characteristics implies distinct approaches to not just apply the results but also to interpret the network outcomes.

Network projection is also an important concept to reduce the network size according to some specific characteristics and features. As we saw in this chapter, the network projection reduces a multi-partition network into a single-partition network. By doing this, we can select a subset of nodes of interest, segregate them into a small network, and keep their relationships characteristics for a more specific and narrowed network analysis. The network projection can be used from customers-items relationships to relationships between metabolites and enzymes, and from authorships relations to genes-diseases correlations, among many other practical applications.

Nodes or graph similarity also has multiple practical applications in science and business. The concept of nodes similarity can be used to compare biological or social networks, and can be used in web searching and to find match in chemical structures. It can also be used to find similar groups of customers from a large network, or more specifically, to identify similar actors practicing any suspicious activity throughout the network over time.

Finally, pattern match has been shown to be useful for many applications. Pattern match perfectly represents the concept of subgraph isomorphism, where the problem is to find a subgraph G' of graph G that is isomorphic to graph H. Both graphs G' and H must have the same topology. Pattern matching addresses the same problem by searching similar subgraphs within the network based on a defined set of nodes and links attributes. Pattern matching is used to detect fraud in financial institutions, to match DNA sequences in biology applications, or to recognize voice, text, and images in computer science problems. There is an endless number of practical applications that can involve the pattern matching algorithm, whether in full or partially.

The concept of subgraphs is important and will be used in the further chapters, narrowing network centralities computation, guiding network analysis, or determining specific solutions in network optimization.

3

Network Centralities

3.1 Introduction

One of the most common objectives of network analysis is to understand the relationships within an interconnected structure, composed of customers, companies, agencies, countries, employees, banks accounts, students, or any set of entities that are correlated. The main goal is to identify how members of the network can influence others. Influence is a form of power, suggestion, or domination. Depending on the type of interconnected structure, influence can take many different shapes and have different strength levels.

Influence is the capacity that one customer has to induce others to follow them in a specific business event, like churn or product adoption. In social structures such as universities and schools, influence can take many forms. In these networks, power is often the strongest form of influence. One particular member of this social structure may be able to affect others in different ways in terms of opinions or actions, depending on the type of subject.

In business scenarios, influence is usually recognized in the context of particular events. One customer can influence another customer in a particular business event. Let us say that first customer decides to make churn. Perhaps, they can influence their peers to also make churn afterwards. Indeed, that would be the worst type of business influence for a company. On the other hand, another customer may purchase or start consuming a brand-new product and later they can influence their peers to also purchase or start consuming. This would definitely be the best type of cascade event based on influence. The main question here is how to identify these customers who can influence others to make churn or lead their peers to purchase and consume more? How can we measure the level of influence of these customers? In most of the cases, customers have a level of influence for a very specific business event. Customers can be influencers for churn but not for purchase. Customers can be influencers for product adoption but not for cross-sell.

The process to identify the influential customers in a specific network is actually similar to identifying the probability of a customer to make churn or purchase something. In supervised models, whether using statistical or machine learning algorithms, we use historical events aiming to find a set of rules based on thresholds, a set of coefficients, or even a combination of weights and predictors that somehow describe the propensity of each customer in making churn or purchasing a product in the future. The method to identify influencers in network analysis is quite similar. In order to highlight the influential customers within a particular network, we look at the historical events. Based on these past events, we build a set of graphs throughout the timeframe used to analyze the data. With this sequence of graphs in time, we analyze what happens with some key customers at a particular point in time and what happens to their peers going forward. For example, let us assume a business event like churn. To analyze the influence, we collect historical transactions between customers, such as text messages and calls in a telecommunications scenario. Based on that set of historical transactions, we build a series of graphs in time, representing the customers' behavior during the timeframe of analysis, for instance, on a weekly or monthly basis. At some specific point in time, let us say at time t, we identify all customers who made churn. The graphs at this point in time going forward, let us say at time $t + 1$, $t + 2$, and so on, will tell us which customers made churn afterward or remain at the company. The graphs before time t, let us say $t - 1$, $t - 2$, and so on, will tell us how the relationships between the customers were before the first event of churn took place. Were the customers who made churn in the second wave at time $t + 1$ or $t + 2$ highly connected at time $t - 1$ or $t - 2$ to the customers who made churn at time t? Let us suppose we can find that all customers who made churn in the second wave were heavily connected to the customers who made churn in the first wave. Of course, we cannot possibly know what customers text to each other or the content of their conversations. But we do know that some key customers made churn at some point in time, and some highly connected customers made churn afterward. Some less connected customers remained at the company. We can therefore assume that a sort of influence took place here

Network Science: Analysis and Optimization Algorithms for Real-World Applications, First Edition. Carlos Andre Reis Pinheiro.
© 2023 John Wiley & Sons, Inc. Published 2023 by John Wiley & Sons, Inc.

and then we start from there. We set time *t* as event time, and we analyze the relationships before time *t*. We can compute all network metrics based on that set of graphs and correlated to the event of churn. We can also analyze which network centrality is more relevant to the event of churn, and then create a influence factor that describes the probability of churn, or to be more precise, the probability of a customer's peers to make churn if that particular customer decides to make churn in the future. We can definitely use all network centralities as input for statistical and machine learning models to predict the event of churn. The possibilities of using network analysis and machine learning is endless.

3.2 Network Metrics of Power and Influence

This chapter covers a great variety of network metrics in terms of network analysis. Most of these network metrics compute and describe the importance and the role of the nodes and the links within the network. These network metrics consider the overall distance and the path among the nodes within the network, the strength of the links connecting the nodes, and the strength or value of the nodes within the network.

This chapter summarizes this theory in relation to the measure of power and influence in network analysis. The measure of power is closely related to the graph theory. Power is calculated by a particular node's connections, the strength of its links, and the length of its path to the other nodes within the network, among other factors. The measure of power is a network metric, or a set of network centralities, that indicates which nodes have higher values within the interconnected structure.

The measurement of power and influence in terms of network analysis is usually assigned to the analysis of social structures. Power is a property of social structures, and hence it relies on the shape of the structure being analyzed. There are distinct types of social networks, and, therefore, there are variations in the process and concept of computing power.

All social structures are based on relationships. Therefore, the measure of power is completely correlated to the relationships within the social network. A particular member of the social structure, or a node within the graph, has power solely because of its relations within the network.

Each network measure has an appropriate interpretation relative to the business goals. These metrics are used to describe the main characteristics of the network, like its topology, and the individual importance for nodes and links. Each network centrality aims to highlight some relevant characteristic of the network, both in terms of nodes and links.

For example, the degree centrality describes how connected a node is to other nodes. The influence centrality describes how the nodes are connected to each other, but it also considers the adjacent links, which means it computes the importance not just based on the number of connections but also the strength of these connections. The closeness centrality measures the average shortest path assigned to each node. It describes how close each node is to the rest of the nodes within the network. This network metric ultimately indices how fast a message can flow throughout the network if it starts from a particular node. The betweenness centrality measures how many shortest paths each node within the network partakes. This network metric indicates how much control a node has over the information flowing throughout the network. The hub centrality measures how many nodes a node refers to. It may also consider the strength of the connections. The authority centrality measures how many nodes it is referred by other nodes. It also considers the strength of the connections existing among the nodes within the network. The centrality eigenvector shows how likely some nodes are to be influencers, or how important they are considering all distances among the nodes. PageRank centrality indicates a similar importance to eigenvector but for directed graphs. Cluster coefficient measures how the node's neighbors are connected.

Network analysis aims to analyze relationships between entities. These relationships are typically defined by using a graph. A graph $G = (N, L)$ is defined over a set N of nodes (or vertices) and a set L of links (edges or arcs). A node is an abstract representation of some entity (or object), and an arc defines some relationship (or connection) between two nodes.

The following graph presented in Figure 3.1 defines some relationships between nodes. This graph will be used to elaborate on the network centralities and how to compute them using proc network on the following sections.

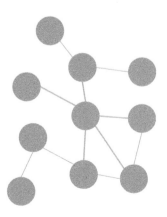

Figure 3.1 Network graph containing nodes and links.

This graph may be slightly modified in the following sections based on the network centrality analyzed, either to add weights for the links, to define directions to the links, or even to select just a subset of the graph and consider a few nodes and links to compute the network centrality.

3.3 Degree Centrality

Degree centrality is a network measure that describes the number of connections that a node v has. It is the number of links incident to that node v. In directed graphs, there is a differentiation in the direction of the connection. Out-degree describes the amount of outgoing links incident to node v. In-degree describes the amount of incoming links incident to node v. In directed graphs, degree is the sum of out-degree and in-degree. In undirected graphs, degree is the out-degree.

Degree centrality represents the number of connections a node has to other nodes. In a directed graph, in which the direction of the node is relevant, there is a differentiation between in-degree, the number of links a node receives, and the out-degree, the number of links a node sends. These directional elements are particularly important in telecommunications and other industries that have an inherent succession in network behavior. The sum of both in-degree and out-degree nodes determines the degree centrality measure.

This network metric is quite useful for understanding how popular or connected a node is to others. For example, in the telecommunications industry, a high degree of centrality for a node means that a subscriber is well connected, probably making and receiving both calls and texts to and from many other subscribers. The degree centrality is typically one of the first possible network measures that describes a formula for customer influence. Usually, this influence factor is described by a set of network metrics formalized in an expression, and most often degree centrality is one metric in this influence formula. The Figure 3.2 shows an example of an undirected graph and the degree centrality.

The degree of node A in the previous graph is 5. Node A is connected to 5 other nodes in the graph. There is no distinction between the direction of the connection. They can be friends, coworkers, coauthors, etc. The relationship between do not hold any direction or causality. The Figure 3.3 shows a directed graph and an example of degree centrality.

The degree of node A in the previous graph is still 5, summarizing the degree-out, which is 2, and degree-in, which is 3. An important note here is the way proc network computes the degree in directed graph. If A has a bi-directional link to another node, which means, it points to this node and it is pointed by it, proc network accounts for 1degree-in and 1degree-out. Summarizing degree-in and degree-out will give us a number greater than the actual number of connections. For example, suppose node A has a bi-directional link with any of the other nodes it is currently connected. Node A would have a degree-out of 4 and a degree-in of 2, or a degree-out of 3 and a degree-in of 3. In any case, the total degree would be 6 (4 + 2 or 3 + 3), even though node A would still have just five distinct connections (five unique nodes connected to it). This is important when trying to describe how connected a node is. The degree centrality in an undirected network also informs the number of distinct connections a particular node has. However, the degree centrality in a directed network informs the number of connections incident to a node, incoming and outgoing connections independently. It does not necessarily inform the number of distinct connections a particular node has.

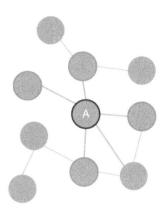

Figure 3.2 Undirected graph representing the connections of node A.

3.3.1 Computing Degree Centrality

In general terms, the degree centrality is defined by the following formulas.

For undirected graphs, degree centrality is defined as:

$$C_{\mathrm{d}}(i) = \sum_{e \in \delta_i} w_l^L$$

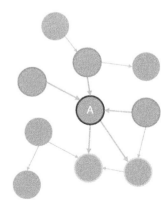

Figure 3.3 Directed graph representing the connections of node A, considering the direction of the relationships between A and the other nodes.

where $C_d(i)$ is the degree centrality of node i, δ_i represents the set of links that are incident to node i, and w_l^L is the weight of link l.

For directed graphs, we have two separate degree metrics summarizing the total degree, the degree-out and the degree-in. The out-degree centrality of a node i is:

$$C_d^{out}(i) = \sum_{e \in \delta_i^{out}} w_l^L$$

where δ_i^{out} represents the set of outgoing links that are connected to node i, and w_l^L is the weight of link l.

Similarly, the in-degree centrality of a node i is:

$$C_d^{in}(i) = \sum_{e \in \delta_i^{in}} w_l^L$$

where δ_i^{in} represents the set of incoming links that are connected to node i, and w_l^L is the weight of link l.

The total degree centrality of a node i for a directed graph is calculated as:

$$C_d(i) = C_d^{out}(i) + C_d^{in}(i)$$

Proc network allows us to compute the degree centrality for both directed and undirected graphs. Also, it allows us to compute the degree centrality using the link weights or not. In this way, the degree centrality will be weighted or unweighted. Proc network can compute both degree centralities though at the same time, weighted and unweighted.

We can define how proc network will calculate the degree centrality by using the option DEGREE=WEIGHT| UNWEIGHT|BOTH.

The following code describes how to compute the degree centrality using SAS proc network. The graph built by this code is unweighted (links with no weights), but it can be used to compute the degree centrality (or any other network centrality) for both directed and undirected graphs.

```
data mycas.links;
    input from $ to $ @@;
datalines;
J B  B C  B A  I A  D A  A E  A F  G F  G H  E F  D E
;
run;

proc network
    direction = undirected
    links = mycas.links
    outlinks = mycas.outlinks
    outnodes = mycas.outnodes
    ;
    linksvar
        from = from
        to = to
    ;
    centrality
        degree
    ;
run;
```

As a result of running proc network, we will get a set of messages summarizing the process. The Figure 3.4 shows the summary results for proc network.

Proc network also generates two output datasets, one for the output links (OUTLINKS) and one for the output nodes (OUTNODES), which includes the degree centrality results assigned to the column centr_degree_out. The Figure 3.5 shows the output table for the nodes.

The NETWORK Procedure

Problem Summary

Number of Nodes	10
Number of Links	11
Graph Direction	Undirected

The NETWORK Procedure

Solution Summary

Problem Type	Centrality
Solution Status	OK
CPU Time	0.00
Real Time	0.00

Output CAS Tables

CAS Library	Name	Number of Rows	Number of Columns
CASUSER(Carlos.Pinheiro@sas.com)	OUTLINKS	11	2
CASUSER(Carlos.Pinheiro@sas.com)	OUTNODES	10	2

Figure 3.4 Output results by proc network.

« OUTNODES Table Rows: 10 | Columns: 2 of 2 | Rows 1 to 10

	node	centr_degree_out
1	J	1
2	B	3
3	C	1
4	A	5
5	I	1
6	D	2
7	E	3
8	F	3
9	G	2
10	H	1

Figure 3.5 Output dataset for the nodes based on an undirected graph with unweighted links.

Notice that in the output dataset OUTNODES we can see the centr_degree_out for node A as 5, like we saw before when we were looking at the graph. This degree centrality is calculated based on an undirected graph with unweighted links, even though the centrality column name is centr_degree_out.

The output links dataset looks pretty much like the original links dataset, particularly because for the degree centrality, there is no calculation for the links. Nevertheless, proc network still produces the output for the links dataset. The Figure 3.6 shows the output table for the links.

« OUTLINKS   Table Rows: 11 | Columns: 2 of 2 | Rows 1 to 11 | ⬆ ↑ ↓ ⬇ ⋮

▽ | Enter expression   ⌕

	⚠ from	⚠ to
1	J	B
2	B	C
3	B	A
4	I	A
5	D	A
6	A	E
7	A	F
8	G	F
9	G	H
10	E	F
11	D	E

Figure 3.6 Output dataset for the links.

For a directed graph, we just need to change the attribute direction in the proc network statement, as shown in the following code (DIRECTION=DIRECTED).

```
proc network
    direction = directed
    links = mycas.links
    outlinks = mycas.outlinks
    outnodes = mycas.outnodes
    ;
    linksvar
        from = from
        to = to
    ;
    centrality
        degree
    ;
run;
```

The following figure shows the output dataset produced by proc network when computing the degree centrality based on a directed graph. Now the nodes dataset has three degree centrality metrics, centr_degree_in, centr_degree_out, and centr_degree. The Figure 3.7 shows the output table for the nodes.

Notice that node A in the OUTNODES dataset has centr_degree_in as 3, centr_degree_out as 2, summing up centr_degree to 5. This centralities metrics represent the graph we saw before, where node A has three incoming connections and two outgoing connections, coming to five connections in total.

Proc network produces the output dataset for the links, but it will be exactly the same as presented before for the undirected graph with unweighted links, and very similar to the original links dataset.

Now, let us take a look at how proc network address graphs with weighted links. This changes completely the way proc network computes the degree centrality, and also produces a small change in the output dataset for the links.

Let us consider the following directed graph with link weights presented in the Figure 3.8.

« OUTNODES   Table Rows: 10 │ Columns: 4 of 4 │ Rows 1 to 10 │ ⤒ ↑ ↓ ⤓ │ ↻ ▾ │ ⋮

▽ | Enter expression   ⌕

	⚠ node	⊞ centr_degree_in	⊞ centr_degree_out	⊞ centr_degree
1	J	0	1	1
2	B	1	2	3
3	C	1	0	1
4	A	3	2	5
5	I	0	1	1
6	D	0	2	2
7	E	2	1	3
8	F	3	0	3
9	G	0	2	2
10	H	1	0	1

Figure 3.7 Output dataset for the nodes based on a directed graph with unweighted links.

Based on weighted links, proc network can compute the degree centrality weighted, which takes into account the weight of the links instead of just the number of connections, both coming into a node and going out from it.

The following code computes both degree centralities, weighted and unweighted, by using the option DEGREE=BOTH.

In this code we are also recreating the links dataset by adding weights to the links, so proc network can use weighted links when computing the degree centrality.

```
data mycas.links;
    input from $ to $ weight @@;
datalines;
J B 2 B C 1 B A 5 I A 3 D A 2 A E 2 A F 1 G F 2 G H 1 E F 3 D E 4
;
run;
```

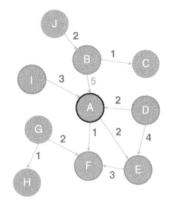

Figure 3.8 Directed graph with weighted links.

```
proc network
    direction = directed
    links = mycas.links
    outlinks = mycas.outlinks
    outnodes = mycas.outnodes
    ;
    linksvar
        from = from
        to = to
    ;
    centrality
        degree = both
    ;
run;
```

The Figure 3.9 shows the results in the output nodes dataset.

	node	⊛ centr_degree_in	⊛ centr_degree_out	⊛ centr_degree	⊛ centr_degree_in_wt	⊛ centr_degree_out_wt	⊛ centr_degree_wt
1	J	0	1	1	0	2	2
2	B	1	2	3	2	6	8
3	C	1	0	1	1	0	1
4	A	3	2	5	10	3	13
5	I	0	1	1	0	3	3
6	D	0	2	2	0	6	6
7	E	2	1	3	6	3	9
8	F	3	0	3	6	0	6
9	G	0	2	2	0	3	3
10	H	1	0	1	1	0	1

Figure 3.9 Output dataset for the nodes considering weighted and unweighted links.

Notice the links weights in the definition of the links dataset and the graph shown before. As we requested proc network to compute the degree centrality weighted and unweighted, now the nodes dataset has six degree centralities metrics, centr_degree_in, centr_degree_out, centr_degree, centr_degree_in_wt, centr_degree_out_wt, centr_degree_wt, centr. We can see that the unweighted metrics accounts for the number of connections, as seen before. For the node A, degree in equals to 3 (three incoming links), degree out equals to 2 (two outgoing links), and total degree equals to 5. However, the weighted degree centrality accounts for the link weights rather than the number of connections. Then, for the same node A, the degree in equals to 10 (one incoming link from B with weight 5, one incoming link from I with weight 3, and one incoming link from D with weight 2). Analogously, the weighted degree out equals to 3 (one outgoing link to F with weight 1, and one outgoing link to E with weight 2).

The weighted degree is a good way to consider not just the number of connections the node has, but mostly the strength of their connections, or if their relationships are frequently or important. The weight for the link can vary according to the problem. For example, in communications, we can value more calls than text messages. We can also value more long calls than short calls. The link weights can then be based on the call duration and/or the call frequency. We can add it up the frequency of the text messages, or the size of them if this information is available. We can consider different types of messages, like SMS and MMS, and then give them different values. As we can easily realize, the way we are going to create the network is very open, and the approach we are going to use to weight nodes and links really depend on the business problems we are trying to solve by using network analysis.

To finalize the degree centrality, let us take a look at the undirected graph again, but now based on weighted links. It also changes the way proc network computes the degree centrality and the outcomes for nodes and links datasets.

The following code executes proc network to compute the degree centrality for the undirected graph based on weighted links.

```
proc network
    direction = undirected
    links = mycas.links
    outlinks = mycas.outlinks
    outnodes = mycas.outnodes
    ;
    linksvar
        from = from
        to = to
    ;
    centrality
        degree = weight
    ;
run;
```

The output table OUTNODES presented in the Figure 3.10 contains the results for the degree centrality for the nodes within the undirected graph considering the links weights.

« OUTNODES   Table Rows: 10   Columns: 2 of 2   Rows 1 to 10   ⊤ ↑ ↓   ⋮

Enter expression

	node	centr_degree_out_wt
1	J	2
2	B	8
3	C	1
4	A	13
5	I	3
6	D	6
7	E	9
8	F	6
9	G	3
10	H	1

Figure 3.10 Output dataset for the nodes based on an undirected graph with weighted links.

« OUTLINKS   Table Rows: 11   Columns: 3 of 3   Rows 1 to 11   ⊤ ↑ ↓ ↧   ⋮

Enter expression

	from	to	weight
1	J	B	2
2	B	C	1
3	B	A	5
4	I	A	3
5	D	A	2
6	A	E	2
7	A	F	1
8	G	F	2
9	G	H	1
10	E	F	3
11	D	E	4

Figure 3.11 Output dataset for the links based on an undirected graph with weighted links.

Proc network also produces the output links dataset showed in the Figure 3.11 as OUTLINKS. But here, for the weighted links, we can see a small difference.

As we can see, the links dataset has now weight for the links. This weights can represent the importance of the link, or the frequency of the connections between the nodes. The new graph is shown below.

In terms of practical applications, the degree centrality is useful to tell us how many connections an individual has, or a customer has. This information may highlight a popular person, an influential one, or even a non-expected well-connected entity in a particular business scenario. For marketing purposes, the degree centrality has a straightforward application spotting individuals who can effectively perform as promoters or detractors, as they are popular in the network holding up many connections. In both cases, reaching out to these customers can represent spread a good message about the company, or product or service, or retain not just these possible detractors but all the individuals are connected to them.

Outlier analysis can also benefit from the degree centrality. In some business problems, the average number of expected connections can be an indicator of normal behavior. Entities with a huge number of connections can be suspicious. However, it is important to notice that depending on the business scenario under analysis, it is legit to have nodes with great number of connections and then a possible filter needs to be deployed. For example, when creating a network based on calls and text messages from a telecommunications provider. We would be interested in the relationships among the subscribers, the customers. However, the provider also uses calls and text messages to communicate to customers and of course, prospects. When looking at all connections over time, it is easy to notice that some numbers have a huge degree centrality. These number can be eventually service numbers for different companies, not just the carrier. For example, call center numbers, contact center numbers, help desk, etc. all these numbers may have a very high degree centrality (both in and out) but they are not important within the network. They are not telling us much about customers relationships. And even worse, they can create groups of customers that in fact do not exist. These numbers can connect customers together just by the fact the call to the contact center to request some information about a product or a service. In most of the cases, these service numbers need to be removed from the network, allowing just customers to be connected to each other. Same thing happens in different networks, in different business scenarios and industries. It is important to evaluate the data describing the network and proceed all possible adjustments before creating the network and start computing the centrality metrics.

3.3.2 Visualizing a Network

So far, we are seeing all the results from proc network in terms of tables and datasets. What if we want to see the real network, allowing us to investigate the actual network by looking at nodes and links and how they are related?

A big question in any network analysis project is always about the visualization. SAS allows us to visualize the network in Visual Analytics. We will need to promote the links dataset first in order to reference it in Visual Analytics. The network analysis objects in Visual Analytics has a small drawback. All nodes plotted should be in the origin/from list. For example, plotting the original **links** dataset we have been working, straightforward in Visual Analytics, would give us the following network diagram, presented in the Figure 3.12.

Carefully looking at this graph, we can notice that there are some connections missing, as well as some nodes. This happens because there are destination (to) nodes not in the origin (from) column in the links dataset. As all nodes in the graph needs to be origin to some connections, we are missing the only destination nodes and their connections. A method to overcome this issue is to create dummy connections containing the destination nodes as origin (from), pointing them to nowhere, with missing values in the destination (to) column.

The following code creates this scenario.

```
data mycas._tmp_linksext;
   set mycas.links(keep=to rename=(to=from));
   to='';
run;

data mycas.linksext(promote=yes);
   set mycas.links mycas._tmp_linksext;
run;
```

Once we recreate the links dataset just for the Visual Analytics, with all nodes in the origin (from) column, now we can see the true graph we have been working on. The new graph is showed in the Figure 3.13.

The graph is not exactly in the same layout, but all nodes and links are there, in both example graph we saw before and the network objective in Visual Analytics.

An alternative way to visually analyze network and plot graphs is by integrating SAS code in SAS Studio to open-source packages. We can do this with any package, to output results of graphs either using maps or not. We will see examples of network plotting on maps once we have coordinates for the nodes. This approach gives us a smooth workflow when analyzing networks, considering the common steps of coding, assessing results, plotting

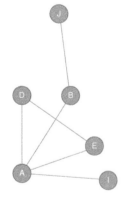

Figure 3.12 Network analysis object in visual analytics with the original links dataset.

graphs, and back to coding. This back-and-forth approach is pretty much what we do in data science, and it is not much different when doing network analysis and network optimization. The reason this process is smooth is because we can go back and forth in the coding-assessment-plotting-coding process without leaving SAS Studio. We can use SAS data steps to create HTML files with the network results (including maps), download and open them up using the same browser we are working on SAS Studio (in a different tab) and keep coding and analyzing the network problem continuously. Here is an example using a network analysis package called **vis.js**. For the network maps, mostly during the network optimization topics, we are going to use the packaged called **Leaflet**.

As we can plot multiple networks along the process, going back and forth during the network outcomes evaluation, we are going to use a macro to write the HTML file to invoke vis.js and plot the graph. Here is a very simple example where we can plot exactly the same graph we have been working on.

The following SAS code create the macro. There are only three parameters, the directory to create the HTML file, the nodes dataset, and the links dataset. Of course, this macro can be much more complex and complete to include the multiple options provided by vis.js to plot a graph. The idea here is just to be as simple as possible.

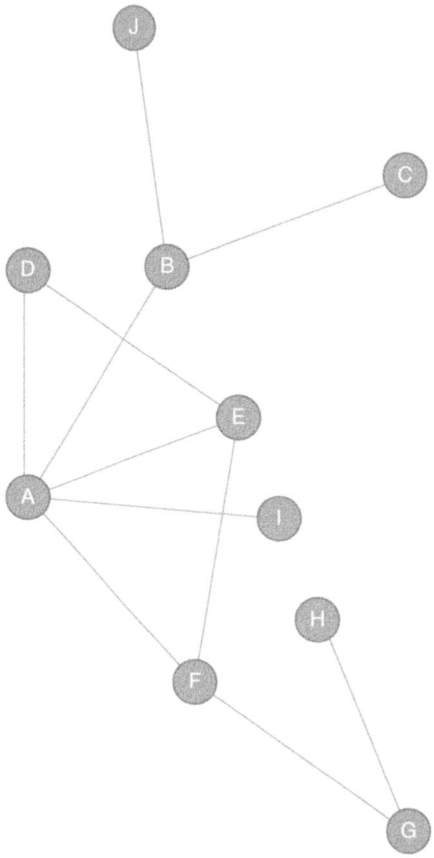

```
%macro cpnetvis(dir=,nodes=,links=);

   data linesnodes;
      length line $32767.;
      set nodes end=eof;
      k+1;
      if k=1 then
         do;
            line='var nodes=new
vis.DataSet([{id:"'||compress(n)||'",label:"'||compress(l)||'",group:"'||compress
(g)||'",value:"'||compress(v)||'"}';
            output;
         end;
      else
         do;
         line=',{id:"'||compress(n)||'",label:"'||compress(l)||'",group:"'||compress
(g)||'",value:"'||compress(v)||'"}';
            output;
         end;
      if eof then
         do;
            line=']);';
            output;
         end;
   run;

   data lineslinks;
      length line $32767.;
      set links end=eof;
      k+1;
      if k=1 then
```

Figure 3.13 Network analysis objects in visual analytics with links dataset modified.

```
            do;
               line='var edges=new vis.DataSet([{from:"'||compress(f)||'", to:"'||compress
(t)||'",value:"'||compress(v)||'"}';
               output;
            end;
         else
            do;
               line=',{from:"'||compress(f)||'",to:"'||compress(t)||'",value:"'||compress
(v)||'"}';
               output;
            end;
         if eof then
            do;
               line=']);';
               output;
            end;
   run;

   data lines;
      set linesnodes lineslinks;
   run;

   proc sql noprint;
      select case when max(v)=0 then 0 else 15 end as aux into :ml from links;
   quit;

   filename arq "&dir/visnet.htm";

   data _null_;
      set lines end=eof;
      file arq;
      i+1;
      if i=1 then
         do;
            put '<html>';
            put '<head>';
            put '<script type="text/javascript" src="https://unpkg.com/vis-network/
standalone/umd/vis-network.min.js"></script>';
            put '<style type="text/css"> #mynetwork {width:"100%";height:"100%";}
></style>';
            put '</head>';
            put '<body>';
            put '<div id="mynetwork"></div>';
            put '<script type="text/javascript">';
         end;
      put line;
      if eof then
         do;
            put 'var container=document.getElementById("mynetwork");';
            put 'var data={nodes:nodes,edges:edges};';
            line='var
            options={layout:{improvedLayout:true},nodes:{shape:"dot",scaling:{min:10,
```

```
         max:30},font:{size:12,face:"Tahoma"}},edges:{scaling:{min:0,max:'||
         compress(&ml)||'},color:{inherit:"from"},smooth:{type:"continuous"}},
         physics:{forceAtlas2Based:{gravitationalConstant:-50,centralGravity:0.01,
         springLength:250,springConstant:0.1},maxVelocity:50,
         solver:"forceAtlas2Based",timestep:0.5,stabilization:{iterations:100}}};';
         put line;
         put 'var network=new vis.Network(container, data, options);';
         put '</script>';
         put '</body>';
         put '</html>';
      end;
   run;

%mend cpnetvis;
```

At some point during the coding-assessment process, we will have some results to visually evaluate. Once we have the datasets for nodes and links, we just need to pass them as parameters to the macro, as shown in the following code.

```
%CPNetVis(dir=&dm,nodes=nodes,links=links);
```

This macro creates a HTML file called visnet.htm.
The file looks like the one presented in the Figure 3.14.

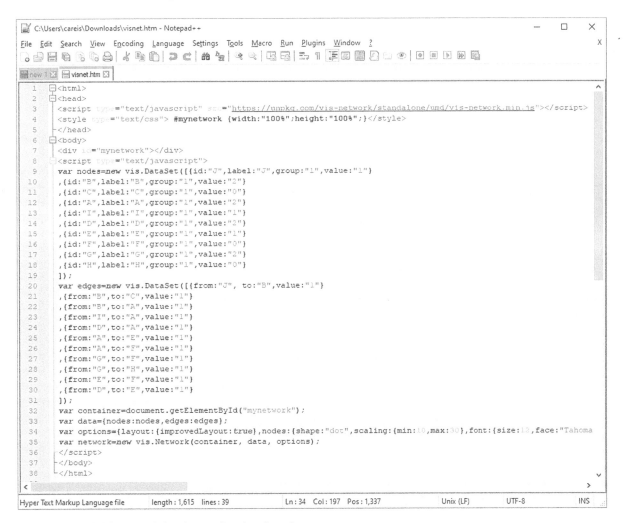

Figure 3.14 HTM file containing the graph to be plotted.

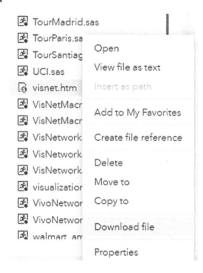

Figure 3.15 Download the visnet.htm file.

File: /enable01-export/enable01-aks/homes/Carlos.P

Figure 3.16 Open the visnet.htm file in a browser.

We can simply download this file from SAS Studio and open it in the same browser we are using to access SAS Studio, as the sequence of pictures describes. The Figure 3.15 shows how to download the file.

Right click on the name of the file (visnet.htm) to open the options menu and select Download file. The Figure 3.16 shows how to open the file in a new tab on the browser.

The HTM file (visnet.htm) will be downloaded to our local computer (Download folder in a Windows machine for example). The file might also appear in the status bar of the browser, like in the Figure 3.16. If so, we just need to click on the file name to open the HTM file with the network in a new tab. The Figure 3.17 shows the resulting graph.

In addition to smooth the network analysis back and forth flow when coding, analyzing, and plotting the datasets and graphs, this packaged allow us to use multiple plotting algorithms and many features to identify relevant characteristics of the network. It allows us to interact dynamically to the graph drag and dropping nodes from one place to another and eventually isolating some key nodes within the network to better investigate them. We can definitely plot larger graphs using this package. The following figure shows the famous Les Miserables network, based on the homonym French movie. The Figure 3.18 shows the graph presenting the Les Miserable network.

3.4 Influence Centrality

This influence centrality measure is a generalization of the degree centrality, considering, in addition to the number of incoming and outgoing links incident to node v, the weights of the nodes and links adjacent to v. The influence centrality can be computed whether we use the weights of the nodes and links.

Let us consider the following undirected graph with weighted links presented in the Figure 3.19.

The centrality measure influence 1 is the first-order centrality for a node. It describes how many other nodes are directly connected to it. This measure is popularly understood as how many friends a node has. It indicates how many nodes can be expected to be directly influenced by a node. The influence 1 for a node A considers the weight of nodes and links adjacent to A.

Sometimes the number of connections does not matter as much as the importance of the connections. The centrality influence 1 considers the relevance of the links (from the weights) and the importance of the node itself (its specific weight). For example, how many customers a subscriber is connected to in a mobile telephone environment is less important than the strength of the connection links and the value of the subscribers themselves.

This centrality is often used to understand the impact of a business event that requires strong relationships between the members, such as churn.

Influence 2 represents the second-order centrality for a node, reflecting how many nodes that a node is connected to are, in turn, connected to other nodes. This measure can be best understood as how many friends my friends have. It describes how many nodes can possibly be influenced by a node. The influence 2 for a node A considers the weights of nodes and links adjacent to A, as well as the weight of the links for the nodes adjacent to the nodes adjacent to A.

The centrality influence 2 extends the concept of centrality to the second order by adding the importance of the links that influence 1 connected nodes have. A subscriber might not directly have important friends, but his friends might have important friends – and thus a second level degree of influence is useful to evaluate event impacts. In terms of business applications, a node with a high value of influence 2 might indicate that the individual is a good candidate to spread a message. In telecommunications, this might be a subscriber who has an average number of connections (with average relevance) to other customers who have a substantial number of connections (and valuable ones). In this case, the average subscribers might be good potential starter candidates for a product-adoption campaign.

Consider this example: a mobile operator identifies two customers, the first one well connected to a few important people, and the second one connected to a few less-important people. In the second instance, the people that she is connected to are well connected to a substantial number of good customers. Thus, the second customer might be in fact a better prospect to initiate the campaign than the first one. That first customer would have an influence 1 centrality higher than the second one, and the second customer would have an influence 2 centrality higher than the first one. On the other hand, for a different business scenario, such as for churn, the mobile company would more likely select the first customer to spread the message for retention, but not for diffusion.

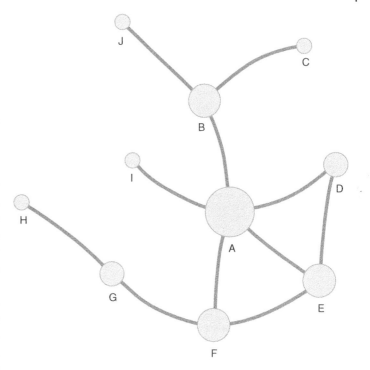

Figure 3.17 The graph plotted by the visnet.htm file containing the links dataset.

3.4.1 Computing the Influence Centrality

For undirected graphs, the neighbors are all nodes connected. The general formulas are as following.

$$C_1(A) = \frac{\sum_{v \in \delta_i} w_l^L}{\sum_{v \in N} w_j^N}$$

$$C_2(A) = \sum_{v \in N_i} C_1(v)$$

where δ_i represents the set of links that are incident to node i, N_i represents the set of neighbors of node i (all the nodes that are connected by links incident with node i), w_l^L is the weight of link l, and w_j^E is the weight of node j.

As the name suggests, this metric gives some indication of potential influence, performance, or ability to transfer knowledge from one node to its network throughout its neighbors.

Considering the graph before, the influence 1 of node A can be computed as follows:

$$C_1(A) = \frac{w_{AB} + w_{AD} + w_{AI} + w_{AE} + w_{AF}}{w_A + w_B + w_C + w_D + w_E + w_F + w_G + w_H + w_I + w_J}$$

$$C_1(A) = \frac{5 + 2 + 3 + 2 + 1}{1 + 1 + 1 + 1 + 1 + 1 + 1 + 1 + 1 + 1} = \frac{13}{10} = 1.3$$

The influence 2 of node A can be computed as follows:

$$C_2(A) = C_1(B) + C_1(D) + C_1(I) + C_1(E) + C_1(F) = 0.8 + 0.6 + 0.3 + 0.9 + 0.6 = 3.2$$

Proc network can compute the influence centrality considering the link weights or not.

We can define how proc network will calculate the influence centrality by using the option INFLUENCE=WEIGHT|UNWEIGHT|BOTH.

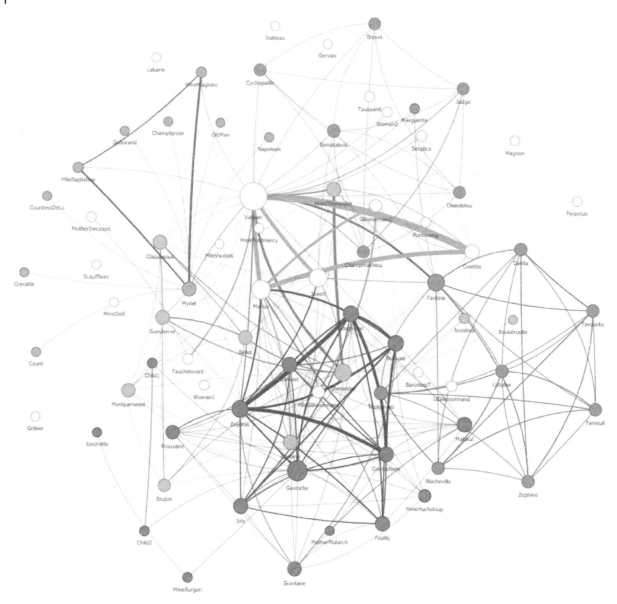

Figure 3.18 Les Miserables dataset plotted in vis.js.

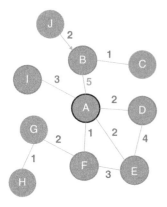

Figure 3.19 Undirected graph with link weights to compute influence centrality.

The following code shows how to compute the influence centrality in proc network. The links dataset, containing the weighted links, created for the degree centrality computation, is used here. This links dataset can be used for both directed and undirected graphs.

```
proc network
   direction = undirected
   links = mycas.links
   outlinks = mycas.outlinks
   outnodes = mycas.outnodes
   ;
   linksvar
      from = from
      to = to
```

```
      weight = weight
   ;
  centrality
      influence = weight
   ;
run;
```

The OUTNODES dataset presented in the Figure 3.20 contains the results for both influence centralities, centr_influence1_wt and centr_influence2_wt.

Notice that the values for node A are 1.3 for influence 1 and 0.3 and for influence 2, as described during the previous formulation. These values are computed based on the weighted links.

For unweighted links the calculation is similar. The links weights are not taken into account, or, as a straightforward method to calculate, all links will have the same weight, equals 1.

The formulation for node A will be as following:

$$C_1(A) = \frac{w_{AB} + w_{AD} + w_{AI} + w_{AE} + w_{AF}}{w_A + w_B + w_C + w_D + w_E + w_F + w_G + w_H + w_I + w_J}$$

$$C_1(A) = \frac{1 + 1 + 1 + 1 + 1}{1 + 1 + 1 + 1 + 1 + 1 + 1 + 1 + 1 + 1} = \frac{5}{10} = 0.5$$

$$C_2(A) = C_1(B) + C_1(D) + C_1(I) + C_1(E) + C_1(F)$$

$$C_2(A) = 0.3 + 0.2 + 0.1 + 0.3 + 0.3 = 1.2$$

In proc network we just need to repeat the previous code and set the option INFLUENCE=UNWEIGHT.

```
proc network
   direction = undirected
   links = mycas.links
   outlinks = mycas.outlinks
   outnodes = mycas.outnodes
   ;
   linksvar
      from = from
```

« OUTNODES   Table Rows: 10 | Columns: 3 of 3 | Rows 1 to 10   ↑ ↑ ↓ ↓   ⋮

	⬧ node	⊕ centr_influence1_wt	⊕ centr_influence2_wt
1	J	0.2	0.8
2	B	0.8	1.6
3	C	0.1	0.8
4	A	1.3	3.2
5	I	0.3	1.3
6	D	0.6	2.2
7	E	0.9	2.5
8	F	0.6	2.5
9	G	0.3	0.7
10	H	0.1	0.3

Figure 3.20   Output nodes dataset containing the influence centrality results for the undirected graph.

<< OUTNODES Table Rows: 10 | Columns: 3 of 3 | Rows 1 to 10   ⟟ ↑ ↓ ⤓ | ↻ ▾ | ⋮

▽ [Enter expression   ⌕]

	△ node	⊕ centr_influence1_unwt	⊕ centr_influence2_unwt
1	J	0.1	0.3
2	B	0.3	0.7
3	C	0.1	0.3
4	A	0.5	1.2
5	I	0.1	0.5
6	D	0.2	0.8
7	E	0.3	1
8	F	0.3	1
9	G	0.2	0.4
10	H	0.1	0.2

Figure 3.21 Output nodes dataset containing the influence centrality results for the undirected graph with unweighted links.

```
      to = to
   ;
   centrality
      influence = unweight
   ;
run;
```

The OUTNODES dataset presented in the Figure 3.21 contains the results for both influence centralities, centr_influence1_unwt and centr_influence2_unwt not considering the links weights.

The values for node A are now 0.5 for influence 1 and 1.2 and for influence 2, as described in the formulation above.

Let us move into directed graphs. For a directed graph the calculation is slightly different. Let us consider the following directed graph with weighted links presented in the Figure 3.22.

For directed graphs, the neighbors are only the out links. The general formula for influence centrality is shown below:

$$C_1(A) = \frac{\sum_{v \in \delta_i^{out}} w_l^L}{\sum_{v \in N} w_j^N}$$

$$C_2(A) = \sum_{v \in N_i^{out}} C_1(v)$$

where δ_i^{out} represents the set of outgoing links that are incident to node i, N_i^{out} represents the set of neighbors of node i (all the nodes that are connected by links incident with node i), w_l^L is the weight of link l, and w_j^E is the weight of node j.

The influence 1 of node A can be computed as follows:

$$C_1(A) = \frac{w_{AF} + w_{AE}}{w_A + w_B + w_C + w_D + w_E + w_F + w_G + w_H + w_I + w_J}$$

$$C_1(A) = \frac{1 + 2}{1 + 1 + 1 + 1 + 1 + 1 + 1 + 1 + 1 + 1} = \frac{3}{10} = 0.3$$

The influence 2 of node A can be computed as follows:

$$C_2(A) = C_1(F) + C_1(E) = 0 + 0.3 = 0.3$$

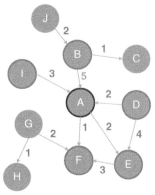

Figure 3.22 Directed graph to compute influence centrality.

The following code shows how to compute the influence centrality for the directed graph considering weighted links.

```
proc network
   direction = directed
   links = mycas.links
   outlinks = mycas.outlinks
   outnodes = mycas.outnodes
   ;
   linksvar
      from = from
      to = to
      weight = weight
   ;
   centrality
      influence = weight
   ;
run;
```

The OUTNODES dataset presented in the Figure 3.23 contains the results for both influence centralities, centr_influence1_wt and centr_influence2_wt, now considering a directed graph with weighted links.

The values for node A are 0.3 for influence 1 and 0.3 for influence 2, as described in the formulation. This values are computed based on a directed graph with weighted links.

We can also compute the influence centralities on a directed graph but not considering the links weights.

The influence 1 and influence 2 of node A would be computed as follows:

$$C_1(A) = \frac{1 + 1}{1 + 1 + 1 + 1 + 1 + 1 + 1 + 1 + 1 + 1} = \frac{2}{10} = 0.2$$

$$C_2(A) = 0 + 0.1 = 0.1$$

The following code shows how to compute the influence centrality for the directed graph not considering the links weights.

```
proc network
   direction = directed
   links = mycas.links
```

	node	⊕ centr_influence1_wt	⊕ centr_influence2_wt
1	J	0.2	0.6
2	B	0.6	0.3
3	C	0	0
4	A	0.3	0.3
5	I	0.3	0.3
6	D	0.6	0.6
7	E	0.3	0
8	F	0	0
9	G	0.3	0
10	H	0	0

Figure 3.23 Output nodes dataset containing the influence centrality results for the directed graph with weighted links.

```
outlinks = mycas.outlinks
outnodes = mycas.outnodes
 ;
 linksvar
    from = from
    to = to
 ;
 centrality
    influence = unweight
 ;
run;
```

The OUTNODES dataset presented in the Figure 3.24 contains the results for both influence centralities, centr_influence1_wt and centr_influence2_wt, now considering a directed graph with unweighted links.

The values for node A are 0.2 for influence 1 and 0.1 for influence 2, as described in the formulation.

As we saw for the degree centrality, we can plot the graph to show the network, using both approaches presented before, by creating and promoting the links dataset and use it in Visual Analytics, and via code invoking vis.js from the links and nodes datasets already created or as results of proc network. In order to make this textbook more concise, from now on, all graphs visualization will be suppressed unless they provide a relevant meaning to the algorithm described.

In business purposes, the influence centrality extends the concept of the degree centrality taking into consideration the strength of the relationships for all nodes within the network. For example, in a telecommunications network, a particular customer may have many connections, represented by a high value in the degree centrality. However, let us say that the strength of each link is very low. For example, that node has 100 connections, but each connection happens one or two times a month. Even though that customer has many connections, it is not exactly popular, or at least, this customer does not hold strong relationships with his or her peers. In a marketing campaign, a message spread out by this customer would probably be ineffective. Let's consider another customer, with just 20 connections. Let us also consider that all the connections for this node happen frequently, given to them a high value weight. Let us say, this customer communicates to his or her peers often, at least once a day. In a marketing campaign, for instance, trying to advertise a new product or service, a communication coming from this customer to his or her peers would be probably more effective than to the first customer, even though the

« OUTNODES Table Rows: 10 | Columns: 3 of 3 | Rows 1 to 10

	node	⊕ centr_influence1_unwt	⊕ centr_influence2_unwt
1	J	0.1	0.2
2	B	0.2	0.2
3	C	0	0
4	A	0.2	0.1
5	I	0.1	0.2
6	D	0.2	0.3
7	E	0.1	0
8	F	0	0
9	G	0.2	0
10	H	0	0

Figure 3.24 Output nodes dataset containing the influence centrality results for the directed graph using unweighted links.

first customer has 100 connections and the second one has only 20. If a message is supposed to be spread out and absorbed, it will probably be more effective in a frequent communication than in a rare one. Then, selecting customers with a reasonable number of connections but more important, with strong relationships (high value weights), would be ideal in a marketing campaign to advertise a product or service.

3.5 Clustering Coefficient

The clustering coefficient for a node v is the number of links between its neighbors divided by the possible number of links between them.

The centrality clustering coefficient computes the number of links between the nodes within its neighborhood, divided by the possible links that could exist among them. It is a metric about how connected a neighborhood of nodes is to each other.

If I am connected to 10 people and these 10 people are connected to each other at some level, the centrality clustering coefficient would measure how connected my neighborhood is by computing the number of connections between all my connected neighbors and all possible connections that can exist among them.

Many different types of networks rely on clustering coefficient to identify good diffusion nodes. If a company needs to target a segment of customers for a specific message (like a product advertisement), customers with high values for the clustering coefficient would be good candidates because they are members of well-connected neighborhoods, encouraging the advertising message to reach further.

3.5.1 Computing the Clustering Coefficient Centrality

The clustering coefficient for a node is the number of links between the nodes within its neighborhood divided by the number of links that could possibly exist between them.

The clustering coefficient centrality can be calculated using the following formulas.

For an undirected graph:

$$C(i) = \frac{\sum_{j \in N_i, k \in N_i: j < k} b_{jk}}{K(N_i)}$$

For a directed graph:

$$C(i) = \frac{\sum_{j \in N_i^{out}, k \in N_i^{out}: j \neq k} b_{jk}}{K(N_i^{out})}$$

where N_i represents the set of neighbors, excluding node i itself (unique nodes that are connected by links excluding self-links), N_i^{out} represents the set of out-neighbors, excluding node i itself (unique nodes that are connected by outgoing links excluding self-links), and b_{jk} specifies whether or not (1 or 0, respectively) a link from node j to node k exists in the link set E, as follows:

$$b_{jk} = \begin{cases} 1, & \text{if there exists } e \in E \text{ such that from } (e) = j \text{ and to}(e) = k \\ 0, & \text{otherwise} \end{cases}$$

For a particular node i, the clustering coefficient determines how close to being a clique (complete subgraph) the subgraph induced by itself, and its neighbor set E are. In social networks, a high clustering coefficient can help predict relationships that might not be known, confirmed, or realized yet. The fact that person A knows person B and person B knows person C does not guarantee that person A knows person C, but it is much more likely that person A knows person C than that person A knows some random person within the network.

Let us use the same undirected graph used to calculate the previous centrality metrics, showed in the Figure 3.25.

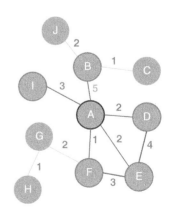

Figure 3.25 Undirected graph to compute the clustering coefficient centrality.

For example, node A in the graph is connected to the nodes B, D, E, F, and I. Then the clustering coefficient for node A is

$$C(A) = \frac{\textit{Number of links between nodes } B, D, E, F, I}{\textit{Number of possible links between them}} = \frac{2}{10} = 0.2$$

The number of links between the nodes in the neighborhood of A can be calculated as follows:

$$\#\textit{possible links} = \frac{N \times (N-1)}{2} = \frac{5 \times 4}{2} = 10$$

We can invoke proc network to calculate the clustering coefficient centrality by using the option CLUSTERINGCOEFFICIENT.

The following code shows how to compute the clustering coefficient centrality using proc network for au undirected graph

```
proc network
    direction = undirected
    links = mycas.links
    outlinks = mycas.outlinks
    outnodes = mycas.outnodes
    ;
    linksvar
        from = from
        to = to
        weight = weight
    ;
centrality
        clusteringcoefficient
    ;
run;
```

The OUTNODES dataset presented in the Figure 3.26 contains the results for the clustering coefficient centrality centr_cluster for an undirected graph.

Figure 3.26 Output nodes dataset containing the clustering coefficient centrality results for the undirected graph.

« OUTNODES Table Rows: 10	Columns: 2 of 2	Rows 1 to 10 ⋮
△ node	⊕ centr_cluster
1	J	0
2	B	0
3	C	0
4	A	0.2
5	I	0
6	D	1
7	E	0.6666666667
8	F	0.3333333333
9	G	0
10	H	0

Notice that the clustering coefficient value for node A is 0.2, as described during the formulation.

For directed graphs, the clustering coefficient computation changes a little bit. It considers as the neighborhood of A only the nodes pointed out by A.

Let us consider now the directed graph showed in the Figure 3.27.

In the example above, the neighborhood of A would be just E and F (instead of I, B, D, E, and F produced by undirected graph).

Then the clustering coefficient centrality would be calculated as follows:

$$C(A) = \frac{Number\ of\ links\ between\ nodes\ E, F}{Number\ of\ possible\ links\ between\ them} = \frac{1}{2} = 0.5$$

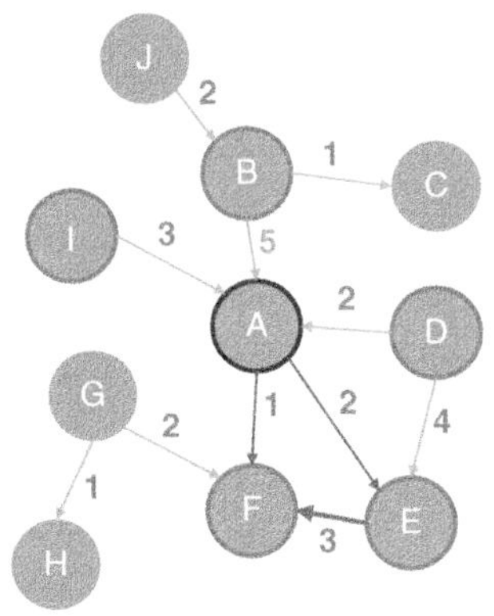

The number of links between the nodes in the neighborhood of A considering a directed graph includes both directions and can be calculated as follows:

$$\#possible\ links = N \times (N-1) = 2 \times (2-1) = 2 \times 1 = 2$$

Figure 3.27 Directed graph to compute the clustering coefficient centrality.

The following code shows how to compute the clustering coefficient centrality using proc network for a directed graph.

```
proc network
    direction = directed
    links = mycas.links
    outlinks = mycas.outlinks
    outnodes = mycas.outnodes
    ;
    linksvar
        from = from
        to = to
        weight = weight
    ;
centrality
    clusteringcoefficient
    ;
run;
```

The OUTNODES dataset presented in the Figure 3.28 contains the results for the clustering coefficient centrality centr_cluster when computed on a directed graph.

Notice that the clustering coefficient value for node A is 0.5, as described during the formulation.

In terms of business, the clustering coefficient centrality can be used to reinforce a message throughout the network. A node, or a person, with high clustering coefficient means that its neighbors are well connected. If this node spreads out a message through its neighbors, there is a great chance that the message will go back and forth throughout the neighborhood. As all neighbors are connected, they will probably communicate the same message between them, emphasizing the message and allowing the message to be faster absorbed and most likely more effectively. A quote attributed to Joseph Goebbels, a German Nazi politician, characterizes that. "A lie told once remains a lie, but a lie told a thousand times becomes the truth." Of course, no matter how many times a lie is told, it is still a lie. Maybe this is not a good example, but the concept here is the following. A message replicated just once does not take effect. But a message replicated many times throughout the same senders and receivers, may be absorbed more effectively. That type of message may occur many times in Congress around the world, no matter if lobby is allowed or not. A message does not need to be spread out widely, but strongly, step by step.

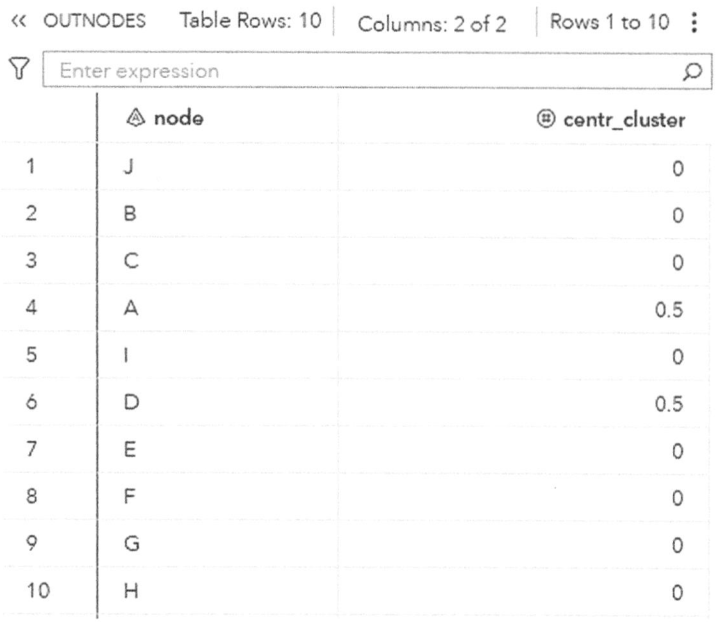

« OUTNODES   Table Rows: 10 | Columns: 2 of 2 | Rows 1 to 10 ⋮

▽ | Enter expression   🔍

	△ node	⊕ centr_cluster
1	J	0
2	B	0
3	C	0
4	A	0.5
5	I	0
6	D	0.5
7	E	0
8	F	0
9	G	0
10	H	0

Figure 3.28 Output nodes dataset containing the clustering coefficient centrality results for the directed graph.

3.6 Closeness Centrality

Closeness centrality is a measure of distance, proximity. It is calculated as the average of the shortest paths from one node to all of the others. This measure indicates how long it would take some message to spread from a particular node to all of the others. Spreading a message throughout the network is the viral effect in many networks (such as social).

The closeness measure indicates the mean of the geodesic distances (that is, the shortest path in the network) between a node and all other nodes connected with it. In other words, this closeness describes the average distances between one node and all other connected nodes. It can be used to understand how long a message will take to spread throughout the network from a node, n, essentially describing the speed of a message within social structures.

The closeness centrality is very often used to indicate what constitutes a central node – that is, a point within the network that is close to other points, measured by short paths to reach other vertices from its position. In some business applications, such as in telecommunications, it is a good measure to describe subscribers who are closely knit to others and therefore are more likely to influence them during an event in which the strength of relationship is important, like churn or portability.

Closeness centrality is the reciprocal of the average of the shortest paths (geodesic distances) to all other nodes. Closeness can be thought of as a measure of how long it would take information to spread from a given node to other nodes in the network.

3.6.1 Computing the Closeness Centrality

The general formula for closeness centrality is shown below:

$$C_C(i) = \frac{|N| - 1}{\sum_{j \in N \setminus u} d_{ij}^N + |N \setminus R(i)| \, p}$$

where $d_{ij}^N = \begin{cases} d_{ij}, & \text{if } d_{ij} < \infty \\ p, & \text{otherwise} \end{cases}$

and d_{ij} is the shortest path between i and j.

For undirected graphs, closeness centrality is calculated considering all nodes and links at once.

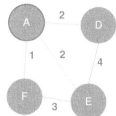

Figure 3.29 Undirected graph to compute closeness.

Let us use just a subset of the previous graph, showed in the Figure 3.29.

It is important to notice here that the shortest path has an optimization concept to minimize an objective function, like minimize the cost or the distance between the nodes. However, in social science studies, this objective function should be maximized instead. The straightforward method to do this is to inverse the weights for the links between the nodes. If they represent for example, the time any two persons spend together over the phone, less time will give the shortest path. But again, in social studies, we want to emphasize the time they spend together, so bigger is better, or shorter. By inversing the weights, the objective function remains the same, minimizing the distance or the time between any two nodes. But we know that by inversing weights, we are focusing on higher weights rather than lower ones. The Figure 3.30 shows the graph with inversed weights.

For undirected graphs the formula to calculate the closeness centrality is as following.

$$C_c(i) = s(i)\left(\frac{n(i)}{\sum_{j \in R(i)} d_{ij} + |N \backslash R(i)| p}\right)$$

where $R(i) = \{j \in N : d_{ij} < \infty\}$ is the set of reachable nodes from node i. The set of unreachable nodes from node i is $N \backslash R(i) = \{j \in N : d_{ij} = \infty\}$.

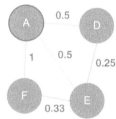

Figure 3.30 Undirected graph with inversed weights.

Considering the undirected subgraph above, the closeness centrality for node A will be as following.

$$C_C(A) = \frac{4-1}{d_{AF} + d_{AE} + d_{AD}} = \frac{3}{0.83 + 0.5 + 0.5} = \frac{3}{1.83} = 1.63$$

For direct graphs, it is possible to compute the closeness centrality considering the direction of the links. In that case, we have three different versions of the closeness centrality: the out centrality, the in centrality, and the total centrality.

$$C_c^{out}(i) = s^{out}(i)\left(\frac{n^{out}(i)}{\sum_{j \in R^{out}(i)} d_{ij} + |N \backslash R^{out}(i)| p}\right)$$

$$C_c^{in}(i) = s^{in}(i)\left(\frac{n^{in}(i)}{\sum_{j \in R^{in}(i)} d_{ji} + |N \backslash R^{in}(i)| p}\right)$$

where $R^{out}(i) = \{j \in N : d_{ij} < \infty\}$ is the set of reachable nodes from node i, whereas $R^{in}(i) = \{j \in N : d_{ji} < \infty\}$ is the set of nodes from which there is a finite path to node i. The set of unreachable nodes from node i is $N \backslash R^{out}(i) = \{j \in N : d_{ij} = \infty\}$, whereas $N \backslash R^{in}(i) = \{j \in N : d_{ji} = \infty\}$ is the set of nodes from which there is no finite path to node i.

The closeness centrality can be computed based on weighted links, unweighted links, or both.

We can define how proc network will calculate the closeness centrality by using the option CLOSE=WEIGHT| UNWEIGHT|BOTH.

Proc network also consider different methods to account for the missing links, or the links that does not exist between some pairs of nodes. When calculating the closeness centrality, proc network accounts for the shortest path between all pairs of nodes within the network. Some of these pairs may not have a link between them. For these cases, there are few approaches to define the "distance" between the pairs of nodes that are not connected. The first approach considers the diameter of the network when accounting for the unconnected nodes (diameter plus one). For each pair of node where a connections does not exist, proc network defines the distance between those nodes as the diameter plus one. The diameter is defined as the shortest path between the most distant nodes within the network. The second approach is to use the harmonic formula to compute the closeness centrality. The harmonic centrality is a variant of the closeness centrality that calculates the average of the reciprocal of the shortest path distances from a specific node to all other nodes within the network. The reciprocal is defined as the sum of the inverted distances instead of the invested sum of the distances, as normally calculated for closeness. The harmonic mean avoids the cases where the infinite distances (unconnected nodes) become more important than the other distances (connected nodes). The third method uses the number of nodes as the shortest path distance between disconnected nodes. Finally, the fourth and last method uses zero as the shortest path distance between the pairs of unconnected nodes.

We can define how proc network will account for the missing links and the shortest path distance between disconnected nodes by using the option CLOSENOPATH=DIAMETER|HARMONIC|NNODES|ZERO.

The following code shows how to compute the closeness centrality using proc network for an undirected graph. The method used to compute the shortest path distance for disconnected nodes is by using the diameter of the network.

```
data mycas.links;
    input from $ to $ weight @@;
datalines;
D A 2 A E 2 A F 1 E F 3 D E 4
;
run;

proc network
    direction = undirected
    links = mycas.links
    outlinks = mycas.outlinks
    outnodes = mycas.outnodes
    ;
    linksvar
        from = from
        to = to
        weight = weight
    ;
    centrality
        close = weight
            closenopath = diameter
    ;
run;
```

The OUTNODES dataset presented in the Figure 3.31 contains the results for the closeness centrality centr_close_wt when computed based on an undirected graph.

Notice that the closeness centrality value for the node A is 1.63, as described during the formulation.

The calculation for an undirected graph considering unweighted links is similar to the process we just saw for the weighted links. For the unweighted links, all links within the graph are set to 1.

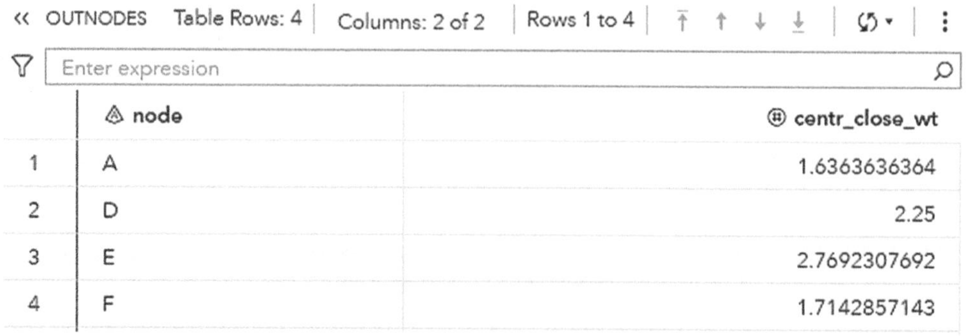

	node	centr_close_wt
1	A	1.6363636364
2	D	2.25
3	E	2.7692307692
4	F	1.7142857143

OUTNODES · Table Rows: 4 · Columns: 2 of 2 · Rows 1 to 4

Figure 3.31 Output nodes dataset containing the closeness centrality results for the undirected graph.

	⬣ node	⊞ centr_close_unwt
1	A	1
2	D	0.75
3	E	1
4	F	0.75

« OUTNODES Table Rows: 4 | Columns: 2 of 2 | Rows 1 to 4 ⊤ ↑ ↓ ⊥ ⋮

▽ Enter expression 🔍

Figure 3.32 Output nodes dataset containing the closeness centrality results for the undirected graph with unweighted links.

$$C_C(A) = \frac{4-1}{d_{AF} + d_{AE} + d_{AD}} = \frac{3}{1+1+1} = \frac{3}{3} = 1$$

The following code invokes proc network to compute the closeness centrality based on an undirected graph with unweighted links.

```
proc network
   direction = undirected
   links = mycas.links
   outlinks = mycas.outlinks
   outnodes = mycas.outnodes
   ;
   linksvar
      from = from
      to = to
   ;
   centrality
      close = unweight
         closenopath = diameter
   ;
run;
```

The OUTNODES dataset presented in the Figure 3.32 contains the results for the closeness centrality centr_close_unwt when computed based on an undirected graph for unweighted links.

Notice that the closeness centrality value for the node A is 1, as described during the formulation.

Let us consider the same subset of nodes from the previous network, but now as a directed graph. The links here are weighted. The Figure 3.33 shows the graph with the original weights.

The Figure 3.34 shows the graph with the inversed weights.

Considering a directed graph, the closeness centrality for node A is calculated as follows:

$$C_C^{out}(A) = \frac{4-1}{d_{AF} + d_{AE} + p} = \frac{3}{0.83 + 0.5 + (0.83 + 1)} = \frac{3}{3.16} = 0.94$$

$$C_C^{in}(A) = \frac{4-1}{0.5 + p + p} = \frac{3}{0.5 + (0.83 + 1) + (0.83 + 1)} = \frac{3}{4.16} = 0.72$$

$$C_C(A) = \frac{0.94 + 0.72}{2} = 0.83$$

Notice that there are two paths from A to F. Straight A–F with weight 1 and the combination of A–E with weight 0.5 and E–F with weight 0.33. The shortest path in this case is A–E–F, with weight 0.83.

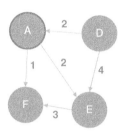

Figure 3.33 Directed graph with original weights.

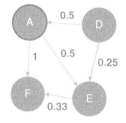

Figure 3.34 Directed graph with inversed weights.

Also, there is no link from A to D. The network procedure uses the diameter + 1. The diameter is the longest shortest path in the network. There are the following shortest paths in the network: A–F: 0.83, D–E: 0.25 (here there are D–E = 0.25 and D–A–E = 1.0), E–F: 0.33, and D–A: 0.5. A–F is the longest shortest path in the network, and it is used to weight the non-existing links.

The following code shows how to compute closeness centrality using proc network. The definition of the links dataset is exactly the same as the one used for the undirected graph before. Here we see just the proc network code.

```
proc network
    direction = directed
    links = mycas.links
    outlinks = mycas.outlinks
    outnodes = mycas.outnodes
    ;
    linksvar
        from = from
        to = to
        weight = weight
    ;
    centrality
        close = weight
            closenopath = diameter
    ;
run;
```

The OUTNODES dataset presented in the Figure 3.35 contains the results for the closeness centrality centr_close_wt, centr_close_in_wt and centr_close_out_wt when computed based on a directed graph.

As described during the closeness centrality formulation for a directed graph, proc network computes the closeness in and out for each node, and averaging them to compute the total closeness centrality.

The closeness centrality in for the node A is 0.72 and the closeness out is 0.94, as described during the formulation. The total closeness centrality is the average of the in and out values, 0.83.

Analogously to the undirected graph, the unweighted links calculation for a directed graph is very similar. All links weights are set to 1.

$$C_C^{out}(A) = \frac{4-1}{d_{AF} + d_{AE} + p} = \frac{3}{1 + 1 + (2+1)} = \frac{3}{5} = 0.6$$

$$C_C^{in}(A) = \frac{4-1}{1 + p + p} = \frac{3}{1 + (2+1) + (2+1)} = \frac{3}{7} = 0.42$$

$$C_C(A) = \frac{0.6 + 0.42}{2} = 0.51$$

« OUTNODES   Table Rows: 4 | Columns: 4 of 4 | Rows 1 to 4

	node	centr_close_wt	centr_close_in_wt	centr_close_out_wt
1	A	0.8336842105	0.72	0.9473684211
2	D	1.125	0	2.25
3	E	0.9556451613	1.1612903226	0.75
4	F	0.8571428571	1.7142857143	0

Figure 3.35 Output nodes dataset containing the closeness centrality results for the directed graph.

« OUTNODES Table Rows: 4 | Columns: 4 of 4 | Rows 1 to 4

Enter expression

	node	centr_close_unwt	centr_close_in_unwt	centr_close_out_unwt
1	A	0.5142857143	0.4285714286	0.6
2	D	0.375	0	0.75
3	E	0.5142857143	0.6	0.4285714286
4	F	0.375	0.75	0

Figure 3.36 Output nodes dataset containing the closeness centrality results for the directed graph with unweighted links.

The following code invokes proc network to compute the closeness centrality based on a directed graph using unweighted links.

```
proc network
   direction = directed
   links = mycas.links
   outlinks = mycas.outlinks
   outnodes = mycas.outnodes
   ;
   linksvar
      from = from
      to = to
   ;
   centrality
      close = unweight
         closenopath = diameter
   ;
run;
```

The OUTNODES dataset presented in the Figure 3.36 contains the results for the closeness centrality centr_close_unwt, centr_close_in_unwt and centr_close_out_unwt when computed based on a directed graph with unweighted links.

Notice that the closeness centrality values for the node A, of 0.42 for the closeness in, 0.6 for the closeness out, and 0.51 for the total closeness centrality, as described during the formulation.

In terms of business, the closeness centrality measure is often used in marketing campaigns to faster spread out a message. All customers with high closeness centrality are the ones able to spread a message faster than anyone else. These customers are in the middle of the network, or at least in the middle of some subnetworks, allowing them to spread out a message faster. These customers are the ones who can spread a rumor pretty fast. They might be the target of a marketing campaign. They will probably replicate that marketing campaign message throughout their networks faster than the regular nodes. They will work for the company spreading out the marketing campaign message, reaching out to the other customers in great speed. Normally, that message spread accounts for several waves of communication. For example, suppose a customer with the highest closeness centrality measure within a subnetwork. If a message reaches out to this customer, he or she will be able to quickly reach out other customer within that subnetwork, who will reach out other peers, and so on. The message will cascade faster throughout the subnetwork, making the marketing campaign more effective.

3.7 Betweenness Centrality

The network metric betweenness centrality counts the number of times a particular node occurs on the shortest paths between other nodes. This measure indicates how much a particular node controls the communication flow within the network. Nodes with high betweenness are known as gatekeepers of information, because of their relative position in the network. The information travels through them several times when flowing.

Betweenness represents how many shortest paths a node is involved in or is related to. Nodes that occur on many shortest paths relative to other nodes have higher betweenness than those that do not. Betweenness can help clarify how central a node is in relation to the entire network and all connections that exist. It also represents how far a message could reach within a network from any node, n. It is a measure that describes the extent or span of a message within social structures.

The centrality betweenness becomes a good measure to indicate who might be a better candidate to diffuse a message throughout the network. A node with high betweenness (inversely computed) describes a point within the graph where most of the shortest paths pass through. If a node like this (one with high betweenness) can control the information flow, it can also diffuse a message the company wants to advertise, such as a product launch or adoption campaign.

Betweenness centrality counts the number of times a node occurs on shortest paths between other nodes. Betweenness can be thought of as a measure of the control a node has over the communication flow among the rest of the network. In this sense, the nodes with high betweenness are the gatekeepers of information, because of their relative location in the network.

3.7.1 Computing the Between Centrality

Betweenness centrality can be calculated for directed and undirected graphs, considering weighted and unweighted links, and can also be computed for nodes and links.

For an undirected graph, the formula for node betweenness centrality is shown below:

$$C_b(i) = \sum_{(s,t) \in N \times N : s < t, \sigma_{st} > 0} \frac{\sigma_{st}(i)}{\sigma_{st}}$$

where σ_{st} is the number of shortest paths from node s to node t, and $\sigma_{st}(i)$ is the number of shortest paths from node s to node t that pass-through node i.

For directed graphs, the formula changes slightly:

$$C_b(i) = \sum_{(s,t) \in N \times N : s \neq t, \sigma_{st} > 0} \frac{\sigma_{st}(i)}{\sigma_{st}}$$

Proc network also computes the betweenness centrality for the links. For undirected graphs, the formula for the link betweenness centrality is as following.

$$C_b(i,j) = \sum_{(s,t) \in N \times N : s < t, \sigma_{st} > 0} \frac{\sigma_{st}(i,j)}{\sigma_{st}}$$

where $\sigma_{st}(i,j)$ is the number of shortest paths from node s to node t that pass-through link (i,j).

For directed graphs, the formula for the link betweenness changes just a little bit:

$$C_b(i,j) = \sum_{(s,t) \in N \times N : s \neq t, \sigma_{st} > 0} \frac{\sigma_{st}(i,j)}{\sigma_{st}}$$

Let us consider the graph showed in the Figure 3.37 to exemplify the betweenness centrality:

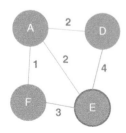

Figure 3.37 Undirected graph with original weights.

The option AUXWEIGHT= can be used in the LINKSVAR statement to determine the links weights that will be used to compute the betweenness centrality. This option defines the auxiliary weights just for closeness and betweenness centralities calculation. If the auxiliary weight is not defined in the LINKSVAR statement, proc network produces the inverse weight (1/w) based on the link weighted defined in the LINKSVAR statement by the option WEIGHT=.

By default, especially in social network analysis, higher values for link weights implies stronger relationships between nodes. This is exactly what we want to emphasize in social studies, strong relationships rather than weak ones. As closeness and betweenness are based on shortest paths, the default shortest path algorithm will look at the link weight in order to minimize the cost and then, the lower the link weights, the closer the nodes. In an optimization problem, this makes all sense. But in a social network analysis, higher values mean closer relations. For that reason, proc network inverse the link weight to turn a stronger relation into lower cost path and then making those nodes closer to each other.

If we want to for instance, keep the original link weights, and not use the inverse values of them, we just need to define the option AUXWEIGHT= in the LINKSVAR statement with the same original values for the link weights.

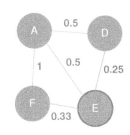

Figure 3.38 Undirected graph with inversed weights to compute the betweenness centrality.

Here is how proc network will inverse of the weights to calculate closeness and betweenness centralities. From the original graph above, we will get the graph showed in the Figure 3.38 with the inversed weights.

Let us consider now node E. Node E participates in the four shortest paths, from D to F and the reciprocal direction, from F to D, and from A to F and the reciprocal direction from F to A.

Let us consider one single direction, from D to F. There are two possible paths, D–A = 0.5 and A–F = 1, which produce D–F = 1.5. There are also D–E = 0.25 and E–F = 0.33, which produce D–F = 0.58. So, the shortest path from D to F is passing by E (not A). As an undirected graph, the direction from F to D would have the same calculation. The shortest path from F to D would be by passing through E (not A again).

Similarly, from A to F, there are two paths possible, A–F = 1, and A–E = 0.5 and E–F = 0.33, which produces D–F = 0.83. So, the shortest path from A to F is passing by E. Even though there is a direct link between A and F, it costs more to take it than go through E.

Then node E has a betweenness centrality equal to 4. This is the unnormalized metric. The network procedure can also produce normalized metrics for the betweenness centrality. The number of pair of nodes is $(N − 1) \times (N − 2)$, or $3 \times 2 = 6$ (A–D, A–E, A–F, D–E, D–F, E–F). The normalized betweenness for node E would be 4/6 = 0.66.

Betweenness centrality can be computed based on weighted links, unweighted links, or both.

We can define how proc network will calculate the betweenness centrality by using the option BETWEEN=WEIGHT| UNWEIGHT|BOTH.

Proc network can also normalize the between centrality metric, varying the results from 0 to 1.

We can define if proc network will normalize or not the betweenness centrality by using the option BETWEENNORM= TRUE|FALSE. By default, proc network normalizes the betweenness centrality results.

Finally, we can define the percentage of the source nodes will be sampled to approximate the betweenness calculation. This value must be a positive number between 1 and 100. The default percentage is 100, which means all the source nodes. To define the percentage of source nodes we can use the option SAMPLEPERCENT=.

The following code shows how to compute the betweenness centrality using proc network for an undirected graph. For this case we are normalizing the results for the betweenness centrality. We are using exactly the same subgraph used to compute the closeness centrality. Then, we are not showing the links dataset definition, just the code with the proc network.

```
proc network
    direction = undirected
    links = mycas.links
    outlinks = mycas.outlinks
    outnodes = mycas.outnodes
    ;
    linksvar
        from = from
        to = to
        weight = weight
    ;
    centrality
        between = weight
            betweennorm = true
    ;
run;
```

The OUTNODES dataset presented in the Figure 3.39 contains the results for the betweenness centrality centr_between_wt when computed based on an undirected graph.

« OUTNODES Table Rows: 4 | Columns: 2 of 2 | Rows 1 to 4

Enter expression

	node	centr_between_wt
1	A	0
2	D	0
3	E	0.6666666667
4	F	0

Figure 3.39 Output nodes dataset containing the betweenness centrality results for the undirected graph.

« OUTLINKS Table Rows: 5 | Columns: 4 of 4 | Rows 1 to 5

Enter expression

	from	to	weight	centr_between_wt
1	A	E	2	0.3333333333
2	A	F	1	0
3	D	A	2	0.1666666667
4	D	E	4	0.3333333333
5	E	F	3	0.5

Figure 3.40 Output links dataset containing the link betweenness centrality results for the undirected graph.

Notice that the betweenness centrality value for the node E is 0.66, as described during the formulation. Here, the betweenness centrality is normalized.

The link betweenness is calculated similarly. Let us consider the possible links between node A and node E. There are three possible links, A–F–E (A–F and F–E), A–D–E (A–D and D–E) and A–E. A–E is the shortest path and then is has a link betweenness of 1/3 = 0.33.

The OUTLINKS dataset presented in the Figure 3.40 contains the results for the link betweenness centrality centr_between_wt when computed based on an undirected graph.

Notice that the link betweenness centrality value for the link A–E is 0.33, as described in the formulation above.

The calculation for the unweighted links is very similar, setting the value of 1 for all links within the graph. For example, now node E partakes the shortest path from node D to node F, as well as node A, as both D–E–F and D–A–F paths have total weight of 2. Then nodes A and E have betweenness of 1, or normalized, 1/6 = 0.16.

The following code invokes proc network to calculate the betweenness centrality based on an undirected graph with unweighted links.

```
proc network
   direction = undirected
   links = mycas.links
   outlinks = mycas.outlinks
   outnodes = mycas.outnodes
   ;
   linksvar
      from = from
      to = to
   ;
```

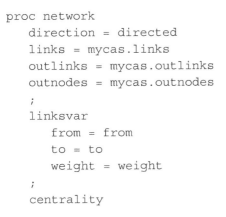

	⬠ node	⊞ centr_between_unwt
1	A	0.1666666667
2	D	0
3	E	0.1666666667
4	F	0

Figure 3.41 Output nodes dataset containing the betweenness centrality results for the undirected graph with unweighted links.

```
centrality
   between = unweight
      betweennorm = true
   ;
run;
```

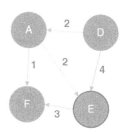

Figure 3.42 Directed graph with original weights.

The OUTNODES dataset presented in the Figure 3.41 contains the results for the betweenness centrality centr_between_wt when computed based on an undirected graph.

For directed graphs, the process is very similar.

Let us consider the graph showed in the Figure 3.42 to exemplify the betweenness centrality:

As we saw before, the network procedure produces the inverse of the weights to compute the betweenness centralities. We then have the graph presented in the Figure 3.43.

Let us consider again node E. Node E participates in the two shortest paths. From D to F, there are two paths possible, D–A = 0.5 and A–F = 1, which produce D–F = 1.5. There are also D–E = 0.25 and E–F = 0.33, which produce D–F = 0.58. So, the shortest path from D to F is passing by E.

Similarly, from A to F, there are two paths possible, A–F = 1, and A–E = 0.5 and E–F = 0.33, which produces D–F = 0.83. So, the shortest path from A to F is passing by E. Even though there is a direct link between A and F, it costs more to take it than go through E.

Then node E has a betweenness centrality equals 2. This is the unnormalized metric. Again, the Network procedure can also produce normalized metrics for the betweenness centrality. The possible number of possible shortest paths in this graph, considering it as undirected, is 6 (A–D, A–E, A–F, D–E, D–F, E–F). The normalized betweenness for node E would be 2/6 = 0.33.

The following code shows how to compute the betweenness centrality using proc network for a directed graph based on weighted links. Here we are going to compute the betweenness centrality first with no normalization, just the counting of partaking shortest paths for each node. Later on, we are going to compute the betweenness centrality normalized.

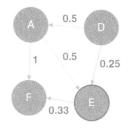

Figure 3.43 Directed graph with inversed weights to compute the betweenness centrality.

```
proc network
   direction = directed
   links = mycas.links
   outlinks = mycas.outlinks
   outnodes = mycas.outnodes
   ;
   linksvar
      from = from
      to = to
      weight = weight
   ;
   centrality
```

	⚠ node	⊕ centr_between_wt
1	A	0
2	D	0
3	E	2
4	F	0

« OUTNODES   Table Rows: 4 | Columns: 2 of 2 | Rows 1 to 4 | ↑̄ ↑ ↓ ↓̲ ⋮

Figure 3.44 Output nodes dataset containing the betweenness centrality results for the directed graph with weighted links unnormalized.

```
      between = weight
         betweennorm = false
   ;
run;
```

The OUTNODES dataset presented in the Figure 3.44 contains the results for the betweenness centrality centr_between_wt when computed based on a directed graph, considering weighted links. Notice that the option BETWEENNORM=FALSE is set to just count the number of shortest paths partaken by each node.

Here node E partakes two shortest paths as showed during the formulation.

To compute the betweenness centrality normalized, we just need to set the option BETWEENNORM=TRUE. By doing this, proc network will divide the number of shortest paths partaken by each node by the number of possible links between all nodes.

The following OUTNODES dataset presented in the Figure 3.45 contains the results for the betweenness centrality centr_between_wt when computed based on a directed graph, considering weighted links, with betweenness centrality normalized.

Notice that the betweenness centrality value for the node E is 0.33 (normalized), as described during the formulation.

Let us take a look at the link betweenness centrality for the directed graph.

Considering the directed graph above, there are eight possible paths within the network, D–A, D–E, D–A–F, D–E–F, E–F, A–F, A–E, and A–E–F. E–F partakes two shortest paths, A–E–F and A–E–F. Then, the link betweenness centrality for E–F is $2/8 = 0.25$.

The OUTLINKS dataset presented in the Figure 3.46 contains the results for the link betweenness centrality centr_between_wt when computed based on a directed graph.

Notice that the link betweenness centrality value for the link E–F is 0.25, as described during the formulation.

As described for the undirected graph, the calculation for the unweighted links in directed graphs is very similar, setting the value of 1 for all links within the graph. For example, now node E partakes the shortest path from node D to node F, as well as node A, as both D–E–F and D–A–F paths have total weight of 2. However, there is small detail here in the

	⚠ node	⊕ centr_between_wt
1	A	0
2	D	0
3	E	0.3333333333
4	F	0

« OUTNODES   Table Rows: 4 | Columns: 2 of 2 | Rows 1 to 4 | ↑̄ ↑ ↓ ↓̲ ⋮

Figure 3.45 Output nodes dataset containing the betweenness centrality results for the directed graph with weighted links normalized.

OUTLINKS Table Rows: 5 Columns: 4 of 4 Rows 1 to 5

	from	to	weight	centr_between_wt
1	A	E	2	0.1666666667
2	A	F	1	0
3	D	A	2	0.0833333333
4	D	E	4	0.1666666667
5	E	F	3	0.25

Figure 3.46 Output links dataset containing the link betweenness centrality results for the directed graph.

calculation. As nodes A and E share the shortest path between nodes D and F, proc network counts just half shortest path for each node. Then, both nodes A and E have betweenness of 0.5. Normalizing the results, both will have betweenness of 0.5/6 = 0.083.

```
proc network
   direction = directed
   links = mycas.links
   outlinks = mycas.outlinks
   outnodes = mycas.outnodes
   ;
   linksvar
      from = from
      to = to
   ;
   centrality
      between = unweight
         betweennorm = false
   ;
run;
```

The OUTNODES dataset presented in the Figure 3.47 contains the results for the betweenness centrality centr_between_unwt when computed based on a directed graph, considering unweighted links. Notice here we are calculating the betweenness centrality with no normalization, just counting the number of shortest paths partaken by each node. We are using the option BETWEENNORM=FALSE.

OUTNODES Table Rows: 4 Columns: 2 of 2 Rows 1 to 4

	node	centr_between_unwt
1	A	0.5
2	D	0
3	E	0.5
4	F	0

Figure 3.47 Output nodes dataset containing the betweenness centrality results for the directed graph with unweighted links unnormalized.

	⚠ node	⊕ centr_between_unwt
1	A	0.0833333333
2	D	0
3	E	0.0833333333
4	F	0

« OUTNODES Table Rows: 4 Columns: 2 of 2 Rows 1 to 4 ⊤ ↑ ↓ ↧ ⋮

▽ Enter expression

Figure 3.48 Output nodes dataset containing the betweenness centrality results for the directed graph with unweighted links normalized.

Notice the betweenness centralities for nodes A and E as 0.5, as they split the shortest path from node D to node F, as described during the formulation.

Now let us compute the betweenness centrality normalize. The only change in the previous code is to set the option BETWEENNORM=TRUE.

The OUTNODES dataset presented in the Figure 3.48 contains the results for the normalized betweenness centrality centr_between_unwt when computed based on a directed graph, considering unweighted links.

Notice that the betweenness centrality value for the nodes A and E is 0.083 (0.5 unnormalized), as described during the formulation. Remember they split the shortest path from node D to node F.

In business scenarios, the betweenness centrality can be used similarly to the closeness centrality, with a particular distinction on the concept. The closeness centrality can be interpreted as the gossip or the rumor factor. A node with a high value closeness centrality is able to spread out a message throughout the network faster than anyone else. The betweenness centrality can be interpreted as the gatekeeper factor. A node with a high betweenness centrality is able to control the information flow within the network. It means that a node with high value betweenness centrality is likely to be aware of all or at least most of the messages circulating within the network. For example, if a company wants to know what customers are saying about products, services, prices, quality, etc., that company should use the customers with high value betweenness centrality to disclose these messages. As gatekeepers of the network, they will likely know what is going on within the network. Companies can use customers like that as promoters, or at least, as informers (not to say spies) of how other customers envision the company, or its products and services.

3.8 Eigenvector Centrality

Eigenvector centrality is an extension to the degree centrality. In this metric, the centrality points are awarded for each neighbor. Not all neighbors are equally important. A connection to an important node contributes more to the centrality score than a connection to a less important node. An eigenvector is defined to be proportional to the sum of the scores of all nodes that are connected to it. Proc network implements two algorithms to compute an eigenvector, the power method, and the Jacobi–Davidson algorithm.

An eigenvector is a nonzero vector that multiplies a square matrix resulting in a multiple of it (vector). The multiplier of the vector is called the eigenvalue.

Let us consider the following square matrix:

$$A = \begin{bmatrix} 3 & 1 \\ 1 & 3 \end{bmatrix}$$

From the definition of the eigenvector v corresponding to the eigenvalue λ we have the following formulation:

$$Av = \lambda v$$

where A represents the adjacent matrix, v represents the eigenvector, and λ represents the eigenvalue.

Then, the equation, with I as the identity matrix:

$$Av - \lambda v = (A - \lambda I) \cdot v = 0$$

has a nonzero solution if and only if:

$$|A - \lambda I| = 0$$

$$|A - \lambda I| = \begin{vmatrix} 3 - \lambda & 1 \\ 1 & 3 - \lambda \end{vmatrix} = \lambda^2 - 6\lambda + 8 = (\lambda - 2) \cdot (\lambda - 4) = 0$$

That gives us:

$$\lambda_1 = 2$$

$$\lambda_2 = 4$$

For every λ we can find its own vectors:
For $\lambda_1 = 2$ we will have:

$$v_1 = \begin{pmatrix} -1 \\ 1 \end{pmatrix}$$

And for $\lambda_2 = 4$ we will have:

$$v_2 = \begin{pmatrix} 1 \\ 1 \end{pmatrix}$$

Given a square matrix A, the eigenvector of the matrix is those nonzero vectors that remain proportional to the original vector after being multiplied by A. Upon multiplication, an eigenvector changes magnitude, but not direction. The corresponding amount that the vector changes in magnitude is the eigenvalue.

3.8.1 Computing the Eigenvector Centrality

The eigenvector centrality measures the importance of a particular node inside the network. Relative scores for all nodes are computed based on their connections, considering both the frequency and strength of the relationship. Eigenvector is assigned to a recursive algorithm that calculates the importance of a particular node considering the importance of all nodes and all connections within the network.

$$x_i = \frac{1}{\lambda} \sum_{j \in N} w_{ij} x_j$$

where x_j is the eigenvector of node i, N is the set of nodes connected to i, w_{ij} is the weight of the link from node i to node i. λ is a constant.

The centrality eigenvector combines the hub and centrality measures, and it can be performed in both undirected and directed graphs.

Eigenvector centrality is also normalized, and the node with the highest eigenvector receives the value of 1 and all other nodes receive a proportion of its value.

Let us consider the same graph we used to compute closeness and betweenness centralities to demonstrate the eigenvector centrality calculation. The graph is showed in the Figure 3.49.

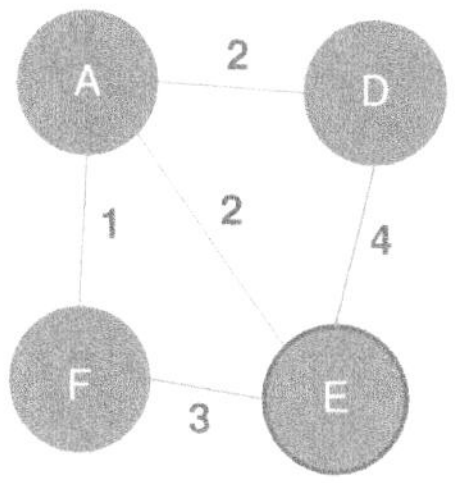

Here, A connects to D, E, and F ([A–D|A–E|A–F] = 1), D connects to A and E ([D–A|D–E] = 1), E connects to A, D, and F ([E–A|E–D|E–F] = 1) and F connects to A and E ([F–A|F–E] = 1). This graph then creates the following adjacent matrix:

Figure 3.49 Undirected graph to shows eigenvector centrality.

$$
\begin{bmatrix} A & D & E & F \end{bmatrix}
$$

$$
\begin{bmatrix} A \\ D \\ E \\ F \end{bmatrix}
\begin{bmatrix}
0 & 1 & 1 & 1 \\
1 & 0 & 1 & 0 \\
1 & 1 & 0 & 1 \\
1 & 0 & 1 & 0
\end{bmatrix}
$$

This creates

$$
|A - \lambda I| =
\begin{bmatrix}
0-\lambda & 1 & 1 & 1 \\
1 & 0-\lambda & 1 & 0 \\
1 & 1 & 0-\lambda & 1 \\
1 & 0 & 1 & 0-\lambda
\end{bmatrix}
=
$$

$$
\lambda^4 - 5\lambda^2 - 4\lambda =
$$

$$
\lambda \cdot \left(\lambda^3 - 5\lambda - 4 \right) =
$$

$$
\lambda \cdot (\lambda + 1) \cdot \left(\lambda^2 - \lambda - 4 \right) =
$$

$$
\lambda \cdot (\lambda + 1) \cdot \left(\lambda + \frac{\sqrt{17} - 1}{2} \right) \cdot \left(\lambda - \frac{\sqrt{17} + 1}{2} \right) = 0
$$

We have the following eigenvalues:

$$
\lambda_1 = 0
$$

$$
\lambda_2 = -1
$$

$$
\lambda_3 = \frac{-\sqrt{17} + 1}{2}
$$

$$
\lambda_4 = \frac{\sqrt{17} + 1}{2}
$$

These eigenvalues give us the following eigenvectors:

$$
v_1 =
\begin{bmatrix}
0 \\
-1 \\
0 \\
1
\end{bmatrix}
$$

$$
v_2 =
\begin{bmatrix}
-1 \\
0 \\
1 \\
0
\end{bmatrix}
$$

$$
v_3 =
\begin{bmatrix}
\dfrac{-\sqrt{17} + 1}{4} \\
1 \\
\dfrac{-\sqrt{17} + 1}{4} \\
1
\end{bmatrix}
$$

$$v_4 = \begin{bmatrix} \dfrac{\sqrt{17}+1}{4} \\ 1 \\ \dfrac{\sqrt{17}+1}{4} \\ 1 \end{bmatrix}$$

The greatest eigenvector is v_4, which give us the following results:

$$\begin{bmatrix} A \\ D \\ E \\ F \end{bmatrix} = \begin{bmatrix} 1.28 \\ 1 \\ 1.28 \\ 1 \end{bmatrix}$$

Normalizing the results by dividing the outcomes by the maximum value, we reach the following numbers:

$$\begin{bmatrix} A \\ D \\ E \\ F \end{bmatrix} = \begin{bmatrix} 1.28/1.28 \\ 1/1.28 \\ 1.28/1.28 \\ 1/1.28 \end{bmatrix} = \begin{bmatrix} 1 \\ 0.78 \\ 1 \\ 0.78 \end{bmatrix}$$

Let us see now how to do it by using proc network.

The eigenvector centrality can be computed based on weighted links, unweighted links, or both. We can define how proc network will calculate the eigenvector centrality by using the option EIGEN=WEIGHT|UNWEIGHT|BOTH.

The eigenvector can be calculated for both directed and undirected graphs. We can define how proc network will compute the eigenvector by using the option DIRECTION=DIRECTED|UNDIRECTED.

We can also define the eigenvector algorithm to be used when calculating the eigenvector centrality by using the option EIGENALGORITHM=AUTOMATIC|JACOBDAVIDSON|POWER. Finally, we can also define the maximum number of iterations performed by the algorithm when computing the eigenvector centrality. The option EIGENMAXITERS= limit the amount of computation time spent when convergence is slow. The default method to select the algorithm is AUTOMATIC. The default number of iterations to compute the eigenvector centrality is 10 000.

The following code shows how to compute the eigenvector centrality using proc network.

```
proc network
    direction = undirected
    links = mycas.links
    outlinks = mycas.outlinks
    outnodes = mycas.outnodes
    ;
    linksvar
        from = from
        to = to
    ;
    centrality
        eigen = unweight
            eigenalgorithm = jacobidavidson
            eigenmaxiters = 100000
    ;
run;
```

« OUTNODES   Table Rows: 4 | Columns: 2 of 2 | Rows 1 to 4 | ⊼ ↑ ↓ ⤓ ⋮

▽ | Enter expression | 🔍

	⬙ node	⊕ centr_eigen_unwt
1	A	1
2	D	0.7807764064
3	E	1
4	F	0.7807764064

Figure 3.50   Output nodes dataset containing the eigenvector centrality based on an undirected graph using unweighted links.

The OUTNODES dataset presented in the Figure 3.50 contains the results for the eigenvector centrality centr_eigen_-unwt. The eigenvector centrality here is computed based on an undirected graph with unweighted links.

Notice that the eigenvector centrality for the nodes A and E are 1, and for the nodes D and F are 0.78, as we observed during the formulation.

We can compute the eigenvector considering the links weights. The computation process is very similar, but the adjacent matrix will be slightly different, resulting in a different outcome.

Considering the same graph with nodes A, D, E, and F, we will replace the values to represent the existence of links between the nodes (1 if exist and 0 otherwise) by the real links weights between them. This will produce the following adjacent matrix:

$$\begin{bmatrix} 0 & 2 & 2 & 1 \\ 2 & 0 & 4 & 0 \\ 2 & 4 & 0 & 3 \\ 1 & 0 & 3 & 0 \end{bmatrix}$$

We have the following eigenvalues:

$$\lambda_1 = -5.005$$

$$\lambda_2 = -1.467$$

$$\lambda_3 = 0.085$$

$$\lambda_4 = 6.387$$

This produces four eigenvectors and the greatest one will be as the following:

$$v_4 = \begin{bmatrix} 1.158 \\ 1.454 \\ 1.743 \\ 1 \end{bmatrix}$$

Normalizing the results by dividing the outcomes by the maximum value, we reach the following numbers:

$$\begin{bmatrix} A \\ D \\ E \\ F \end{bmatrix} = \begin{bmatrix} 1.158/1.743 \\ 1.454/1.743 \\ 1.743/1.743 \\ 1/1.743 \end{bmatrix} = \begin{bmatrix} 0.66 \\ 0.83 \\ 1 \\ 0.57 \end{bmatrix}$$

Let us see this calculation in proc network. The code below invokes proc network to compute the eigenvector centrality still based on undirected graphs but now on weighted links. Notice that the only difference in the code is EIGEN=WEIGHT.

```
proc network
   direction = undirected
   links = mycas.links
   outlinks = mycas.outlinks
   outnodes = mycas.outnodes
   ;
   linksvar
      from = from
      to = to
      weight = weight
   ;
   centrality
      eigen = weight
         eigenalgorithm = jacobidavidson
         eigenmaxiters = 100000
   ;
run;
```

The OUTNODES dataset presented in the Figure 3.51 contains the results for the eigenvector centrality centr_eigen_wt. computed based on an undirected graph using weighted links.

Notice that the eigenvector centrality for the node A is 0.66, B is 0.83, E is 1, and F is 0.57, exactly like we saw during the formulation.

For the directed graph, the calculation is very similar. Once again, the difference will be the adjacent matrix. Now we consider just the incident links from one node to another, holding the direction of the graph.

Let us assume the graph showed in the Figure 3.52, the same we used for closeness and betweenness centralities.

In the directed graph, D connects to A, but A does not connect to D. The same happens for all other relations. This will produce a different adjacent matrix:

$$
\begin{bmatrix} A \\ D \\ E \\ F \end{bmatrix}
\begin{matrix} [A & D & E & F] \end{matrix}
\begin{bmatrix} 0 & 0 & 1 & 1 \\ 1 & 0 & 1 & 0 \\ 0 & 0 & 0 & 1 \\ 0 & 0 & 0 & 0 \end{bmatrix}
$$

	node	centr_eigen_wt
1	A	0.664230909
2	D	0.8342935079
3	E	1
4	F	0.5737198019

≪ OUTNODES Table Rows: 4 | Columns: 2 of 2 | Rows 1 to 4

Figure 3.51 Output nodes dataset containing the eigenvector centrality based on an undirected graph using weighted links.

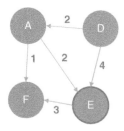

Figure 3.52 Directed graph to shows eigenvector centrality.

This creates

$$|A - \lambda I| = \begin{bmatrix} 0-\lambda & 0 & 1 & 1 \\ 1 & 0-\lambda & 1 & 0 \\ 0 & 0 & 0-\lambda & 1 \\ 0 & 0 & 0 & 0-\lambda \end{bmatrix} = \lambda^4 = 0$$

Approximating the results, we basically have one eigenvalue:

$$\lambda_1 = 0$$

That single eigenvalue produces one single eigenvectors:

$$v_1 = \begin{bmatrix} 0 \\ 1 \\ 0 \\ 1 \end{bmatrix}$$

Proc network considers other factors when computing the eigenvalues and eigenvectors, especially to improve performance in sparse matrices. The results for very small network like this one we used in the example might be slightly different, but the main figures will match. The final eigenvector normalized will give us the following results:

$$\begin{bmatrix} A \\ B \\ C \\ D \end{bmatrix} = \begin{bmatrix} \approx 0 \\ 1 \\ \approx 0 \\ \approx 0 \end{bmatrix}$$

Let us see now how to do it in proc network. Notice the options DIRECTION=DIRECTED and EIGEN=UNWEIGHT.

```
proc network
   direction = directed
   links = mycas.links
   outlinks = mycas.outlinks
   outnodes = mycas.outnodes
   ;
   linksvar
      from = from
      to = to
   ;
   centrality
      eigen = unweight
         eigenalgorithm = jacobidavidson
         eigenmaxiters = 100000
   ;
run;
```

The OUTNODES dataset presented in the Figure 3.53 contains the results for the eigenvector centrality centr_eigen_unwt. computed based on a directed graph using unweighted links.

Notice the values for nodes A, E, and F approximately to 0 and for node D equals 1, as we saw during the formulation.

The calculation for directed graphs based on weighted links will be very similar to the previous one, producing a similar adjacent matrix. We will replace the ones in the matrix by the actual weights. This will still produce one single eigenvalue, and then one eigenvector. The approximate results will be 0 for all nodes except D with eigenvector equals to 1.

In proc network, let us just reuse the previous code and change the option EIGEN=WEIGHT.

The OUTNODES dataset presented in the Figure 3.54 contains the results for the eigenvector centrality centr_eigen_wt. computed based on a directed graph using weighted links.

As showed in the formulation, and similarly to the calculation for unweighted links, nodes A, E, and F equal to approximately 0 and node D equals 1.

In terms of business, the eigenvector centrality is one of the most complete network metrics. It can be interpreted as the possible influence of a node within the network. For this, it can be used to target customers in marketing campaigns, sales events and retention and loyalty programs. There is a magic number in the market that measures customer acquisition and customers retention. It says that it is eight times more expensive to acquire new customers than to keep them in the company. Of course, this number varies depending on the industry and the company, but there is a common sense that it is much more expensive to acquire than to retain. Also, it is easier to sell to existing customer than to prospects. According to marketing metrics, the probability of converting existing customers is around 60–70%, and the probability of converting a new prospect on the other hand is around 5–20%. Based on that, losing customers is bad for the business. But losing good customers can be far worse. Customers with high eigenvector centrality are the ones who are likely to be influential, have a substantial number of strong connections, are usually central to the network and close to their peers. All that turn them into key nodes and a target for close monitoring and good relationship management. Keep these customers in the company are key to eventually retain many other customers within the network.

	node	centr_eigen_unwt
1	A	0.031645987
2	D	1
3	E	0.0010004991
4	F	0.0000316311

Figure 3.53 Output nodes dataset containing the eigenvector centrality based on a directed graph using unweighted links.

	node	centr_eigen_wt
1	A	0.0285919801
2	D	1
3	E	0.0008575474
4	F	0.0000169836

Figure 3.54 Output nodes dataset containing the eigenvector centrality based on a directed graph using weighted links.

3.9 PageRank Centrality

PageRank centrality is a variant of eigenvector centrality, but more effective on directed graphs. This network centrality measure was proposed by Google's cofounder Larry Page. The PageRank models the stationary distribution of a Markov process, assuming each node as a web page and each link as a hyperlink from one page to another. A web surfer may choose a random link on every page or may jump to a random page on the entire web with some probability. The PageRank centrality of a web page is the percentage of time the surfer spends on the page.

When dealing with undirected graphs, PageRank treats each undirected link as two directed links going in both directions. The damping factor is an important concept for the PageRank centrality computation used to avoid pages without any incoming links having PageRank equal zero. The damping factor can vary from 0 to 1. The default damping factor in proc network is 0.85. this threshold means that a random web surfer has a 15% of chance of jumping to any other web page in the web at any time. The PageRank algorithm takes more iterations to converge as the damping factor gets close to 1.

3.9.1 Computing the PageRank Centrality

The mathematical formulation to the PageRank centrality can be established as following.

$$x_i = \frac{1-d}{|N|} + d \left(\sum_{j \in N: w_j^{out} \neq 0} \frac{w_{ij}}{w_j^{out}} x_j + \sum_{j \in N: w_j^{out} = 0} \frac{1}{|N|} x_j \right)$$

$$w_i^{out} = \sum_{j \in N} w_{ij}$$

where x_i is the PageRank score of node i, N is the set of nodes that link to node i, w_{ij} is the sum of the outbound link weights from node i to node j, d is a damping factor that ranges from 0 to 1, and $1 - d$ is the probability of jumping to a random page.

In terms of social relations, the PageRank centrality describes the percentage of possible time that other nodes within the network might spend with a particular node n. In networks other than the Internet, this measure can infer how a node may be relevant or influent to the overall network based on the frequency of relationships (such as websites visited) with other nodes.

Networks representing corporate relationships can use PageRank centrality as a way to predict contacts among employees, vendors, consultants, providers, and so on. The PageRank of a particular node, considering its historical relationships, might also indicate the probability of it being contacted by the other nodes in the network or in its own subnetwork.

As computation of a distribution of a Markov process, the PageRank algorithm is iterative, where the PageRank of a node relies on the PageRank of its connected nodes, and the PageRank of its connected nodes relies on its own PageRank. The whole process may take a huge number of iterations in order to converge to the final results, particularly in large network, when the need to update the PageRank values for all nodes along the process might be exhausted.

In order to exemplify the calculation, and make it easier at the same time, here we are going to see basically two distinct approaches. The first one is based on a simple linear algebra point of view. The second one is based on a dynamical systems point of view, describing the iterative process.

Let us consider the following graph to demonstrate the PageRank centrality. Notice this graph is slightly different than the one we have been using to demonstrate the other centrality measures. The reason is to make the process simpler. In PageRank calculation, some nodes can be considered dangling nodes, when they do not have any outgoing links. In our previous subgraph, node F receives links from nodes A and E, but it does not link to any other node. Node F is a dangling node as it has no outgoing links. The other problem with the previous subgraph to exemplify the PageRank calculation is the nodes with no incoming links. Node D in that subgraph points to nodes A and E, but it has no incoming links. Nodes with no incoming links would have a PageRank score of zero. These nodes cannot be distinguished by the PageRank scores.

Due to these small issues in the previous subgraph, and to make the calculation simpler, let us change a little bit the previous graph, adding outgoing links to node F and incoming links to node D. The new graph is shown in the Figure 3.55.

Now, all nodes in the graph have incoming and outgoing links. This will make easier the description of the steps during the PageRank calculation on the linear algebra point of view. This new graph produces the following adjacent matrix:

$$\begin{bmatrix} 0 & 0 & 1 & 1/2 \\ 1/3 & 0 & 0 & 0 \\ 1/3 & 1/2 & 0 & 1/2 \\ 1/3 & 1/2 & 0 & 0 \end{bmatrix}$$

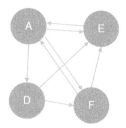

Figure 3.55 Directed graph with unweighted links to shows PageRank centrality.

It is important to emphasize that here we are really simplifying the linear algebra behind the PageRank calculation, as a method to make it more understandable. The way proc network computes the PageRank centrality is more complex and it accounts for large graphs. It uses multiple iterations in both eigenvector algorithms (power method and Jacobi–Davidson).

In PageRank centrality calculation, we assume that each node transfer evenly its importance to the other nodes it links to. For example, node A links to nodes D, E, and F, so it transfers its importance evenly to them, 1/3 for each. Node E links to D and A so it transfers evenly its importance to them, 1/2 for each. Same for F and D.

Considering the evenly distribution of importance, the new graph looks like the one showed in the Figure 3.56.

Again, proc network performs a series of iterations to converge the final results. Here, to exemplify the calculation, we are showing the linear algebra point of view.

Let us assume x_1, x_2, x_3, and x_4 the importance of the four nodes A, D, E, and F.

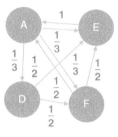

Figure 3.56 Directed graph with link weights evenly distributed.

$$\begin{cases} x_1 = 1 \cdot x_3 + \dfrac{1}{2} \cdot x_4 \\ x_2 = \dfrac{1}{3} \cdot x_1 \\ x_3 = \dfrac{1}{3} \cdot x_1 + \dfrac{1}{2} \cdot x_2 + \dfrac{1}{2} \cdot x_4 \\ x_4 = \dfrac{1}{3} \cdot x_1 + \dfrac{1}{2} \cdot x_2 \end{cases}$$

This is equivalent to ask for solutions to the equation:

$$A \cdot \begin{bmatrix} x_1 \\ x_2 \\ x_3 \\ x_4 \end{bmatrix} = \begin{bmatrix} x_1 \\ x_2 \\ x_3 \\ x_4 \end{bmatrix}$$

The eigenvector for this adjacent matrix corresponding to the eigenvalue 1 is:

$$c \cdot \begin{bmatrix} 2 \\ \dfrac{2}{3} \\ \dfrac{3}{2} \\ 1 \end{bmatrix} = c \cdot \begin{bmatrix} 12 \\ 4 \\ 9 \\ 6 \end{bmatrix}$$

Since the PageRank centrality should reflect the relative importance of the nodes, we can select c to be the unique eigenvector with the sum of all entries equal to 1. The eigenvector:

$$\frac{1}{31} \cdot \begin{bmatrix} 12 \\ 4 \\ 9 \\ 6 \end{bmatrix} \sim \begin{bmatrix} 0.38 \\ 0.12 \\ 0.29 \\ 0.19 \end{bmatrix}$$

The PageRank centrality can be computed based on weighted links, unweighted links, or both. It can also be computed for both directed and undirected graphs.

We can define how proc network will calculate the PageRank centrality based on the links by using the option PAGERANK=WEIGHT|UNWEIGHT|BOTH.

We can also define the damping factor to use in the PageRank algorithm. The damping value must be between 0 and 1. The default value is 0.85. This number indicates the probability a web surfer will randomly jump to any another web page within the network at any time. For example, the default value of 0.85 means that a random node has 15% of chance to be connected to or point to any other node within the network at any time. As closer the dumping factor gets to 1, more iterations the algorithm might take to converge. Eventually the algorithm does not converge at all. To define the dumping factor, we use the option PAGERANKALPHA=.

Proc network allow us to define the convergence tolerance for the PageRank algorithm. This value must be a positive number. The default is 1E-9. The PageRank algorithm stops iterating when the difference between the PageRank score of the current iteration and the previous iteration is less than or equal to the tolerance defined. To define the convergence tolerance, we can use the option PAGERANKTOLERANCE=.

The following code shows how to compute the PageRank centrality using proc network based on a directed graph using unweighted links.

```
data mycas.links;
    input from $ to $ @@;
datalines;
A D  A E  A F  D E  D F  E A  F A  F E
;
run;


proc network
    direction = directed
    links = mycas.links
    outlinks = mycas.outlinks
    outnodes = mycas.outnodes
    ;
    linksvar
        from = from
        to = to
    ;
    centrality
        pagerank = unweight
    ;
run;
```

The OUTNODES dataset presented in the Figure 3.57 contains the results for the PageRank centrality centr_pagerank_ unwt based on a directed graph using unweighted links.

Notice the values for nodes A, D, E, and F as 0.37, 0.14, 0.29, and 0.20, very similar to the values found during the formulation, as 0.38, 0.12, 0.29, and 0.19. The difference here is due to the iterative method used by proc network to converge the PageRank centrality, considering the dumping factor, and mostly the distribution of the Markov process.

« OUTNODES   Table Rows: 4 | Columns: 2 of 2 | Rows 1 to 4 | ⤒ ↑ ↓ ⤓ ⋮

▽ | Enter expression | 🔍

	⚠ node	⊕ centr_pagerank_unwt
1	A	0.3681506771
2	D	0.1418093586
3	E	0.2879616285
4	F	0.2020783358

Figure 3.57 Output nodes dataset containing the PageRank centrality based on a directed graph using unweighted links.

Let us take a look at the second approach, the dynamical systems point of view, which consider the multiple iterations to converge to the final result. Let us reuse the subgraph we have been using for the other centralities like closeness, betweenness and eigenvector. Let us consider this new example as an undirected graph with weighted links. This graph will produce the same adjacent matrix we saw during the eigenvector demonstration. Here we will show the approximated steps based on the iterative method. The graph is showed in the Figure 3.58.

The adjacency matrix for this undirected graph based on weighted links is shown below.

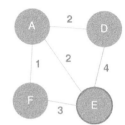

$$\begin{bmatrix} 0 & 2 & 2 & 1 \\ 2 & 0 & 4 & 0 \\ 2 & 4 & 0 & 3 \\ 1 & 0 & 3 & 0 \end{bmatrix}$$

Figure 3.58 Undirected graph with weighted links to show the PageRank centrality calculation.

Analogously to performed before, the PageRank centrality supposed to evenly distribute the importance to the connected nodes. Then, let us normalize the adjacency matrix based on the evenly degree. To compute the degree-normalized adjacent matrix, we divide each matrix column by the sum of the column.

$$\begin{bmatrix} 0/5 & 2/6 & 2/9 & 1/4 \\ 2/5 & 0/6 & 4/9 & 0/4 \\ 2/5 & 4/6 & 0/9 & 3/4 \\ 1/5 & 0/6 & 3/9 & 0/4 \end{bmatrix} = \begin{bmatrix} 0 & 0.3333 & 0.2222 & 0.25 \\ 0.4 & 0 & 0.4444 & 0 \\ 0.4 & 0.6667 & 0 & 0.75 \\ 0.2 & 0 & 0.3333 & 0 \end{bmatrix}$$

Let us see the iterations based on the power method to calculate the eigenvector, here considering the dumping factor α as 0.85 (default) and the four nodes within the graph ($N = 4$).

First, the importance is uniformly distributed among the four nodes, each getting ¼. This will be represented by the initial rank vector x_0 having all entries equal to 1/4. Each incoming link increases the importance of a node, so at step 1, we update the rank of each page by adding to the current value the importance of the incoming links. This is the same as multiplying the adjacent matrix with x_0. At step 1, the new importance vector is $x_1 = Ax_0$. We can iterate the process, thus at step 2, the updated importance vector is $x_2 = A(Ax_0) = A^2x_0$.

The initial state is:

$$x_0 = \begin{bmatrix} 0.25 \\ 0.25 \\ 0.25 \\ 0.25 \end{bmatrix}$$

In the first iteration, we will multiply the degree-normalized adjacent matrix by the evenly probability (1/4 or 0.25):

$$
\begin{bmatrix}
0 & 0.3333 & 0.2222 & 0.25 \\
0.4 & 0 & 0.4444 & 0 \\
0.4 & 0.6667 & 0 & 0.75 \\
0.2 & 0 & 0.3333 & 0
\end{bmatrix}
\times
\begin{bmatrix}
0.25 \\
0.25 \\
0.25 \\
0.25
\end{bmatrix}
=
\begin{bmatrix}
0.2014 \\
0.2111 \\
0.4542 \\
0.1333
\end{bmatrix}
$$

$$
\frac{1-\alpha}{N}
\begin{bmatrix}
1 \\
1 \\
1 \\
1
\end{bmatrix}
+ \alpha
\begin{bmatrix}
0.2014 \\
0.2111 \\
0.4542 \\
0.1333
\end{bmatrix}
=
\begin{bmatrix}
0.2087 \\
0.2169 \\
0.4235 \\
0.1508
\end{bmatrix}
$$

The second iteration follows the same process:

$$
\begin{bmatrix}
0 & 0.3333 & 0.2222 & 0.25 \\
0.4 & 0 & 0.4444 & 0 \\
0.4 & 0.6667 & 0 & 0.75 \\
0.2 & 0 & 0.3333 & 0
\end{bmatrix}
\times
\begin{bmatrix}
0.2087 \\
0.2169 \\
0.4235 \\
0.1508
\end{bmatrix}
=
\begin{bmatrix}
0.2041 \\
0.2717 \\
0.3412 \\
0.1829
\end{bmatrix}
$$

$$
\frac{1-\alpha}{N}
\begin{bmatrix}
1 \\
1 \\
1 \\
1
\end{bmatrix}
+ \alpha
\begin{bmatrix}
0.2041 \\
0.2717 \\
0.3412 \\
0.1829
\end{bmatrix}
=
\begin{bmatrix}
0.2110 \\
0.2685 \\
0.3275 \\
0.1930
\end{bmatrix}
$$

Same for the third iteration:

$$
\begin{bmatrix}
0 & 0.3333 & 0.2222 & 0.25 \\
0.4 & 0 & 0.4444 & 0 \\
0.4 & 0.6667 & 0 & 0.75 \\
0.2 & 0 & 0.3333 & 0
\end{bmatrix}
\times
\begin{bmatrix}
0.2110 \\
0.2685 \\
0.3275 \\
0.1930
\end{bmatrix}
=
\begin{bmatrix}
0.2105 \\
0.2300 \\
0.4081 \\
0.1514
\end{bmatrix}
$$

$$
\frac{1-\alpha}{N}
\begin{bmatrix}
1 \\
1 \\
1 \\
1
\end{bmatrix}
+ \alpha
\begin{bmatrix}
0.2105 \\
0.2300 \\
0.4081 \\
0.1514
\end{bmatrix}
=
\begin{bmatrix}
0.2164 \\
0.2330 \\
0.3844 \\
0.1662
\end{bmatrix}
$$

The process continues until the power method converges to the final result:

$$
\begin{bmatrix}
0.2136 \\
0.2471 \\
0.3627 \\
0.1766
\end{bmatrix}
$$

The following code shows how to compute the PageRank centrality using proc network based on an undirected graph using weighted links. The data step recreates the links dataset.

```
data mycas.links;
   input from $ to $ weight @@;
datalines;
D A 2 A E 2 A F 1 E F 3 D E 4
;
run;

proc network
   direction = undirected
   links = mycas.links
```

OUTNODES   Table Rows: 4   Columns: 2 of 2   Rows 1 to 4

	node	centr_pagerank_wt
1	D	0.2471393607
2	A	0.2135599116
3	E	0.3627237996
4	F	0.1765769281

Figure 3.59 Output nodes dataset containing the PageRank centrality based on an undirected graph using weighted links.

```
outlinks = mycas.outlinks
outnodes = mycas.outnodes
;
linksvar
    from = from
    to = to
    weight = weight
;
centrality
    pagerank = weight
        eigenalgorithm = power
        pagerankalpha = 0.85
;
run;
```

The OUTNODES dataset presented in the Figure 3.59 contains the results for the PageRank centrality centr_pagerank_wt based on an undirected graph using weighted links.

Notice the values for nodes A, D, E, and F as 0.2135, 0.2471, 0.3627, and 0.1765, exactly the same to the values found during the formulation.

Obviously, we can also compute the PageRank centrality for undirected graph with weighted links and undirected graph with unweighted links.

To simply exemplify these cases, like we did for the other centrality measures, let us take a look at the undirected graph with unweighted links first. Let us use the same graph as before, showed in the Figure 3.60.

Discarding the current links weights and defining all weights to be 1, we will have the following adjacency matrix.

$$\begin{bmatrix} 0 & 1 & 1 & 1 \\ 1 & 0 & 1 & 0 \\ 1 & 1 & 0 & 1 \\ 1 & 0 & 1 & 0 \end{bmatrix}$$

The degree-normalized adjacent matrix will be like below.

$$\begin{bmatrix} 0/3 & 1/2 & 1/3 & 1/2 \\ 1/3 & 0/2 & 1/3 & 0/2 \\ 1/3 & 1/2 & 0/3 & 1/2 \\ 1/3 & 0/2 & 1/3 & 0/2 \end{bmatrix} = \begin{bmatrix} 0 & 0.5 & 0.33 & 0.5 \\ 0.33 & 0 & 0.33 & 0 \\ 0.33 & 0.5 & 0 & 0.5 \\ 0.33 & 0 & 0.33 & 0 \end{bmatrix}$$

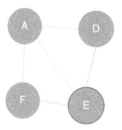

Figure 3.60 Undirected graph with unweighted links to show the PageRank centrality calculation.

« OUTNODES Table Rows: 4 | Columns: 2 of 2 | Rows 1 to 4 | ⊼ ↑ ↓ ↓ ⋮

▽ | Enter expression   ⌕ |

	⚠ node	⊕ centr_pagerank_unwt
1	D	0.2047872341
2	A	0.2952127659
3	E	0.2952127659
4	F	0.2047872341

Figure 3.61 Output nodes dataset containing the PageRank centrality based on an undirected graph using unweighted links.

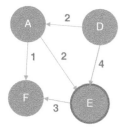

Figure 3.62 Directed graph to shows the PageRank centrality calculation.

After few iterations, the process continues until the power method converges to the final result:

$$\begin{bmatrix} 0.29 \\ 0.20 \\ 0.29 \\ 0.20 \end{bmatrix}$$

We change the previous code to consider the PAGERANK=WEIGHT option.

The OUTNODES dataset presented in the Figure 3.61 contains the results for the PageRank centrality centr_pagerank_unwt computed based on an undirected graph using unweighted links.

Notice that based on an undirected graph with unweighted links, the PageRank centrality values are symmetric. Nodes A and E have the same PageRank, 0.2952, and nodes D and F have the same value as well, 0.2047, as showed during the formulation.

To simplify the demonstration, let us use the same graph to present the results for the directed graph with weighted links. The graph is showed in the Figure 3.62.

This graph produces the following adjacency matrix.

$$\begin{bmatrix} 0 & 0 & 2 & 1 \\ 2 & 0 & 4 & 0 \\ 0 & 0 & 0 & 3 \\ 0 & 0 & 0 & 0 \end{bmatrix}$$

The degree-normalized adjacent matrix will be like below.

$$\begin{bmatrix} 0 & 0 & 0.33 & 0.25 \\ 1 & 0 & 0.66 & 0 \\ 0 & 0 & 0 & 0.75 \\ 0 & 0 & 0 & 0 \end{bmatrix}$$

After few iterations, the process continues until the power method converges to the final result:

$$\begin{bmatrix} 0.1626 \\ 0.1267 \\ 0.2907 \\ 0.4199 \end{bmatrix}$$

We change the previous code to consider the DIRECTION=DIRECT and PAGERANK=WEIGHT options.

The OUTNODES dataset presented in the Figure 3.63 contains the results for the PageRank centrality centr_pagerank_wt computed based on a directed graph using weighted links.

	△ node	⊕ centr_pagerank_wt
1	D	0.1267324502
2	A	0.1626399777
3	E	0.2907101593
4	F	0.4199174128

(Table header: « OUTNODES   Table Rows: 4 | Columns: 2 of 2 | Rows 1 to 4   ⊤ ↑ ↓ ⊥   ⋮)

(Filter: ▽ | Enter expression   ⌕)

Figure 3.63 Output nodes dataset containing the PageRank centrality based on a directed graph using weighted links.

Notice that based on a directed graph with weighted links, the PageRank centrality values are as presented during the formulation, 0.16, 0.12, 0.29, and 0.41 for nodes A, D, E, and F, respectively.

In terms of business actions, the PageRank centrality has a similar interpretation as the eigenvector centrality. The PageRank also gives a sense of the importance of the node within the network. It describes how central the node is, the number of connections, the strength of these connections, and ultimately, how influential a node can be. The major difference between the PageRank and the eigenvector centralities is the type of the network, or the direction of the links in the network. PageRank is tailored to directed graphs, where the direction of the nodes plays an important role. The eigenvector centrality can be computed for both directed and undirected graphs. It might be more suitable to undirected graphs but still can be calculated for both. Therefore, the application of PageRank centrality in business would be very similar to the application of eigenvector. The PageRank centrality would give companies the list of important or influential customers to keep satisfied, to closely monitor and to use in marketing campaigns. These customers can naturally diffuse marketing and corporate messages throughout the network as well as trigger other customers in consuming products and services along with them.

3.10 Hub and Authority

The network metrics hub and authority centralities were first developed to rank the importance of web pages. Some web pages are important because they point to many other important pages. This is the hub centrality. Some web pages are important because they are linked by many other important pages. This is the authority centrality. A good hub points to many good authorities. A good authority is pointed to by many good hubs. In network analysis, other types of entities (such as customers) might be considered as web pages and have their hub and authority computed.

The hub measure represents the number of important nodes a node n, points to. This measure describes how a node refers to other important nodes. The higher the number of important nodes that it refers to, the more important the node is.

The centrality hub considers only the important nodes in the subgraph that it belongs to. If this node points to several important nodes, no matter its own importance, it constitutes a hub.

For example, in an airline traffic network, a node would be the airport and the links could refer to the flights. If most of the international flights for Delta Airlines depart from Atlanta, the Hartsfield–Jackson Atlanta International Airport would be a hub when describing the traffic network. Most of the Delta international flights departs from Atlanta. If I live in a different city in U.S., I most likely need to fly from that city to Atlanta and then fly to my international destination. Atlanta is a hub given it connects many other important airports.

Authority represents the number of important nodes that point to a node n. Like the hub, this measure describes how this node is referenced (versus what nodes it refers to as with the hub). The more important nodes that refer to it, the more important it is.

If many important nodes in the network refer to a node, this node should therefore be even more important. The concept of centrality authority is quite applicable, for example, in authorship networks. Authors can often refer to other authors in their papers. An author, referenced many times by other important authors, is definitely an authority in a particular subject. This concept is straightforward and might be applied in distinct types of networks, such as telecommunications, insurance, and banking.

Hub and authority centralities were originally developed by Kleinberg to rank the importance of web pages. Certain web pages are important in the sense that they point to many important pages (these are called hubs). On the other hand, some web pages are important because they are linked by many important pages (called authorities). In other words, a good hub node is one that points to many good authorities, and a good authority node is one that is pointed to by many good hub nodes. This idea can be applied to many other types of graphs besides web pages. For example, it can be applied to a citation network for journal articles. A review article that cites many good authority papers has a high hub score, whereas a paper that is referenced by many other papers has a high authority score.

The authority centrality of a node is proportional to the sum of the hub centrality of nodes that point to it. Similarly, the hub centrality of a node is proportional to the sum of the authorities of nodes that it points to.

3.10.1 Computing the Hub and Authority Centralities

The overall formulas of hub and authority are presented as following.

$$x_i = \alpha + \sum_{j \in N} w_{ji} y_j$$

$$y_j = \beta + \sum_{j \in N} w_{ij} x_i$$

where x_i is the authority of node i, y_j is the hub of node j, w_{ij} is the sum of weights of the links from i to j, w_{ij} is the sum of weights of the links from j to i, and α and β are constants.

The hub and authority metrics are normalized. The node with the highest hub receives a hub centrality of 1, and the node with the highest authority receives the authority centrality of 1. All other values are in proportion to 1.

Let us consider the graph showed in the Figure 3.64 to demonstrate both hub and authority centralities.

In the graph above, node B has a hub centrality equal to 1. Notice that node B and D have both the sum of outgoing links as 6. Both are connected to A, which is the authority in the network, receiving incoming links summing up as 10. However, the link between B and A has a weight of 5, and the link between D and A has a weight of 2. Then, node B is identified with the highest hub centrality.

Hub and authority centralities can be computed based on weighted links, unweighted links, or both.

We can define how proc network will calculate the hub and authority centralities by using the option AUTH=WEIGHT| UNWEIGHT|BOTH and HUB=WEIGHT|UNWEIGHT|BOTH.

As a matrix operation, we can define the eigenvector algorithm to compute hub and authority centralities by using the option EIGENALGORITHM=AUTOMATIC|JACOBDAVIDSON|POWER.

The following code shows how to compute hub and authority centralities using proc network based on a directed graph using link weights.

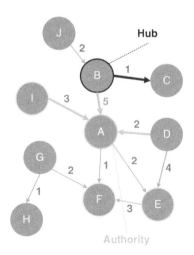

Figure 3.64 Directed graph to shows hub and authority centralities.

```
proc network
    direction = directed
    links = mycas.links
    outlinks = mycas.outlinks
    outnodes = mycas.outnodes
    ;
    linksvar
        from = from
        to = to
        weight = weight
    ;
    centrality
        hub = weight
        auth = weight
    ;
run;
```

« OUTNODES   Table Rows: 10 | Columns: 3 of 3 | Rows 1 to 10 | ⤒ ↑ ↓ ⤓ ⋮

▽ | Enter expression   🔍

	⬠ node	⊕ centr_hub_wt	⊕ centr_auth_wt
1	A	0.150867526	1
2	B	1	0
3	C	0	0.1231558261
4	D	0.6815324015	0
5	E	0.0158805427	0.3728991733
6	F	0	0.0271194982
7	G	0.0108477993	0
8	H	0	0.0013359697
9	I	0.5855765668	0
10	J	0	0

Figure 3.65 Output nodes dataset containing the hub and authority centralities results for the directed graph.

The OUTNODES dataset presented in the Figure 3.65 contains the results for the hub and authority centralities centr_hub_wt and centr_auth_wt. Hub and authority centralities are computed only for directed graphs.

Notice that the hub centrality value for the node B is 1, and the authority centrality for the node A is also 1, indicating these nodes have the highest hub and authority centralities, respectively, as demonstrated during the formulation. The hub centrality for the node D, with the same sum of weights for the outgoing links (6) is not 1 as B, but a proportional value to 1, because node B has a link weight of 5 to node A (the highest authority within the network) and node D has a link weight of 2 to node A.

However, if we compute the hub and authority centralities not accounting for the link weights, the results will be slightly different.

The following code shows how to compute hub and authority centralities using proc network for unweighted links.

```
proc network
   direction = directed
   links = mycas.links
   outlinks = mycas.outlinks
   outnodes = mycas.outnodes
   ;
   linksvar
      from = from
      to = to
      weight = weight
   ;
   centrality
      hub = unweight
      auth = unweight
   ;
run;
```

The OUTNODES dataset presented in the Figure 3.66 contains the results for the hub and authority centralities centr_hub_unwt and centr_auth_unwt. The hub and authority centralities here are computed based on unweighted links.

Notice that the hub centrality value for the node B is 1, and the hub centrality for the node D is also 1. As the weight links are not considered here, both have two outgoing links and then they have the highest hub centrality within the network. However, nodes A and G also have both two outgoing links. Why do not they have the highest hub centrality like nodes B and D? Node

	« OUTNODES	Table Rows: 10	Columns: 3 of 3	Rows 1 to 10	

	⚠ node	⊕ centr_hub_unwt	⊕ centr_auth_unwt
1	A	1	1
2	B	0.6888921825	0
3	C	0	0.3111078175
4	D	1	0
5	E	0.5254275608	0.9032119259
6	F	0	1
7	G	0.6888921825	0
8	H	0	0.3111078175
9	I	0.5254275608	0
10	J	0	0

Figure 3.66 Output nodes dataset containing the hub and authority centralities results for the directed graph and unweighted links.

A has three incoming links, and then it has the highest authority centrality. Because both nodes B and D have outgoing links to A, the highest authority within the network, and nodes A and G do not have outgoing links to node A (considering possible auto-relationships in node A), nodes B and D get the highest hub centrality. Something important to notice although is that node F has also three incoming links, just like node A. So, the same question is valid. Why node F has not the highest authority centrality too? Because node A is connected to nodes B and D, the highest hubs within the network, and node F is not connected to the highest hubs. Then, node A has the highest authority centrality and node F has not.

Here we can observe that hubs and authorities present what is called as mutually reinforced relationship. That means, a good hub usually points to many good authorities, and a good authority usually points to many good hubs. Due to this mutually relationships, to calculate hubs and authorities within the network we need a method to break this circularity.

An iterative algorithm updates the weights for hubs and authorities and normalize their values, so the squares sum to one. This iterative algorithm expresses the mutually relationship between hubs and authorities. If a node n points to many nodes with large x-values (authorities), then it should receive a large y-value (hub). On the same way, if n is pointed to by many nodes with large y-values (hubs), then it should receive a large x-value (authority). That means hubs and authorities reinforce one another. This iterative method of calculation can be achieved by linear algebra, considering the network a symmetric $n \times n$ matrix M. The eigenvector of M will be used to compute both hub and authority.

Similar to eigenvector centrality, the definitions of authority and hub centrality can be written in matrix form as

$$x = \alpha A^T y$$
$$y = \beta A x$$

where A represents the adjacency matrix that corresponds to the outgoing link weights in L and A^T represents the adjacency matrix that corresponds to the incoming link weights in L. In constructing the matrices A and A^T, the weights of multiple links are aggregated.

Combining the two equations results in

$$A^T A x = \lambda x$$
$$A A^T y = \lambda y$$

where $\lambda = (\alpha\beta)^{-1}$. Thus, the authority and hub centralities are the principal eigenvectors of $A^T A$ and $A A^T$, respectively.

Let us take a look at this process a little deeper. To simplify the demonstration, let us use the same subgraph we used to demonstrate the eigenvector and PageRank centralities. We are going to use the subgraph based on directed and weighted links. The graph is showed in the Figure 3.67.

As we saw before, this graph produces the following adjacency matrix.

$$\begin{bmatrix} 0 & 0 & 2 & 1 \\ 2 & 0 & 4 & 0 \\ 0 & 0 & 0 & 3 \\ 0 & 0 & 0 & 0 \end{bmatrix}$$

The transpose matrix A^T looks like the following.

$$\begin{bmatrix} 0 & 2 & 0 & 0 \\ 0 & 0 & 0 & 0 \\ 2 & 4 & 0 & 0 \\ 1 & 0 & 3 & 0 \end{bmatrix}$$

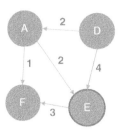

Figure 3.67 Directed graph with weighted links to demonstrate the hub and authority centralities.

By multiplying the adjacent matrix A by its transpose matrix A^T, we get the following result.

$$AA^T = \begin{bmatrix} 0 & 0 & 2 & 1 \\ 2 & 0 & 4 & 0 \\ 0 & 0 & 0 & 3 \\ 0 & 0 & 0 & 0 \end{bmatrix} \cdot \begin{bmatrix} 0 & 2 & 0 & 0 \\ 0 & 0 & 0 & 0 \\ 2 & 4 & 0 & 0 \\ 1 & 0 & 3 & 0 \end{bmatrix} = \begin{bmatrix} 5 & 8 & 3 & 0 \\ 8 & 20 & 0 & 0 \\ 3 & 0 & 9 & 0 \\ 0 & 0 & 0 & 0 \end{bmatrix}$$

As we saw before, the authority and hub centralities are the principal eigenvectors of A^TA and AA^T, respectively.

The eigenvector corresponding to the largest eigenvalue for the AA^T matrix is presented below. The eigenvalue is approximately 23.566.

$$\begin{bmatrix} 4.855 \\ 10.893 \\ 1 \\ 0 \end{bmatrix}$$

Normalizing this eigenvector (dividing all values by the maximum), we get this result:

$$\begin{bmatrix} 4.855/10.893 \\ 10.893/10.893 \\ 1/10.893 \\ 0/10.893 \end{bmatrix} = \begin{bmatrix} 0.4457 \\ 1 \\ 0.0918 \\ 0 \end{bmatrix}$$

This final normalized eigenvector contains the hub scores for the subgraph.

Similar approach can be performed to compute the authority centrality. By multiplying the transpose matrix A^T by the adjacent matrix A, we get the following result.

$$A^TA = \begin{bmatrix} 0 & 2 & 0 & 0 \\ 0 & 0 & 0 & 0 \\ 2 & 4 & 0 & 0 \\ 1 & 0 & 3 & 0 \end{bmatrix} \cdot \begin{bmatrix} 0 & 0 & 2 & 1 \\ 2 & 0 & 4 & 0 \\ 0 & 0 & 0 & 3 \\ 0 & 0 & 0 & 0 \end{bmatrix} = \begin{bmatrix} 4 & 0 & 8 & 0 \\ 0 & 0 & 0 & 0 \\ 8 & 0 & 20 & 2 \\ 0 & 0 & 2 & 10 \end{bmatrix}$$

The eigenvector corresponding to the largest eigenvalue for the A^TA matrix is presented below. The eigenvalue is approximately 23.566.

$$\begin{bmatrix} 2.773 \\ 0 \\ 6.783 \\ 1 \end{bmatrix}$$

Normalizing this eigenvector (dividing all values by the maximum), we get this result:

$$\begin{bmatrix} 2.773/6.783 \\ 0/6.783 \\ 6.783/6.783 \\ 1/6.783 \end{bmatrix} = \begin{bmatrix} 0.4488 \\ 0 \\ 1 \\ 0.1474 \end{bmatrix}$$

This final normalized eigenvector contains the authority scores for the subgraph.

To solve this eigenvector problem, proc network provides two algorithms, the Jacobi–Davidson algorithm, and the power method. As described before, we can select the algorithm to use when computing hub and authority centrality by using the option EIGENALGORITHM=JACOBIDAVIDSON|POWER. The Jacobi–Davidson algorithm is the default, and it is usually used for solving large-scale eigenvalue problems. The power method is one of the standard algorithms for solving eigenvalue problems, but it converges slowly for certain problems.

Let us call proc network to compute the hub and authority centralities considering the previous subgraph. The following code shows how to do that.

```
data mycas.links;
    input from $ to $ weight @@;
datalines;
D A 2 A E 2 A F 1 E F 3 D E 4
;
run;

proc network
    direction = directed
    links = mycas.links
    outlinks = mycas.outlinks
    outnodes = mycas.outnodes
    ;
    linksvar
        from = from
        to = to
        weight = weight
    ;
    centrality
        hub = weight
        auth = weight
            eigenalgorithm = power
    ;
run;
```

The OUTNODES dataset presented in the Figure 3.68 contains the results for the hub and authority centralities centr_hub_wt and centr_auth_wt based on weighted links.

Notice the values for the hub centralities for the nodes A, D, E, and F as 0.4457, 1, 0.0918, and 0, exactly as demonstrated during the formulation. The same for the authority centrality values, as 0.4088, 0, 1, and 0.1474 for nodes A, D, E, and F, respectively. Same values presented during the formulation.

Hub and authority centralities have a straightforward application in business. They clearly describe the role of the nodes within the network. Nodes with high value authority centrality are the nodes frequently reached out by other nodes. Customers in that case are often consulted by other customers and hence can be used to consolidate a corporate message. For example, customers with great authority values can strength the corporate image every time they are reached out by the other customers. The other customers in the network most likely trust on the authorities to build up their opinion about a particular product, a service, or about the company in overall. On the other hand, nodes with high value hub centrality are the nodes frequently reaching out a substantial number of other nodes within the network. Customers in that situation usually have a great level of communication with other customers, reaching out to them quite often. Companies can

	⚠ node	⊕ centr_hub_wt	⊕ centr_auth_wt
1	A	0.4457326581	0.4088754332
2	D	1	0
3	E	0.0918035638	1
4	F	0	0.1474288997

Figure 3.68 Output nodes dataset containing the hub and authority centralities results for the directed graph on weighted links.

use these customers to widely spread corporate and institutional messages. Reaching out a hub means to a company possibly reaching out many other customers indirectly. Because these customers are not exactly powerful customers within the network, or at least they are not the authorities that other customers rely on, the type of the message can be sensitive. Corporate and institutional messages fit perfectly to that case. Imagine a situation where there are many people reaching out to you for questions about machine learning models. Perhaps you are a professor or a senior member in a group. You are definitely an authority in that scenario. Suppose that some of your peers come to you very frequently to ask you about technical details in machine learning models. As they do this with you, they also do with other members of your group, or even different groups. They always ask questions about professional career, technology, trends, etc. They often go for multiple members within different groups in and out the company. They are probably hubs within the network, as they have much more outgoing connections than incoming ones. If they come to you to share some sensitive information, perhaps we may get suspicious and will validate that information before doing anything with that. They may be good to spread out a non-sensitive information, but not a sensitive one. If we translate that scenario to marketing or sales, the outcome should be similar.

3.11 Network Centralities Calculation by Group

All network metrics describe in this chapter were demonstrated individually. For each network centrality, we looked at the theory behind the math, the business concepts assigned to them, how to use them to solve a business problem or to better understand a specific scenario, the step-by-step calculations, the proc network code to compute these metrics and finally the outcomes, the datasets containing the results.

However, we can run all these network centralities calculation at once, in a single execution of proc network, saving a lot of time, not just running the code, but also writing it.

Let us see how to do it in proc network. The following code creates the links dataset we have been using and invokes proc network to compute all network centralities.

```
data mycas.links;
   input from $ to $ weight @@;
datalines;
J B 2 B C 1 B A 5 I A 3 D A 2 A E 2 A F 1 G F 2 G H 1 E F 3 D E 4
;
run;

proc network
   direction = directed
   links = mycas.links
   outlinks = mycas.outlinks
   outnodes = mycas.outnodes
   ;
   linksvar
      from = from
```

```
      to = to
      weight = weight
   ;
   centrality
      degree
      influence = weight
      clusteringcoef
      close = weight
         closenopath = diameter
      between = weight
         betweennorm = true
      eigen = weight
      pagerank = weight
         pagerankalpha = 0.85
      hub = weight
      auth = weight
   ;
run;
```

The OUTNODES dataset presented in the Figure 3.69 contains the results for all the network centralities computed by proc network in a single execution. This dataset contains 15 columns. In addition to the node's identification, it contains columns for all network metrics computed, centr_auth_wt, centr_between_wt, centr_close_in_wt, centr_close_out_wt, centr_close_wt, centr_cluster, centr_degree, centr_degree_in, centr_degree_out, centr_eigen_wt, centr_hub_wt, centr_influence1_wt, centr_influence2_wt, and centr_pagerank_wt.

Notice that we computed the network metrics based on a directed graph with weighted links. Here we still can compute the network centralities based on an undirected graph (be sure to select the centralities accordingly – some of them are calculated just for directed graphs), and using weighted links, unweighted links, or both.

The OUTLINKS dataset presented in the Figure 3.70 contains only the links, defined by the columns from, to and weight, and the network centrality computed for links, the link betweenness, in the column centr_between_wt.

3.11.1 By Group Network Centralities

Proc network enables the calculation of all network metrics by separate groups or subgraphs. This separate analysis is performed when the BY statement is used followed by a variable that uniquely identifies the groups. These groups can be identified by network subgraph detection, like cores, communities, connected components, reach, etc. Theoretically, these groups can be determined by any rule, for example, customer segmentation, geographic locations, business segments, etc. It really depends on the problem we are trying to solve or the business scenario we are trying to better understand.

We can specify a BY statement in proc network to obtain separate analyses of observations in groups that are defined by the values of the BY variables. If you specify more than one BY statement, only the last one specified is used.

The BY statement in proc network is not supported when a node or a nodes subset dataset is used. The BY variable must come from the links dataset.

For example, when running community detection, we can define the output links dataset by using the statement OUTLINKS=. The links dataset specified here contains all the links and the IDs for the communities detected, even when

	node	centr_degree_in	centr_degree_out	centr_degree	centr_eigen_wt	centr_hub_wt	centr_auth_wt	centr_close_wt	centr_close_in_wt	centr_close_out_wt	cent
1	A	3	2	5	0.0022358273	0.150867526	1	0.548513986	0.625	0.472027972	(
2	B	1	2	3	0.0747928394	1	0	0.5051549574	0.4333868379	0.5769230769	(
3	C	1	0	1	0.0000077311558	0	0.1231558261	0.2224052718	0.4448105437	0	
4	D	0	2	2	0.0342539266	0.6815324015	0	0.2721774194	0	0.5443548387	
5	E	2	1	3	0.0001627813	0.0158805427	0.3728991733	0.5489239581	0.6609547124	0.4368932039	(
6	F	3	0	3	0.0000077311558	0	0.0271194982	0.407239819	0.814479638	0	
7	G	0	2	2	0.0001627813	0.0108477993	0	0.2339688042	0	0.4679376083	
8	H	1	0	1	0.0000077311558	0	0.0013359697	0.2115987461	0.4231974922	0	
9	I	0	1	1	0.0448477861	0.5855765668	0	0.2566539924	0	0.5133079848	
10	J	0	1	1	1	0	0	0.2890792291	0	0.5781584582	

Figure 3.69 Output nodes dataset containing all the network centralities results for the directed graph on weighted links.

‹‹ OUTLINKS Table Rows: 11 | Columns: 4 of 4 | Rows 1 to 11 ⊼ ↑ ↓ ↓ | ↺ ▾ | ⋮

▽ [Enter expression   ○]

	△ from	△ to	⊞ weight	⊞ centr_between_wt
1	J	B	2	0.0555555556
2	B	C	1	0.0222222222
3	B	A	5	0.0666666667
4	I	A	3	0.0333333333
5	D	A	2	0.0111111111
6	A	E	2	0.0888888889
7	A	F	1	0
8	G	F	2	0.0111111111
9	G	H	1	0.0111111111
10	E	F	3	0.0666666667
11	D	E	4	0.0222222222

Figure 3.70 Output links dataset containing the link betweenness centrality results for the directed graph on weighted links.

multiple communities are detected during the same execution. In that case, we might have more than one community. This is the case when we run the community detection based on a resolution list. In this scenario, the links dataset contains the from and to identification for the links, the link weight if used, and all the community IDs, such as community_1, community_2, and so on. One of these IDs will be defined in the BY statement to allow proc network to compute the network centralities within each group separately.

Proc network applies all parameter settings to each individual group independently. These parameters are not applied to the entire process as a whole. For example, when a stopping criterion such as the MAXTIME= option is specified for a particular algorithm, this limit pertains to each individual group as it is processed, rather than to the entire graph or all subgraphs.

Let us see how to do it in proc network. Let us use the same graph we have been using for all network centralities so we can compare the results based between the entire network and the subgraphs (groups). The graph is showed in the Figure 3.71.

The following code shows how to invoke proc network to compute network centralities based on multiple groups. First, we need to create these groups. A simple community detection is executed to split the original graph into two distinct communities. Then, proc network is used again to compute the network centralities for each community separately.

```
proc network
   direction = undirected
   links = mycas.links
   outlinks = mycas.outlinkscomm
   outnodes = mycas.outnodescomm
   ;
   linksvar
      from = from
      to = to
      weight = weight
   ;
   community
      algorithm = louvain
      resolution_list = 1
   ;
run;
```

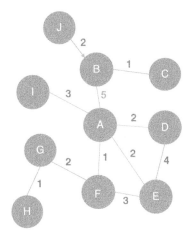

Figure 3.71 Directed graph used to compute the network centralities.

	⬓ node	⊕ community_1
	« OUTNODESCOMM Table Rows: 10 ǀ Columns: 2 of 2 ǀ Rows 1 to 10	
	▽ Enter expression	🔍
1	A	1
2	B	1
3	C	1
4	D	2
5	E	2
6	F	2
7	G	2
8	H	2
9	I	1
10	J	1

Figure 3.72 Output nodes dataset containing the community identification for the undirected graph on weighted links.

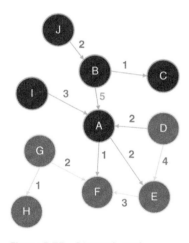

Figure 3.73 Directed graph comprising two communities.

The OUTNODESCOM dataset presented in the Figure 3.72 contains the results for the community detection. The column community_1 determines the community for each one of the nodes within the network.

Notice here that we used the algorithm LOUVAIN to identify the communities within the network. This algorithm is supported only for undirected graph. Proc network also provide an algorithm to detect communities on directed graphs, the PARALLELLABELPROP. This algorithm finds the communities based on the information flow along the directed links, rather than considering the bidirectional relationship between the nodes. For example, if we run the community detection using the PARALLELLABELPROP algorithm, we find three communities, two communities with a single node, and the third community with eight nodes (the rest of the network). This is not appropriate here as we want to investigate the network centralities calculation by groups. It would be nice to have groups with similar size. On the other hand, running the community detection by using the LOUVAIN algorithm we find just two communities, each one of them with five nodes, as we can see in the OUTNODESCOMM dataset above.

The new structure for the original network would be divided by two groups, as shown in the graph presented in the Figure 3.73.

The OUTLINKSCOM dataset presented in the Figure 3.74 contains the results for the community detection. The column community_1 identifies the links pertained to each community identified.

Notice that there are three links with no community identification, between nodes A and E, nodes A and F, and finally between nodes A and D. These are the links within the network that were broken to create the two communities. This small graph is a connected component, where each node can reach out to any other node in the network by some particular path. The nodes are all connected somehow. Based on the average link weights between them, they were split into two groups, and these links mentioned before do not make part of any group now. They are exactly the links that will be ignored when computing the network centralities by group.

Now we have two distinct groups in the network, and therefore we can compute the network centralities based on them.

	⬳ from	⬳ to	⊕ weight	⊕ community_1
1	A	E	2	.
2	A	F	1	.
3	B	A	5	1
4	B	C	1	1
5	D	A	2	.
6	D	E	4	2
7	E	F	3	2
8	G	F	2	2
9	G	H	1	2
10	I	A	3	1
11	J	B	2	1

OUTLINKSCOMM   Table Rows: 11   Columns: 4 of 4   Rows 1 to 11

Figure 3.74 Output nodes dataset containing the community identification for the undirected graph on weighted links.

The following code invokes proc network to compute the network centralities based on the groups identified by the community_1 identification, generated in the previous step (community detection).

```
proc network
    direction = directed
    links = mycas.outlinkscomm(where=(community_1 ne .))
    outlinks = mycas.outlinks
    outnodes = mycas.outnodes
    ;
    linksvar
        from = from
        to = to
        weight = weight
    ;
    centrality
        degree
        influence = weight
        clusteringcoef
        close = weight
            closenopath = diameter
        between = weight
            betweennorm = true
        eigen = weight
        pagerank = weight
            pagerankalpha = 0.85
        hub = weight
        auth = weight
    ;
    by community_1
    ;
run;
```

	⊞ community_1	△ node	⊞ centr_degree_in	⊞ centr_degree_out	⊞ centr_degree
1	1	A	2	0	2
2	1	B	1	2	3
3	1	C	1	0	1
4	1	I	0	1	1
5	1	J	0	1	1
6	2	D	0	1	1
7	2	E	1	1	2
8	2	F	2	0	2
9	2	G	0	2	2
10	2	H	1	0	1

Table Rows: 10 | Columns: 16 of 16 | Rows 1 to 10

Figure 3.75 Output nodes dataset containing the network centralities by group.

The OUTNODES dataset presented in the Figure 3.75 contains the results for the network centralities computed by groups (communities).

Let us take a look at the results for some of the network centralities and compare them to the results when considering the whole graph. The degree centrality is probably the easiest one to consider. Let us take node A as an example. Considering the entire network, node A has degree centrality of 5, 3 of degree in and 2 of degree out, as we can see in the outcome dataset below. Now, based on the groups (communities), node A has connections just to nodes B and I. Then, when proc network computes the degree centrality for node A, it considers just the nodes within the same group (A, I, B, J, and C). In other others, proc network discards exactly the links we mentioned before (A–D, A–E, and A–F). These links would add up three more connections to node A, summing the degree to 5. As they were eliminated from the links dataset (community_1 ne .), they are not counted as connections for node A.

The OUTNODES dataset presented in the Figure 3.76 contains the results for the centralities considering the entire network.

Analogously to the degree centrality computed here for node A, considering the entire network, and based on only its own community, all other network centralities will be affected by the computation by groups. The computation by group may be relevant in several business scenarios, where huge networks just do not make sense, such as telecommunications, banking, and insurance.

Another practical use of the computation by group is to allow proc network to calculate some specific centralities for very large networks, particularly for the network centralities based on distances or matrices multiplication. For example, computing degree centrality, influence, and clustering coefficient, is relatively simple and straightforward even when considering a very large network. These centrality metrics relies on the connections of each node, or on the connections of the nodes' neighbors. For network centralities based on distances, like closeness and betweenness, proc network needs to determine all shortest paths for the entire network in order to compute the average shortest paths for each node, when computing the closeness centrality, or how many shortest paths each node partakes, when computing the betweenness centrality. For network centralities based on matrices multiplications, like eigenvector, PageRank, hub, and authority, proc network needs to create the adjacent matrices and perform several multiplications or iterations to finally achieve the results. In both cases, if the network is very large, the process to calculate all distances (which is basically a matrix as well) or to multiple extremely large matrices, can simply be unfeasible. For these cases, the by group calculation turns out to be a good solution, or a feasible approach. Imagine a network created based on telecommunications

	⚠ node	⊞ centr_degree_in	⊞ centr_degree_out	⊞ centr_degree
1	A	3	2	5
2	B	1	2	3
3	C	1	0	1
4	D	0	2	2
5	E	2	1	3
6	F	3	0	3
7	G	0	2	2
8	H	1	0	1
9	I	0	1	1
10	J	0	1	1

Figure 3.76 Output nodes dataset containing the original degree centrality for the whole network.

transactions, like calls and text messages. Imagine the largest mobile carrier in United States, one of the largest mobile carriers in the world. This carrier has around 180 million subscribers. If we need to compute network centralities based on distances or matrices multiplication, we might end up creating a matrix of 180 million columns by 180 lines, which has 3.24e+16 entries, or 32 400 000 000 000 000 entries, or thirty-two quadrillion entries! That size of matrix would make unfeasible to compute most of the network centralities.

However, suppose we perform first a community detection. Thinking about an average number of members for each community as around 60 subscribers, we might end up with three million communities. Now, we need to compute network centralities for three million communities, but each subgraph would have on average a matrix with 3600 entries. This is completely feasible to proc network. Of course, there are also business approaches we can apply here, considering the same business scenario. For that mobile carrier, we can reduce the size of the original graph by dividing the network into geographies. Based on that, we would have multiple smaller networks based on geographies. Then, we would perform community detection on each geo-network. Finally, we would compute the network centralities based on each community, within each geo-network. The entire process would be divided into multiple smaller parallel processes. First computation on geographies, and then computation on communities by geographies. This type of approach is common in large network like telecommunications. Of course, there are customers between states who talk to each other. But the majority of customers talk to other customers within their local community. That makes more sense. We tend to communicate more with people are close to us, geographically close. For example, suppose someone who lives in the state of North Carolina, U.S. He can eventually talk to someone who live in the state of Florida, U.S. for sure. But most likely, the majority of his communications will be with people who live in North Carolina, probably with people who live in the same geographic area, like neighborhood, county, or city. A major telecommunications company in United States, who cover the whole country, can definitely build a network comprising the entire country, upon all communications from its customers. However, that would be a huge network, described by a very sparse matrix. Most of the customers talk to a small number of other customers. Then, it makes all sense to eventually divide the entire network into states, and then, for each state, to identify the multiple communities within the network. The network centrality metrics would be then computed based on the communities, instead of the entire network.

Finally, in some cases, we can have a hybrid approach. Based on a specific network, we can identify subnetworks like communities, but still compute the network centralities upon the whole network. The communities identified would be important to evaluate how the nodes are grouped together based on their relations, but every link between nodes would be accounted to compute the network metrics. For example, social networks in particular environments like companies or universities can be targeted for this approach. We can use community detection to identify groups of employees that commonly communicate to each other and form groups, but when computing the network metrics like degree, influence, or even closeness and betweenness, we will consider the entire network, not the multiple groups identified.

3.12 Summary

Network analysis is an important tool for exploratory analysis, deeper investigation, and mostly to raise very useful knowledge about not just entities' behavior, but more important, about entities' relationships. How these entities behave within interconnected structures. General and common knowledge about groups of friends, students, employees, or customers, can be easily identified by using clustering methods, which usually recognize similarities among the members of a particular group by their individual attributes. These clusters are, therefore, created based on individual characteristics, whether we talk about people or organizations. As long as the individuals or entities within an interconnected structure have similar attributes, they are fitted into the same cluster.

Network analysis allows us to define a distinguish method to understand interconnected structures, or groups. Instead of looking at individual characteristics, network analysis investigates the relationships among the members of those interconnected structures. This approach emphasizes how the members are related rather than their individual attributes. However, all these individual characteristics can be used to assign value to their members, as nodes, and to their relations, as links. That makes the focus of the network analysis on the relationships rather than on the individualities.

This chapter presented concepts about network centralities, including those related to the measurement of power and influence. Power within a social structure is assigned to the individual nodes, but it is calculated based on the position of these nodes within the graph, which makes the links the crucial information in network analysis. The nodes position within the network is determined by the relationships among them. Thus, the relations within the interconnected structure or groups define the measurement of power and influence for the nodes.

Fundamental measurements of power and influence were discussed in this chapter, such as the ones accounting for the degree of connections, the importance of the connections, or the distance of the connections. The degree centrality gauges the power of a particular node according to the number of connections it holds. These connections may or may not be directional. In directed graphs, where the direction of the relation matters, the measurement of power based on the degree centrality can be split into two distinct measures, the in-degree, for incoming links, and the out-degree, for outgoing links. Also, the degree centrality can be measured based on different levels of connections, such as first order, second order, and so on. The strength of the connections also can be considered. The First order centrality is the number of direct connections among the nodes, weighted by the strength of their connections. This is the influence centrality 1. The second order centrality is the number of direct connections that the nodes connected to the original node have, also weighted by the strength of their connections. This the influence centrality 2. The first order centrality is commonly referred to by the question, "How many friends do I have?" Second order centrality can be addressed by asking, "How many friends do my friends have?"

Clustering coefficient measures how the nodes' neighborhood is connected. Neighbors that are well connected may emphasize the information flow among the members of communities. Start spreading a message from a node with neighbors well connected should be more effective than start spreading a message from an isolated node or a node with neighbors not well connected.

Closeness centrality is also based on the relationships among the nodes. This measurement gauges how close a particular node is to its related nodes. It can be understood as the average number of steps that a particular node needs in order to reach its correlated nodes. This is known as the average length of path needed to diffuse information through its social network. The closeness centrality indicates how far information can spread through the network. Betweenness centrality is also based on relations among the nodes. It gauges how central a particular node is within the social structure. Short distance between two nodes is the smallest path connecting them. Considering different levels of connections, node A might reach node B by a distinct path, passing through several different nodes on the way. The smallest distance between those two nodes is the short path between them. If a node partakes of several short paths within the social structure, it is considered to be more central, and its betweenness is high. If not, the node's betweenness level, and thus its position within the network, is lower. The betweenness centrality indicates nodes which can control the information flow throughout the network.

PageRank and eigenvector centralities mix nodes attributes and distances. It measures the overall importance of the nodes based on their relationships, the strength of their relations, and how close nodes are to the others within the network, highlighting the overall proximity of the nodes to their peers, and then how likely they can be randomly accessed by other in the graph.

Finally, hub and authority centralities consider the importance of the nodes in addition to the number of connections they have. These metrics describe the importance of the nodes within a network as they have many connections going out or coming in. clustering coefficient centrality measures the level of interconnectivity between the nodes's neighbors. How well connected the connections of a particular node are? It helps define how a message can navigate throughout the network once it reaches a particular node.

4

Network Optimization

4.1 Introduction

The mathematical science behind network analysis and network optimization is the graph theory. Several concepts and centrality measures associated with network analysis and network optimization come from graph theory. One of the greatest advantages of graph theory is the mathematical formulation, the formalism that allows us to develop algorithms to be run on computers in order to solve business problems.

Graphs are considered mathematical structures used to model pairs of relations between objects or entities. The study of graphs considers a set of nodes or vertices (the entities or objectives mentioned before), and a set of links or edges (the relations between those objects or entities). The links are used to represent all the connections between the nodes in the graph. These connections are associated with a pair of nodes. These pairs of nodes can be represented by distinct nodes, or by a single node. A connection based on a single node usually refers to an auto-relationship, which a node connects to itself. For example, if an author refers to a paper they published before, that author refers to themself in an authorship graph. Analogously to social science, when we seek descriptions about people's relationships, the graph theory provides explanations about entities and relations. However, as opposed to social sciences, graph theory uses mathematical formalism to describe these relations. Based on mathematical formulas to describe relations between entities, it is possible to create algorithms and translate them into programming language codes to run them on computers. Algorithms running on computers gives us the necessary scale to apply network analysis and network optimization to solve real-world problems. Social sciences usually involve small networks, considering a limited number of persons (nodes) and relationships (links). However, real-world problems, particularly in some industries such as telecommunications, banking, insurance, and credit-card, may comprise very large networks, based on millions of nodes and billions of links. Imagine a mobile carrier with millions of customers sending messages to each other. Or a financial institution processing credit card transactions for millions of customers daily. These networks can easily evolve to a huge graph, and analyzing all these relations between credit card holders or mobile subscribers will not be an easy task. Because of that, in such industries, we need efficient algorithms to address these business issues. These algorithms need to be based on formal methods to be strictly translated into effective programming language codes and then be executed efficiently on computer systems.

4.1.1 History

One of the first scientists to use graph theory and topology was the Swiss mathematician and physicist Leonhard Euler in 1735 when he proposed a solution to the problem known as The Seven Bridges of Königsberg. In the early eighteenth century, people from Königsberg (now Kaliningrad) in the old Prussia (now Russia), used to walk on the complicated set of bridges across the waters of the Pregel River. Like many cities in Europe, Königsberg evolved near the river. The city was set on both sides of the Pregel River, which included two large islands connected to the mainland by seven bridges (nowadays there are only two original bridges from Euler's time; two were destroyed during World War II, two were demolished and replaced by highways, and one was rebuilt). The problem was formulated to find a way to walk through the city crossing each bridge once and only once to reach every part of the city. The islands and mainland could not be reached by any route other than the bridges. For this particular problem, it was not required that you start and end the tour at the same point (as we will see in other network optimization problems). Figure 4.1 shows the map for the city of Königsberg created by Merian Erben.

Network Science: Analysis and Optimization Algorithms for Real-World Applications, First Edition. Carlos Andre Reis Pinheiro.
© 2023 John Wiley & Sons, Inc. Published 2023 by John Wiley & Sons, Inc.

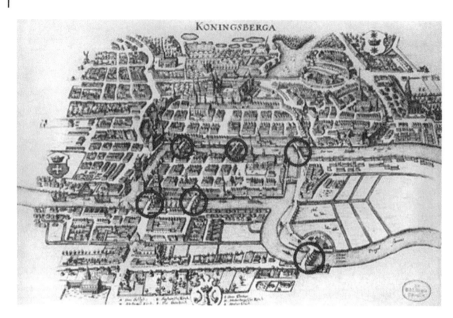

Figure 4.1 Map by Merian-Erben (1652) showing the city of Königsberg. Source: Merian-Erben / wikimedia commons / Public Domain Mark 1.0.

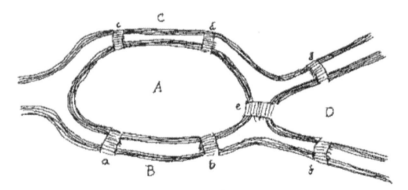

Figure 4.2 Euler's drawing of the Königsberg bridges.

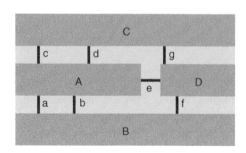

Figure 4.3 Königsberg bridges as links and distinct mainlands as nodes.

Euler indicated that the only important information in this particular problem was the sequence of the bridges to be crossed. Then, it was possible to discard any other information except the number of bridges connecting two land masses. Consequently, the number of bridges was more important than the bridges location. Euler drew the following diagram to formulate the problem. It contains the islands, the mainland, and the bridges. This diagram contains two islands denoted as C and B, and two land masses, denoted as A and D. The seven bridges are denoted as a, b, c, d, e, f, and g. Figure 4.2 shows Euler's drawing of the Königsberg bridges problem.

Extrapolating this problem to build a graph, each land mass is a node (C and D), as well as each island (A and D), and each bridge is a link (a, b, c, d, e, f, and g). Figure 4.3 describes the structure in terms of nodes and links.

As Euler noticed, in this type of problem, only the connections constitute relevant information. The graph can be represented in different shapes without changing the graph itself. The existence or absence of a link between a pair of nodes represents everything in terms of the network. Figure 4.4 describes the problem in terms of a graph, with nodes and links within a connected structure.

The links constitute the only relevant information in this problem and raise a very important concept in network analysis and network optimization. Even when calculating individual network metrics for the nodes, the links are what is considered to compute these centralities. For example, the degree centrality of a node is the number of connections incident to it, or the number of links the node has. No individual information or characteristic of the node itself is relevant, just the links, or particularly here, the number of links in this node. The related nodes are defined by the links between that node and the nodes connected to it. Then, that node's centrality degree for instance is computed by counting the number of links incident to it. Most of the network centralities rely exclusively on the links and the links' characteristics or attributes, such as the links' weight between the nodes. Several network optimization algorithms also rely only on the links and its attributes. Finally, as we saw before, most of the methods to detect subgraphs also rely on the links and the links' characteristics.

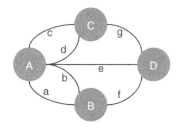

Figure 4.4 Königsberg bridges problem as nodes and links.

Let's go back to the seven bridges problem. Euler observed that whenever a walker gets a bridge (link) and reaches a land mass (node) by a bridge, they leave the node by a link, in a graph perspective. In practical terms, during any walk through the city, the number of times a walker enters a bridge is equal to the number of times they leave a bridge.

In a mathematical formalism, Euler said that the existence of a path in a particular graph depends on the nodes' degrees. We already saw that the degree of a node is the number of links it has, regardless of whether or not the link is arriving to or departing from it. Then, the important information in this problem is the number of links incident to each node. Based on Euler's observation, a necessary condition for a walk through the city is that the graph should be connected and have zero or two nodes with odd degrees. This graph has all four nodes with odd degrees (three connections each). This concept was formulated as the Euler Path. First, the graph must be traversable, which means, we can trace over the links of a graph exactly once without lifting our pencil. Second, the graph needs to have only 2 odd nodes. Third, the walk will start and stop on different odd nodes.

An alternative formulation for the seven bridges problem is to discover a path by which all bridges are crossed and the starting and ending points are the same. This path is known as the Euler Circuit, and this walk exists only if the graph is connected (traversed) and there are no nodes with degrees having odd values.

At the end, Euler proved that the number of the bridges should be even. If a walker wants to cross each bridge once and travel to each part of the city, the number of bridges should be six instead of seven. The solution views each bridge as an endpoint, a link in mathematical terms, connecting the nodes, or the mainland in the city. Euler noticed that only an even number of links produced the correct result of being able to reach every part of the city's mainland without crossing a bridge twice. He mathematically proved it was impossible to cross all seven bridges only once and visit every part of the city.

The Euler solution to this problem evolved to a mathematical field of study called topology. In computing, topology is useful to understand networks, the flow of information throughout the network, or paths within a system. The Seven Bridges of Königsberg is similar to another common problem in network optimization called Traveling Salesman Problem (TSP), where we try to find the most efficient route given a set of places to visit, with some restrictions, just like in the seven bridges problem. We will see the TSP in more detail later in this chapter. For now, it is good to know that we experience the TSP in our daily lives, when we use our cars, we get on a train, bus or even airplanes. We need to figure out the most efficient way to travel from one place to another, particularly when we need to cover multiple destinations. We also face the TSP when we receive our purchases delivered by online vendors. They are definitely the ones that need to account for the TSP in order to find the most efficient way to deliver all the goods purchased by multiple customers located at different addresses. The most efficient way may represent the minimum time or eventually the delivery's minimum cost.

This type of problem is perfect for computing because computers are faster and more efficient than humans in calculating things. But before asking computers to do all these calculations, we need Euler and other authors to formulate the problems and define solutions based on mathematical formulas. Then we can create programs to do all the math for us, based on those formulas to run on computers.

In this chapter we will cover some of those algorithms we experience or face in our daily lives, even if we don't really notice them. Algorithms like the TSP, Shortest Path, Minimum Cost Network Flow, Minimum Cut, Minimum Spanning Tree, and Vehicle Routing Problem (VRP) are just some of them.

4.1.2 Network Optimization in SAS Viya

Proc OptNetwork provides several network optimization algorithms that can augment more generic mathematical optimization approaches. Many practical applications of network optimization depend on an underlying network topology. For example, retailers facing the problem of shipping goods from warehouses to stores in a distribution network to satisfy demand at minimum cost. Commuters choosing routes in a road network to travel from home to work in the shortest time. Retailers searching for optimal routes to reach out to multiple destinations at the minimal time. Drivers looking for specific vehicle routes to minimize the distance traveled based on certain conditions. Industries defining teams to produce products at the maximum profitability or establishing quantities of production to minimize the cost. There are endless examples where network optimization can be applied to solve real-world problems, in virtually any industry.

Networks can be explicit and implicit in different scenarios. Networks are often built based on relationships that occur within those scenarios. For instance, relationships between researchers who coauthor articles, or actors who appear in the same movie, words or topics that occur in the same document, items that appear together in a shopping basket, messages exchanged by subscribers, money wired by customers, these are all explicit relations that create networks. Terrorism suspects who travel together or are seen in the same location, infected people been at the same location at the same time, these are all implicit relations that can also create networks. In both types of relationship, the entities involved in the relations (customers, authors, terrorists) are the nodes within the network. The interactions between them (money, message, flight) are the links within the network. The strength, value, importance, or frequency of these interactions are modeled as weights on the corresponding links of the network. These weights are crucial when running most of the network optimization algorithms.

Most of the network problems described here can actually be solved by using traditional optimization approaches, or general methods, like linear programming or mixed integer linear programming. Nevertheless, proc OptNetwork provides a set of optimization algorithms specifically tailored to network problems. These methods implemented in proc OptNetwork require less coding by the users and offer superior computational performance.

Proc OptNetwork makes no assumption about the context or application considered when building the network. The procedure provides a set of network analysis and network optimization algorithms that take an abstract graph or network as input. The procedure's outcomes help users to better understand the network structure and guide them in solving specific network optimization problems. Depending on the application or business problem, this type of network analysis can stand on its own and provide independent value or final solution to the problem. In some other business scenarios, the network analysis can provide additional input for subsequent work in optimization, or even in other forms of analytics, like supervised and unsupervised machine learning models.

Similar to what we see in Chapter 3 for proc Network, proc OptNetwork requires a graph $G = (N, L)$, where N is defined as a set of nodes and L is defined as a set of links. A node is an abstract representation of some entity or object within the network, and a link defines the relationship or connection between two entities or objects. Often, the terms node and vertex are interchangeable in describing an entity. Analogously, the term link is interchangeable with the terms edge or arc in describing a connection, interaction, or relationship between two nodes. Finally, the terms graph and network are also interchangeable.

4.2 Clique

A clique of a graph is a subgraph that is a complete graph. In a complete graph, every node within the graph is connected to every other node. A clique is then a subgraph within a graph where every node in the clique is connected to each and every other node within the same clique. In large graphs, it is easy to find cliques within the network, and sometimes cliques within cliques. Then, there is an important concept known as the maximal clique. A maximal clique is a clique that is not a subset of the nodes of any larger clique. That is, it is a set C of nodes such that every pair of nodes in C is connected by a link and every node not in C is missing a link to at least one node in C. The number of maximal cliques in a graph can be quite large and can grow exponentially with every node that is added to the network.

Cliques in network analysis and network optimization have multiple applications. Industries like bioinformatics, social sciences, electrical engineering, and chemistry are good candidates for problem solving by using the concepts of cliques. The analysis of transportation systems and traffic routes can also benefit from the concept of cliques. Even analytical models in banking, insurance and money laundering can use outcomes provided by clique analysis.

Proc optnetwork finds the maximal cliques of a graph by using the CLIQUE statement. The clique algorithm works only with undirected graphs (with no self-links). Clique can be seen as a sequence of nodes where, from each of the nodes within the graph, there is a link to the next node in that sequence.

The results of the clique algorithm are written to the output data table that is specified in the OUT= option. The output lists each node of each clique along with the variable clique to identify the clique to which it belongs. The clique identifiers are numbered sequentially, starting from the value of the INDEXOFFSET= option in the proc optnetwork statement. Proc optnetwork can find multiple cliques in the same run. Then, a particular node can appear multiple times in the output dataset if it belongs to multiple cliques.

The INDEXOFFSET= option in proc optnetwork specifies the index offset for identifiers in the log and results output tables. For example, if there are three cliques within the network, the clique enumeration algorithm labels the cliques as 1, 2, and 3. If INDEXOFFSET= 4, the clique enumeration algorithm labels the cliques as 4, 5, and 6. The value assigned to the INDEXOFFSET= option must be an integer greater than or equal to 0. By default, INDEXOFFSET= option is 1.

Proc optnetwork can compute all maximal cliques within the network. For very large graphs, proc optnetwork may not scale well in enumerating all maximal cliques.

If we simply want to count the cliques, we just need to suppress the output option and proc optnetwork will not write the results. If we suppress all options, just invoking the clique algorithm, proc optnetwork will return just the first clique.

In proc optnetwork, the CLIQUE statement invokes the algorithm that finds the maximal cliques in the input graph.

There are many options to be set when running proc optnetwork to find the maximal cliques within a graph.

The MAXCLIQUES= option specifies the maximum number of cliques for clique enumeration to return. We can specify a number (which can be any 32-bit integer greater than or equal to 1) or ALL (which represents the maximum that can be represented by a 32-bit integer). By default, MAXCLIQUES = 1.

The MAXLINKWEIGHT= option specifies the maximum sum of link weights in a clique. Any clique in which the sum of the link weights is greater than the number specified is removed from the results. This option can be used to discard a sequence of very strong relations between the links, and then, as a result, to reduce the number of possible maximal cliques within the network. The default value for this option is the largest number that can be represented by a double. When the default is used, no cliques are removed from the final results, or all possible maximal cliques are output.

The MINLINKWEIGHT= option specifies the minimum sum of link weights in a clique. Any clique in which the sum of the link weights is less than the number specified in the option is removed from the results. This option can be used to discard a sequence of very weak relations between the links, and then, as a result, to reduce the number of possible maximal cliques within the network. The default value for this option is the largest negative number that can be represented by a double. When the default is used, no cliques are removed from the final results.

The MAXNODEWEIGHT= option specifies the maximum sum of node weights in a clique. Any clique in which the sum of the node weights is greater than the number specified is removed from the results. Analogous to the link weights, this option can be used to discard a sequence of relations between very strong nodes, and then, as a result, to reduce the number of possible maximal cliques within the network. The default is the largest number that can be represented by a double. When the default is used, no cliques are removed from the final results.

Similarly, the MINNODEWEIGHT= option specifies the minimum sum of node weights in a clique. Any clique in which the sum of the node weight is less than the number specified in the option is removed from the results. Again, analogous to the link weights, this option can be used to discard a sequence of relations between weak nodes, and then, as a result, to reduce the number of possible maximal cliques within the network. The default value for this option is the largest negative number that can be represented by a double. When the default value is used, no cliques are removed from the final results.

There is a useful option in proc optnetwork when computing cliques. The MAXSIZE= option specifies the maximum number of nodes in a clique. Any clique in which the size is greater than the number specified is removed from the results. This option is useful in well-connected networks, where the maximal cliques can be quite large. The default is the largest number that can be represented by a 32-bit integer. When the default is used, no cliques are removed from the results. For example, if we are trying to find triangles, a common concept in fraud detection or money laundering, we can invoke the clique enumeration algorithm in proc optnetwork to limit the number of nodes in each clique to 3. Then proc optnetwork finds just cliques with a maximum of 3 nodes. We may find cliques with just 2 nodes though. Another option can be used to specifically determine cliques with 3 nodes.

The MINSIZE= option complements the MAXSIZE= option in searching cliques with specific sizes. This option specifies the minimum number of nodes in a clique. Any clique in which the size is less than the number specified in the option is

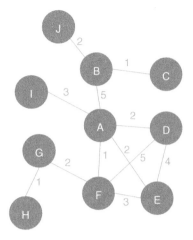

Figure 4.5 Undirected graph with weighted links.

removed from the results. By default, MINSIZE = 1 and no cliques are removed from the results. This option can complete the search for specific size cliques in the fraud detection and money laundering example. If we are looking for cliques with exactly 3 nodes, we can specify MINSIZE = 3 and MAXSIZE = 3. Then proc optnetwork finds only triangles within the network.

As we notice, clique enumeration can be exhaustive, particularly in large graphs. Proc optnetwork allows us to limit the time of running when invoking the network optimization algorithms. The MAXTIME= option specifies the maximum amount of time to spend finding all maximal cliques within the network. The type of time is determined by the value of the TIMETYPE= option. If TIMETYPE=CPU, the restriction of time is applied per processing machine. If TIMETYPE=REAL, the number specified in the MAXTIME= option defines the units of real-time. By default, the time type is real. The default value for the MAXTIME= option is the largest number that can be represented by a double.

The last option is where to output the results. The OUT= option specifies the output data table contains all maximal cliques found by proc optnetwork. The output data table must be a CAS-libref.data-table, where CAS-libref refers to the caslib, and data-table specifies the name of the output data table.

4.2.1 Finding Cliques

Let's consider the undirected graph with weighted links presented in Figure 4.5. This graph is similar to the ones we used to demonstrate the network centralities, except for the additional link between D and F with weight of 5.

The following code describes how to create the new links dataset and search for the maximal cliques within the undirected graph.

```
data mycas.links;
    input from $ to $ weight @@;
datalines;
J B 2  B C 1 B A 5 I A 3 D A 2 A E 2 A F 1 G F 2 G H 1 E F 3 D E 4 D F 5
;
run;


proc optnetwork
    direction = undirected
    links = mycas.links
    ;
    clique
      maxcliques = all
      out = mycas.outclique
    ;
run;
```

A summary result is created for each execution of proc optnetwork. Figure 4.6 shows the results for the clique algorithm.

As a result, proc optnetwork finds seven cliques, including all pairs of nodes (J-B, B-C, B-A, I-A, G-F, and G-H) plus the clique comprising 4 nodes (A-D-F-E). We can see this list checking the output dataset.

Proc optnetwork generates the output dataset OUTCLIQUE presented in Figure 4.7 containing all maximal cliques found. The dataset shows the identification for each clique and the nodes belonging to them.

We can set the minimum and the maximum size for the cliques searched by proc optnetwork. This constraint forces proc optnetwork to produce less cliques within the network as a final result. The following code specifies that all cliques found by proc optnetwork must have at least 3 nodes and the maximum of 10 nodes.

The OPTNETWORK Procedure

Problem Summary

Number of Nodes	10
Number of Links	12
Graph Direction	Undirected

The OPTNETWORK Procedure

Solution Summary

Problem Type	Clique
Solution Status	OK
Number of Cliques	7
CPU Time	0.02
Real Time	0.00

Output CAS Tables

CAS Library	Name	Number of Rows	Number of Columns
CASUSER(Carlos.Pinheiro@sas.com)	OUTCLIQUE	16	2

Figure 4.6 Summary results for clique enumeration using proc optnetwork.

>> OUTCLIQUE   Table Rows: 16   Columns: 2 of 2   Rows 1 to 16

Enter expression

	clique	node
1	1	A
2	1	D
3	1	E
4	1	F
5	2	A
6	2	B
7	3	A
8	3	I
9	4	B
10	4	C
11	5	B
12	5	J
13	6	F
14	6	G
15	7	G
16	7	H

Figure 4.7 Output dataset for the cliques.

```
proc optnetwork
   direction = undirected
   links = mycas.links
   ;
   clique
     maxcliques = all
     minsize = 3
     maxsize = 10
     out = mycas.outclique
   ;
run;
```

With the constraints for the minimum and maximum number of nodes within the clique, proc optnetwork finds only 1 clique. Figure 4.8 shows the summary results for the clique algorithm.

The output dataset OUTCLIQUE presented in Figure 4.9 shows the nodes assigned to the single clique. The single clique founds has 4 nodes, A, D, E, and F.

The OPTNETWORK Procedure

Problem Summary

Number of Nodes	10
Number of Links	12
Graph Direction	Undirected

The OPTNETWORK Procedure

Solution Summary

Problem Type	Clique
Solution Status	OK
Number of Cliques	1
CPU Time	0.01
Real Time	0.00

Output CAS Tables

CAS Library	Name	Number of Rows	Number of Columns
CASUSER(Carlos.Pinheiro@sas.com)	OUTCLIQUE	4	2

Figure 4.8 Output results for proc optnetwork searching for cliques.

| OUTCLIQUE | Table Rows: 4 | Columns: 2 of 2 | Rows 1 to 4 |

	⊕ clique	△ node
1	1	A
2	1	D
3	1	E
4	1	F

Figure 4.9 Output dataset with the nodes within the single clique.

Figure 4.10 highlights the single clique based on the constraints for minimum and maximum number of nodes.

Finally, depending on the constraints specified, proc optnetwork may return no results. For instance, if we increase the minimum number of nodes in the clique to 5, proc optnetwork will not find any clique within the network.

```
proc optnetwork
   direction = undirected
   links = casuser.links
   ;
   clique
     maxcliques = all
     minsize = 5
     out = casuser.outclique
   ;
run;
```

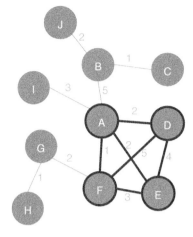

Figure 4.10 Single clique within the network based on a set of constraints.

The summary results presented in Figure 4.11 show that there is no clique found within the network based on the constraints defined.

The OPTNETWORK Procedure

Problem Summary

Number of Nodes	10
Number of Links	12
Graph Direction	Undirected

The OPTNETWORK Procedure

Solution Summary

Problem Type	Clique
Solution Status	OK
Number of Cliques	0
CPU Time	0.00
Real Time	0.00

Output CAS Tables

CAS Library	Name	Number of Rows	Number of Columns
CASUSER(Carlos.Pinheiro@sas.com)	OUTCLIQUE	0	2

Figure 4.11 No cliques found with a minimum of 5 nodes.

As a final note on the clique enumeration, large graphs can produce a huge number of cliques, and make the computational process take too long to run or even be unfeasible. A good practice in searching for cliques in large networks is to define constraints to limit the minimum and maximum sizes for the cliques, the minimum and maximum link weights to form the cliques, the minimum and maximum node weights to compose the cliques, the maximum time to run the algorithm, and finally, the maximum number of cliques returned for the clique enumeration.

4.3 Cycle

A sequence of links where the destination node of each link is the origin node of the next link is known as a path. Cycle is an elementary path in a graph where the starting node is also the ending node. No node within the path can appear more than once in the sequence. For example, a sequence of connections such as $A \rightarrow B \rightarrow C \rightarrow D \rightarrow A$.

Proc optnetwork searches for elementary cycles in a graph. A graph can comprise multiple cycles. Proc optnetwork searches and counts all cycles within that graph.

Cycles in network analysis and network optimization have several business applications. Social sciences, computer networks, and logistics are common candidates for cycle enumeration used when solving real-world problems. Like clique enumeration, cycles can also be used in transportation systems and as part of a broader set of analytical models to find anomaly transactions in banking, insurance, and mostly money laundering.

The results of the cycle enumeration algorithm are written to the output data tables that are specified in options. Proc optnetwork can provide one output for the nodes and one for the links. We can also simply define the OUT= option to produce a single data table as the result of the cycle enumeration. This output lists each node of each cycle along with the variable to identify to which cycle the node belongs and the order of that node in the cycle. Analogous to the clique enumeration algorithm, the cycle identifiers are numbered sequentially, starting from the value of the INDEXOFFSET= option in the proc optnetwork statement. Proc optnetwork can find multiple cycles within the same network at the same run. Therefore, a particular node can appear multiple times in the output dataset if it belongs to multiple cycles. The INDEXOFFSET= option is explained in the clique enumeration algorithm.

Proc optnetwork can search and count all cycles within the network. For very large graphs, proc optnetwork may not scale well in enumerating all cycles existing in the network. We can always define constraints to limit the number of cycles that will be searched and counted along the process.

In proc optnetwork, the CYCLE statement invokes the algorithm that finds the cycles in the input graph.

There are many options to define how proc optnetwork runs when counting cycles within a graph. If we suppress the output options, proc optnetwork simply counts the number of elementary cycles within the graph.

We can simply specify the option OUT= to have a single output data table as the result of the cycle algorithm, or we can specify the options OUTCYCLESNODES= to report the nodes in which cycle, and its order, similarly to the OUT= option, and the option OUTCYCLESLINKS= to report the links in which cycle, and its order.

The cycle enumeration is primarily tailored to directed graphs. It makes more sense when a sequence of nodes connected to each other creates a path where the origin node is the final node too. However, cycles can also be found in undirected graphs. In this case, each link represents two directed links, in both directions. For example, a link $A - B$ represents two links in both directions, $A \rightarrow B$ and $B \rightarrow A$. This can generate a substantial number of cycles in the final report. For this reason, trivial cycles like $A \rightarrow B \rightarrow A$ and duplicate cycles found by traversing a cycle in both directions like $A \rightarrow B \rightarrow C \rightarrow A$ and $A \rightarrow C \rightarrow B \rightarrow A$ are filtered out from the results.

Proc optnetwork uses two different algorithms to enumerate cycles. If the MAXLENGTH= option is greater than 20, the algorithm used is the backtracking (ALGORITHM=BACKTRACK). This can properly scale to large graphs that contain few cycles. However, some graphs can have a large number of cycles, so the algorithm might not scale well. If the MAXLENGTH= option is less than or equal to 20, then the algorithm used is the Build (ALGORITHM=BUILD). This algorithm is usually faster than the backtracking algorithm when the length of the cycles is sufficiently restricted. Those are the default options. Users can always change the default setting by using the option ALGORITHM=BACKTRACK|BUILD.

When searching and counting all cycles in a graph, the output table can become very large. The option MAXCYCLES=ALL requests proc optnetwork to count all cycles in a graph. A good practice before outputting the results is to simply verify the number of cycles within the network by suppressing the OUT= option. If the number of cycles is not too large, the output data table can be used subsequently to report all the cycles.

The option OUTCYCLESNODES= (or OUT=) creates the output table containing the enumerated cycles as a sequence of nodes. This table contains the following columns:

- cycle: the cycle identifier
- order: the order of the node in the cycle
- node: the node label

The option OUTCYCLESLINKS= creates the output table containing the enumerated cycles as a sequence of links. This table contains the following columns:

- cycle: the cycle identifier
- order: the order of the link in the cycle
- from: the from (origin) node label
- to: the to (destination) node label

The MAXCYCLES= option specifies the maximum number of cycles that proc optnetwork returns based on the cycle enumeration algorithm. The number specified ranges from 1 to the greatest number defined by a 32-bit integer. The option ALL returns all cycles existing in the network, limit to a maximum number that can be represented by a 32-bit integer. By default, MAXCYCLES = 1.

The option MAXLENGTH= specifies the maximum number of links in a cycle. Any cycle containing more links than the number specified in the option is removed from the results. The default is the largest number that can be represented by a 32-bit integer. When the default is used, no cycles are removed from the results.

The MINLENGTH= option specifies the minimum number of links in a cycle. Any cycle containing less links than the number specified in the option is removed from the results. By default, MINLENGTH = 1 and no cycles are removed from the results.

The option MAXLINKWEIGHT= specifies the maximum sum of link weights in a cycle. Any cycle with the sum of all its link weights greater than the number specified in the option is removed from the results. The default is the largest number that can be represented by a double. When the default is used, no cycles are removed from the results.

The option MINLINKWEIGHT= works similarly. It specifies the minimum sum of link weights in a cycle. Any cycle with the sum of all link weights less than the number specified in the option is removed from the results. The default is the largest negative number that can be represented by a double. When the default is used, no cycles are removed from the results.

The option MAXNODEWEIGHT= specifies the maximum sum of node weights in a cycle. Any cycle with the sum of all node weights greater than the number specified in the option is removed from the results. The default is the largest number that can be represented by a double. When the default is used, no cycles are removed from the results.

The MINNODEWEIGHT= option specifies the minimum sum of node weights in a cycle. Any cycle with the sum of all node weights less than the number specified in the option is removed from the results. The default is the largest negative number that can be represented by a double. When the default is used, no cycles are removed from the results.

Similar to the clique enumeration, a graph may contain a great number of cycles. The process can be exhaustive and require a long computing time. To avoid proc optnetwork from running for long periods, we can limit the amount of time the procedure will run to search all cycles within the network. The option MAXTIME= specifies the maximum amount of time to spend finding cycles. The type of time is CPU time or real-time. This is determined by the option TIMETYPE= in the proc optnetwork. The default value is the largest number that can be represented by a double.

4.3.1 Finding Cycles

Let's consider the directed graph with unweighted links presented in Figure 4.12. This graph is similar to the previous graph except by one missing link between A and E.

The following code describes how to create the new links dataset and then search for the existing cycles within the directed graph.

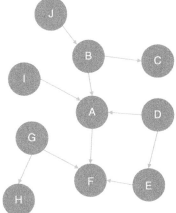

Figure 4.12 Directed graph with unweighted links.

```
data mycas.links;
   input from $ to $ @@;
datalines;
J B  B C  B A  I A  A D  F A  G F  G H  E F  D E
;
run;

proc optnetwork
   direction = directed
   links = mycas.links
   ;
   cycle
     maxcycles = all
     outcyclesnodes = mycas.outcyclenodes
     outcycleslinks = mycas.outcyclelinks
   ;
run;
```

Figure 4.13 shows the summary result created by proc optnetwork after the execution.

As a result, proc optnetwork finds one single cycle, formed by $A \rightarrow D \rightarrow E \rightarrow F \rightarrow A$.

The output dataset OUTCYCLENODE presented in Figure 4.14 contains the cycle found as a sequence of nodes. The dataset contains the identification for the cycle, the nodes within that cycle, and their order within the cycle.

The output dataset OUTCYCLELINK presented in Figure 4.15 contains the cycle found as a sequence of links. The dataset contains the identification for the cycle, the order of the link, the from (origin) node, and the to (destination) node

Analogously to the clique enumeration, the cycle enumeration can produce a high number of cycles, particularly in large networks, making the computational process exhaustive and long. A good practice in searching for cycles, especially in large

The OPTNETWORK Procedure

Problem Summary

Number of Nodes	10
Number of Links	10
Graph Direction	Directed

The OPTNETWORK Procedure

Solution Summary

Problem Type	Cycle
Solution Status	OK
Number of Cycles	1
CPU Time	0.00
Real Time	0.00

Output CAS Tables

CAS Library	Name	Number of Rows	Number of Columns
CASUSER(Carlos.Pinheiro@sas.com)	OUTCYCLENODE	5	3
CASUSER(Carlos.Pinheiro@sas.com)	OUTCYCLELINKS	4	5

Figure 4.13 Summary results for cycle enumeration using proc optnetwork.

| « OUTCYCLENODE | Table Rows: 5 | Columns: 3 of 3 | Rows 1 to 5 | ↑ ↑ ↓ ↓ | ⟳ ▾ | ⋮ |

Enter expression

	⊞ cycle	⊞ order	⚠ node
1	1	1	A
2	1	2	D
3	1	3	E
4	1	4	F
5	1	5	A

Figure 4.14 Output dataset containing the nodes within the cycle.

| « OUTCYCLELINK | Table rows: 4 | Columns: 4 of 4 | Rows 1 to 4 | ↑ ↑ ↓ ↓ | ⟳ ▾ | ⋮ |

Enter expression

	⊞ cycle	⊞ order	⚠ from	⚠ to
1	1	1	A	D
2	1	2	D	E
3	1	3	E	F
4	1	4	F	A

Figure 4.15 Output dataset containing the cycles.

graphs, is to define constraints to limit the possible number of cycles to be found and reported in the output datasets, which includes the size of the cycle, the link weights, the node weights, the time to run, and of course, the maximum number of cycles to be returned.

Figure 4.16 highlights the single cycle found in the input graph.

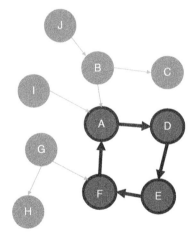

Figure 4.16 Single cycle within the input graph.

4.4 Linear Assignment

The linear assignment problem is a fundamental problem in combinatorial optimization that involves assigning workers to tasks at minimal costs. In graph theoretic terms, linear assignment is equivalent to finding the minimum link weights matching in a weighted bipartite directed graph. In a bipartite graph, the nodes can be divided into two disjointed sets W (workers) and T (tasks). Links connect nodes between both sets W and T, but not nodes within each set W or T. That means, the sets of nodes W and T are independent. There are no connections inside sets W and T, just connections between nodes from W and T. The concept of assigning workers to tasks can be generalized to the assigning of any abstract object from one group to some abstract object to another group.

The linear assignment problem can be formulated as an integer programming optimization problem. The form of the problem depends on the sizes of the two sets of nodes W and T.

Let's assume that A represent the set of possible assignments between the sets of nodes W and T. In a bipartite graph, these assignments are the links between the nodes, or the workers and the tasks.

First, let's define some rules for the optimization problem, like the decision variables, the objective function, and the constraints for the possible solutions.

The decision variable can be defined as

$x_{w,t} = 1$ if the task t is assigned to the worker w.

Otherwise, $x_{w,t} = 0$.
The objective function can be defined as:

$$minimize \sum_{a \in A} c_a x_a$$

The solution is subjected to some constraints, such as each worker is assigned to one single task, each task is assigned to one single worker, and finally, the cost of the worker doing the task is always positive.

If the number of workers is greater than or equal to the number of tasks ($|W| \geq |T|$), then the optimization problem can be solved as follows:

$$subject\ to \sum_{a \in A} c_a x_a$$

$$\sum_{a \in \delta_w^{out}} x_a \leq 1 \quad w \in W$$

$$\sum_{a \in \delta_t^{in}} x_a = 1 \quad t \in T$$

$$x_a \in \{0,1\} \quad a \in A$$

where c_a is the cost of the assignment a associated with x_a, which is the worker w doing the task t. The δ_w^{out} represents the set of outgoing links that are connected from node w (the workers) and δ_t^{in} represents the set of incoming links that are connected to node j (the tasks).

In the case of the number of workers being strictly greater than the number of tasks ($|W| > |T|$), the model allows for some workers to go unassigned.

If the number of workers is less than the number of tasks ($|W| < |T|$), then the optimization problem can be solved as follows:

$$minimize \sum_{a \in A} c_a x_a$$

$$subject\ to \sum_{a \in A} c_a x_a$$

$$\sum_{a \in \delta_w^{out}} x_a = 1 \quad w \in W$$

$$\sum_{a \in \delta_t^{in}} x_a \leq 1 \quad t \in T$$

$$x_a \in \{0,1\} \quad a \in A$$

In this case, the model allows for some tasks to go unassigned.

In proc optnetwork, the LINEARASSIGNMENT statement invokes the algorithm that solves the minimal-cost linear problem. This algorithm is based on the augmentation of shortest paths. It can be applied only to bipartite graphs. For the matching problem, we first define the input graph as a directed network by specifying the DIRECTION= option as directed. Then we use the LINK= option to define how the links dataset will determine the problem. The workers are set in the FROM= option. The tasks are set in the TO= option. Finally, the cost to perform a task by a worker is set in the WEIGHT= option. Internally, the graph is treated as a bipartite graph where the *from* nodes define one set, and the *to* nodes define the other set.

There are few options for the linear assignment algorithm in proc optnetwork.

The resulting assignment is reported in the output data table specified in the OUT= option. The resulting matching table keeps the same from, to, and weight column names. The output table is a two-level name, including the caslib name and the output table name.

The option MAXTIME= specifies the maximum amount of time that the linear assignment algorithm will spend to find the solution. The type of time can be either CPU time or rea- time, and it is determined by the value of the TIMETYPE= option. The default is the largest number that can be represented by a double.

4.4.1 Finding the Minimum Weight Matching in a Worker-Task Problem

Let's consider the bipartite graph containing the relationship between workers and tasks presented in Figure 4.17. Each relation describes the cost to produce each part of a chair by each one of the workers.

The following code describes how to create the bipartite graph containing the cost of each worker to produce each one of the pieces of a chair. We can directly create a table with all the workers repeating as lines for their costs to build each piece of a chair, or we can create a matrix containing workers by pieces containing the respective costs. Let's see the second approach, which seems more elegant.

```
data mycas.chairs;
    input employee $ leg arm seat back;
datalines;
John 14 18 23 27
Clark 12 14 19 22
Megan 15 17 28 25
Paul 21 25 32 19
Beth 16 20 22 28
Lisa 13 21 20 32
Linda 15 19 25 29
Mike 15 16 24 26
;
run;
```

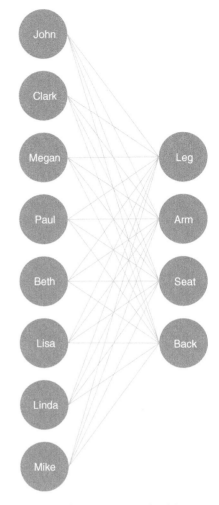

Figure 4.17 Bipartite graph with relations between workers and tasks.

This first step creates the matrix with 8 lines (workers) and 4 columns (pieces of a chair). Figure 4.18 shows the table representing that matrix.

The second step transposes the matrix into a table, repeating the workers based on their costs to produce each piece of the chair. The result is a table with 32 lines (8 workers repeated 4 times – for each piece of the chair) and 3 columns (the worker, the piece of the chair, and the respective cost). This is the table we need to pass as a parameter to proc optnetwork in order to search for the minimum weight or cost to produce the chairs.

```
data mycas.chairslinks(keep=employee part cost);
    set mycas.chairs;
    length part $ 4;
    array a[4] leg arm seat back;
    do i = 1 to dim(a);
        part = vname(a[i]);
        cost = a[i];
        output;
    end;
run;
```

This second step creates the table presented in Figure 4.19.

« CHAIRS   Table Rows: 8 | Columns: 5 of 5 | Rows 1 to 8   ⊤ ↑ ↓ ⊥   ⋮

▽ | Enter expression   ⌕

	△ employee	⊞ leg	⊞ arm	⊞ seat	⊞ back
1	John	14	18	23	27
2	Clark	12	14	19	22
3	Megan	15	17	28	25
4	Paul	21	25	32	19
5	Beth	16	20	22	28
6	Lisa	13	21	20	32
7	Linda	15	19	25	29
8	Mike	15	16	24	26

Figure 4.18   Matrix workers by pieces of a chair containing the respective costs.

« CHAIRSLINKS   Table Rows: 32 | Columns: 3 of 3 | Rows 1 to 32   ⊤ ↑ ↓ ⊥   ⟳ ▾ | ⋮

▽ | Enter expression   ⌕

	△ employee	△ part	⊞ cost
1	John	leg	14
2	John	arm	18
3	John	seat	23
4	John	back	27
5	Clark	leg	12
6	Clark	arm	14
7	Clark	seat	19
8	Clark	back	22
9	Megan	leg	15
10	Megan	arm	17
11	Megan	seat	28
12	Megan	back	25
13	Paul	leg	21
14	Paul	arm	25
15	Paul	seat	32
16	Paul	back	19
17	Beth	leg	16
18	Beth	arm	20
19	Beth	seat	22
20	Beth	back	28

Figure 4.19   Table of workers containing the costs to produce each piece of a chair.

This final table is then passed as a parameter to the proc optnetwork in order to search for the minimum cost to produce chairs. The linear assignment algorithm will search for the combination of workers and pieces of a chair to minimize the cost (the weight).

The following code invokes the linear assignment algorithm in proc optnetwork. Notice that in the linksvar= option the weight receives the cost to produce the piece of a chair by each worker.

```
proc optnetwork
   direction = directed
   links = mycas.chairslinks
   ;
   linksvar
      from = employee
      to = part
      weight = cost
   ;
   linearassignment
      out = mycas.outlap
   ;
run;
```

Figure 4.20 shows the output for the linear assignment algorithm. It says an optimal solution was found and the objective function (the minimal cost) is 67.

Figure 4.21 shows the optimal combination, which means which worker should produce which part of the chair in order to minimize the cost of production.

As a result, Clark will produce the arm at cost 14, John will produce the leg at cost 14, Lisa will produce the seat at cost 20, and Paul will produce the back at cost 19. The total cost will be 67, and this will be the minimal cost to produce a chair

The OPTNETWORK Procedure

Problem Summary

Number of Nodes	12
Number of Links	32
Graph Direction	Directed

The OPTNETWORK Procedure

Solution Summary

Problem Type	Linear Assignment
Solution Status	Optimal
Objective Value	67
CPU Time	0.00
Real Time	0.00

Output CAS Tables

CAS Library	Name	Number of Rows	Number of Columns
CASUSER(Carlos.Pinheiro@sas.com)	OUTLAP	4	3

Figure 4.20 Output results by proc optnetwork.

	⌂ employee	⌂ part	⊞ cost
1	Clark	arm	14
2	John	leg	14
3	Lisa	seat	20
4	Paul	back	19

OUTLAP Table Rows: 4 Columns: 3 of 3 Rows 1 to 4

Figure 4.21 Table of selected workers and the costs to produce each piece of a chair.

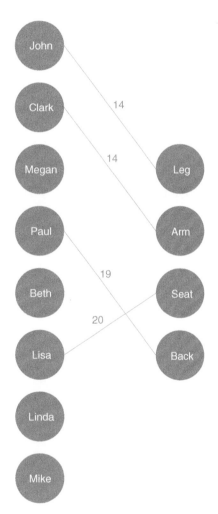

Figure 4.22 Bipartite graph with the optimal match.

considering all different costs of all workers to produce each chair's piece. Notice that Clark is the cheapest for leg, arm, and seat. He was selected for arm so others can be picked to minimize the total cost.

Figure 4.22 shows how the bipartite graph looks like when the minimal cost is achieved.

Suppose we are trying to maximize the objective function. Instead of minimizing the cost, for instance, we need to maximize the profit. Proc optnetwork doesn't maximize the objective function but if we invert the weights and minimize the objective function, it also works.

Just as an example, let's use the same data, but now producing the inverse weight for the cost. The following code describes this approach:

```
data mycas.chairslinks(keep=employee part
profit invwgt);
    set mycas.chairs;
    length part $ 4;
    array a[4] leg arm seat back;
    do i=1 to dim(a);
        part=vname(a[i]);
        profit=a[i];
        invwgt=1/a[i];
        output;
    end;
run;
```

This code produces the table presented in Figure 4.23, switching cost to profit and producing the inverse of the cost as invwgt.

The following code invokes the linear assignment algorithm in proc optnetwork. Here we are using the option vars= (_all_) in the linksvar statement in order to export all original variables to the result table. With that, we will have the inverse weight, used to minimize the objective function, but also the original weight, here considered as the maximal profit.

```
proc optnetwork
    direction = directed
    links = mycas.chairslinks
```

CHAIRSLINKS   Table Rows: 32   Columns: 4 of 4   Rows 1 to 32   ↑   ↑   ⋮

	⚠ employee	⚠ part	⊞ profit	⊞ invwgt
1	John	leg	14	0.0714285714
2	John	arm	18	0.0555555556
3	John	seat	23	0.0434782609
4	John	back	27	0.037037037
5	Clark	leg	12	0.0833333333
6	Clark	arm	14	0.0714285714
7	Clark	seat	19	0.0526315789
8	Clark	back	22	0.0454545455
9	Megan	leg	15	0.0666666667
10	Megan	arm	17	0.0588235294

Figure 4.23 Table containing the inverse of the weights.

```
    ;
    linksvar
        from = employee
        to = part
        weight = invwgt
        vars = (_all_)
    ;
    linearassignment
        out = mycas.outlap
    ;
run;
```

Proc optnetwork produces the output presented in Figure 4.24. Notice that an optimal solution was found when the objective function was minimized to 0.1645.

Figure 4.25 shows the optimal combination, which means which worker should produce which part of the chair in order to maximize the profit. This scenario doesn't make much sense, but it can illustrate the concept. Translate this example for instance to create a bundle of products and services that maximizes the profit for a telecommunications company.

Notice that the final result produces a total profit (originally cost) as 101. If we think about the cost only, when minimizing the objective function, we found a total cost of 67. Here we minimize the inverse of the cost, which works similarly to maximize the cost. The maximal cost then was 101.

4.5 Minimum-Cost Network Flow

The minimum-cost network flow problem is aimed to find the cheapest possible way of sending a certain amount of flow through a network. This problem is useful for a set of real-life situations involving network with associated costs and flows to be sent through. The minimum-cost network flow algorithm is commonly applied in telecommunications networks, energy networks, computer networks, and of course, in supply chains.

The OPTNETWORK Procedure

Problem Summary

Number of Nodes	12
Number of Links	32
Graph Direction	Directed

The OPTNETWORK Procedure

Solution Summary

Problem Type	Linear Assignment
Solution Status	Optimal
Objective Value	0.1645833333
CPU Time	0.00
Real Time	0.00

Output CAS Tables

CAS Library	Name	Number of Rows	Number of Columns
CASUSER(Carlos.Pinheiro@sas.com)	OUTLAP	4	4

Figure 4.24 Output results produced by proc optnetwork.

OUTLAP   Table Rows: 4   Columns: 4 of 4   Rows 1 to 4

	employee	part	invwgt	profit
1	Beth	arm	0.05	20
2	Lisa	back	0.03125	32
3	Megan	seat	0.0357142...	28
4	Paul	leg	0.0476190...	21

Figure 4.25 Table of selected workers and the costs to produce each piece of a chair.

The minimum-cost network flow problem is a fundamental problem in network analysis and network optimization that involves sending flow over a network at a minimal cost. Suppose that $G = (N, L)$ is a directed graph, comprising a set of nodes N and a set of links L. For each link (i, j) of L, we can associate a cost per unit of flowing something through the network (goods for example in a supply chain scenario), designated by c_{ij}. The demand or supply at each node i of N is designated as b_i, where $b_i \geq 0$ denotes a supply node, and $b_i < 0$ denotes a demand node. In some cases, when $b_i = 0$, the node is called a transshipment node. These values must be within lower and upper limits when we define the constraints of the business scenario. In this business scenario, we can also define a decision variable x_{ij}. This decision variable represents the number of units flowing through the network, the specific amount of goods sent from node i to node j. As we may have constraints assigned to the nodes, for both demand and supply, defined by the lower and upper limits, we may also have constraints defined to the links. The amount of flow that can be sent across each link can therefore be bounded within $[l_{ij}, u_{ij}]$. The lower bound l_{ij} represents the minimum flow that must be sent from node i to node j. The upper bound u_{ij} represents the maximum flow that can be sent from node i to node j and these constraints are optional. That means, we can define or not a minimum amount of goods that need to be sent throughout a link, as well as we may or may not define the maximum amount of flow that can be sent through the link.

The minimum-cost network flow problem can be modeled as a linear programming problem that minimizes the sum of the cost per unit of flow, that is, the goods that are sent through all the links (i, j) of L, and the lower and upper limits of flows that must or can be sent throughout the links.

The mathematical formulation is represented by the following equations:

Minimize

$$\sum_{i,j \in L} c_{ij} x_{ij}$$

Subject to

$$b_i^l \leq \sum_{i,j \in \delta_i^{out}} x_{ij} - \sum_{i,j \in \delta_i^{in}} x_{ij} \leq b_i^u \qquad i \in N$$

$$l_{ij} \leq x_{ij} \leq u_{ij} \qquad\qquad ij \in L$$

where δ_i^{out} represents the set of outgoing links that are connected from node i, and δ_i^{in} represents the set of incoming links that are connected to node i. The lower and upper limits for the demand or supply for each node was previously defined as b_i^l and b_i^u respectively. The cost to flow through a link was defined as c_{ij}, where the link was defined as x_{ij}. When the demand or supply for all nodes equals the lower and upper limits, $b_i = b_i^l = b_i^u$, the problem is called a standard network flow problem. For these problems, the sum of the demand and supply values must be equal to 0 to ensure a feasible solution.

In proc optnetwork, the MINCOSTFLOW statement invokes the algorithm that solves the minimum-cost network flow. The algorithm to solve the minimum-cost network flow in proc optnetwork is a variant of the primal network simplex algorithm. If the directed graph is disconnected, which means some nodes cannot be reached by other nodes throughout a sequence of links, the problem is first decomposed into two steps. First, the algorithm finds all connected components within the disconnected graph. Then, for each connected component, the minimum-cost network flow is executed. In this way, the procedure searches for the minimum-cost network flow for each disconnected part of the graph before the output of the final solution.

There are very few options for the minimum-cost network flow algorithm in proc optnetwork. The first option is to control the frequency of displaying the iteration logs. The option LOGFREQUENCY= specifies a number that controls how much information is iteratively added to the log when the procedure calculates the minimum-cost network flow for a directed graph. For example, for directed graphs that contain one single connected component, this option displays progress for every number (the number specified in the LOGFREQUENCY= option) of simplex iterations. The default number is 10 000 iterations. If the graph has multiple connected components, and the LOGLEVEL= option is MODERATE, the procedure displays progress after processing every number of connected components. If the LOGLEVEL= option is AGGRESSIVE, the procedure displays progress for every number of simplex iterations for each connected component within the directed graph.

The option MAXTIME= specifies the maximum amount of time that the minimum-cost network flow algorithm will spend to find the solution. The type of time can be either CPU time or real-time, and it is determined by the value of the TIMETYPE= option. The default is the largest number that can be represented by a double.

The final result for the minimum-cost network flow algorithm is reported and saved in output tables defined by the OUTLINKS= and OUTNODES= options. The optimal flow through the network including the reduced cost of each link is saved on the output table specified in the OUTLINKS= option. The original variables are also saved in that table, the from and to nodes, the cost, and the lower and upper limits to flow on each link. The optimal dual value for each node is saved on the output table specified in the OUTNODES= option. The original variables are also saved in that table, the node, and the lower and upper limits for the supply or demand on each node.

The definition of the links and the nodes in the minimum-cost network flow algorithm is crucial. It defines the resources and constraints for the network flow problem and how the algorithm will search for the optimal solution. The links dataset is defined using the LINKS= option, and the nodes dataset is defined using the NODES= option. The DIRECTION= option must be specified as directed. In the LINKSVAR statement, the variables *from=* and *to=* are used to define the link x_{ij}, the variable *weight=* is used to receive the cost c_{ij}, the variable *lower=* is used to receive the lower bound l_{ij} for the link, and the variable *upper=* is used to receive the upper bound u_{ij} for the link. In the

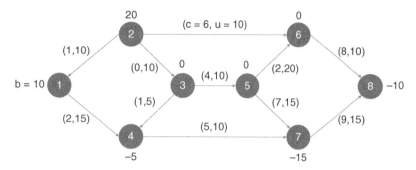

Figure 4.26 Network flow with costs and lower and upper bounds for nodes and links.

NODESVAR= statement, the variable *node=* is used to define the node, the variable *lower=* is used to receive the lower bound b_i^l for the supply or demand of the node, and the variable *upper=* is used to receive the upper bound b_i^u for the supply or demand of the node.

If the lower bound for the link is not defined, the algorithm assumes zero. If the upper bound is not defined, the algorithm assumes infinity. Similarly, for the nodes supply or demand, no value for the lower bound is assumed 0, and no value for the upper bound is assumed ∞.

To define a pure network, or a standard network flow problem, where the node supply must be met exactly, we can use the variable *weight=* only in the node's definition. Also, we don't need to specify all the node supply or demand bounds. For any missing node, the solver in proc optnetwork will use a lower and upper bound of 0.

An explicit upper bound of ∞ can be specified by using the special missing value ".I". To explicitly define a lower bound of −∞, a special missing value ".M" is used.

Some constraints are applied to the algorithm. The flow on a link must be bounded from below. That means, a lower bound $l_{ij} = -∞$ cannot be used. The flow balance constraints cannot be free. That means, the lower bound for supply or demand $b_i^l = -∞$ and the upper bound for supply or demand $b_i^u = ∞$ cannot be used.

4.5.1 Finding the Minimum-Cost Network Flow in a Demand–Supply Problem

Let us consider a very simple example to demonstrate the minimum-cost network flow problem using proc optnetwork. Consider the graph presented in Figure 4.26. It shows a network with 8 nodes, 2 of them supplying goods, 3 of them demanding goods, and the remaining 3 not having any supply or demand constraints. Also notice that this is a pure network, a standard network flow problem. There are 2 nodes supplying 30 units in total (1 = 10 and 2 = 20), and 3 nodes demanding 30 units in total (4 = −5, 7 = −15 and 8 = −10). This balance will produce a perfect match in minimizing the cost to flow goods throughout the network. In this way, we have the nodes definition with the lower bound supply or demand specified. There is no definition for the upper bound, so it is set as infinity as default.

This network also shows the links. The links have the cost specified, and the upper bound definition. There is no lower bound specification, so it is assumed 0 as default.

The following code describes how to create the input datasets for the minimum-cost network flow problem. Two datasets are created. The nodes dataset defines the lower bound for the supply or demand. Positive values set the supply and negative values set the demand. These values describe how much flow a node can send through at maximum and how much flow a node can receive at maximum. The links dataset defines the connections between the nodes, and for each connection, the cost to flow the goods on it and the upper bound, or the maximum flow that can be sent through on that particular link. Once again, there is no upper bound for the nodes, which means, they have infinity capacity, and there is no lower bound for the links, which means they have no minimum flow to be sent through.

```
data mycas.nodes;
   input node supdem @@;
datalines;
1 10  2 20  3 0  4 -5  5 0  6 0  7 -15  8 -10
;
run;
```

```
data mycas.links;
    input from to cost max @@;
datalines;
1 4 2 15   2 1 1 10   2 3 0 10   2 6 6 10   3 4 1 5   3 5 4 10
4 7 5 10   5 6 2 20   5 7 7 15   6 8 8 10   7 8 9 15
;
run;
```

Once the input datasets are defined, we can now invoke the minimum-cost network flow algorithm using proc optnetwork. The following code describes how to do it. Notice the links and nodes definition using the LINKSVAR and NODESVAR statements. For the links, the variable weight receives the cost to flow through that link and the variable upper receives the maximum that can be flowed. For the nodes, the variable lower receives the supply or demand values, which is the minimum that a node can send or receive in terms of flow.

```
proc optnetwork
    direction = directed
    links = mycas.links
    nodes = mycas.nodes
    outlinks = mycas.linksoutmcnf
    outnodes = mycas.nodesoutmcnf
    ;
    linksvar
        from = from
        to = to
        weight = cost
        upper = max
    ;
    nodesvar
        node = node
        lower = supdem
    ;
    mincostflow
    ;
run;
```

Figure 4.27 shows the output for the minimum-cost network flow algorithm. It says an optimal solution was found and the objective function (the minimal cost to flow goods through the network) is 270.

The following pictures show the minimum-cost network results. The LINKSOUTMCNF dataset presented in Figure 4.28 shows the original variables, the *from* and *to,* which defines the link, the *cost* to flow throughout the link, and the upper bound, or the maximum allowed to be flowed throughout the link. The solution comes with the *mcf_flow* variable. It says how much should be flowed on each link to minimize the overall cost to flow all the goods. The variable *mcf_rc* presents the reduced cost for the optimal solution on each link.

Notice the reduced cost on each link. The reduced cost can be understood as the gain obtained by the optimal solution in minimizing the overall cost in flowing the goods throughout the network. Sometimes in pricing optimization, the reduced cost means the cost of buying something at node i, shipping it from i to j, and then selling it at j. For example, here the only way to flow something from node **1** to node **4** is exactly that link, from **1** to **4**. Then, there is no gain by flowing something using the link **1** to **4**. The reduced cost for this link is therefore 0. The cost to flow from **2** to **1** is 1. However, nothing was flowed using this link, so the reduced cost is 1. The cost to flow from **3** to **4** is 1. There is another way to flow things to **4**, through node **1**. But the cost to flow from **1** to **4** is 2. Then, the gain in flowing from **3** to **4** is −1. Let's take another example. The reduced cost for the link **4** to **7** is −4. The cost to flow from **4** to **7** is 5. Node **4** demands 5 and that was fulfilled through **3**. Node **4** also received 10 from node **1** at a cost of 2. Then the total cost to fulfill node **7** using node **4** is 7. There is another way

The OPTNETWORK Procedure

Problem Summary

Number of Nodes	8
Number of Links	11
Graph Direction	Directed

The OPTNETWORK Procedure

Solution Summary

Problem Type	Minimum-Cost Network Flow
Solution Status	Optimal
Objective Value	270
CPU Time	0.00
Real Time	0.00

Output CAS Tables

CAS Library	Name	Number of Rows	Number of Columns
CASUSER(Carlos.Pinheiro@sas.com)	LINKSOUTMCNF	11	6
CASUSER(Carlos.Pinheiro@sas.com)	NODESOUTMCNF	8	3

Figure 4.27 Output results by proc optnetwork running the minimum-cost network flow algorithm.

« LINKSOUTMCNF Table Rows: 11 | Columns: 6 of 6 | Rows 1 to 11 | ⫪ ↑ ↓ ⫫ | ↺ ▾ | ⋮

Enter expression

	⊞ from	⊞ to	⊞ cost	⊞ max	⊞ mcf_flow	⊞ mcf_rc
1	1	4	2	15	10	0
2	2	1	1	10	0	1
3	2	3	0	10	10	0
4	2	6	6	10	10	0
5	3	4	1	5	5	-1
6	3	5	4	10	5	0
7	4	7	5	10	10	-4
8	5	6	2	20	0	0
9	5	7	7	15	5	0
10	6	8	8	10	10	0
11	7	8	9	15	0	6

Figure 4.28 The minimum-cost network flow solution for the links.

to flow things to node *7*. All the way down from node *2* at cost 0, then from *3* to *5* at cost 5, and finally from *5* to *7* at cost 7. The total cost is 11. Then flowing through node 4 represents a gain of −4.

There is a solution report for the nodes as well. The NODESOUTMCNF dataset presented in Figure 4.29 shows the original variables, the *node* identification, and the *supdem* variable, which defines the lower bound for the supply or demand for

	⊕ node	⊕ supdem	⊕ mcf_dual
1	1	10	14
2	2	20	14
3	3	0	14
4	4	-5	12
5	5	0	10
6	6	0	8
7	7	-15	3
8	8	-10	0

Figure 4.29 The minimum-cost network flow solution for the nodes.

each node to flow throughout the link. The variable *mfc_dual* presents the optimal dual value for the minimum cost network flow problem.

Notice the dual. The dual value is commonly used in price optimization. It is known sometimes as dual price. The optimal dual value is reported for each node in network flow problems. It gives the improvement in the objective function if the constraint is relaxed by one unit.

The dual values are computed during the iterations of the algorithm. The dual solution is used when the amount of flow is bounded on links within the network. The links within the network are either basic or non-basic. The non-basic links are limited at upper or lower bounds. Then, if the amount of flow between node i and j $\bar{x}_{ij}$ is basic, the cost is complementary for the dual values, $c_{ij} = \bar{y}_i - \bar{y}_j$. For example, the link between nodes **1** and **4** is basic, as the amount of flow is not limited at any upper or lower bound. The link between nodes **6** and **8** is non-basic as the amount of flow is limited by the upper bound. The dual values are given by the basic links.

Let's set the dual value for node **8** is zero, $\bar{y}_8 = 0$. Then, $\bar{x}_{68}$ implies $\bar{y}_6 - \bar{y}_8 = 0$. As $c_{ij} = \bar{y}_i - \bar{y}_j$, we have for this link the following cost equation: $c_{68} = \bar{y}_6 - \bar{y}_8$, which is $8 = \bar{y}_6 - 0$. Then the dual for the node **6** is 8. Going backwards in the network flow, the cost for the link between nodes **2** and **6** is 6: $c_{26} = \bar{y}_2 - \bar{y}_6$, or $6 = \bar{y}_2 - 8$. Then, the dual for node **2** is 14. The process continues as the optimization algorithms run. There may be multiple sets of dual values as a result of the network simplex algorithm. Proc optnetwork searches for the optimal dual values.

Figure 4.30 shows the solution for the minimum-cost network flow problem, including the amount to flow for each link throughout the network.

The optimal solution defines a flow of 10 from **1** to **4**, 10 from **2** to **3**, 10 from **2** to **6**, 5 from **3** to **4**, 5 from **3** to **5**, 10 from **4** to **7**, 5 from **5** to **7**, and finally 10 from **6** to **8**.

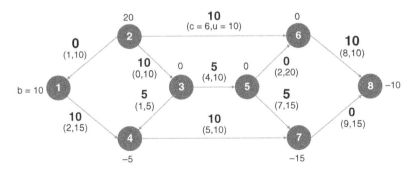

Figure 4.30 Minimum-cost network flow results.

As stated before, this example describes a pure network, where the supply and demand are balanced. Considering all supplying nodes, the total supply is 30. Considering all demanding nodes, the total demand is also 30.

Let's consider now a more flexible example. Nodes may have an infinity or huge capacity or can have a range in the lower and upper bounds. For example, nodes may supply in a particular range, from 10 to 100, or can have a flexible demand, from 10 to 20.

The following code defines a more flexible network considering the exact same topology. Here, some nodes have a range in supply or demand, including infinity supply and demand values. Analogously, the links also have a range in the lower and upper bound for the amount of flow.

```
data mycas.nodes;
    input node min max @@;
datalines;
1 10 .I  2 0 40  3 0 0  4 -10 -5
5 0 0  6 0 0  7 .M -10  8 -50 -20
;
run;

data mycas.links;
    input from to wgt min max @@;
datalines;
1 4 2 5 10  2 1 1 0 10  2 3 0 5 20  2 6 6 10 20  3 4 1 5 5  3 5 4 0 10
4 7 5 10 10  5 6 2 0 20  5 7 7 5 15  6 8 8 0 10  7 8 9 0 15
;
run;
```

The following code describes how to invoke proc optnetwork to search for the minimum-cost network flow considering this new scenario. Notice the definition of links and nodes to specify lower and upper bounds for supply or demand, and amount of flow.

```
proc optnetwork
    direction = directed
    links = mycas.links
    nodes = mycas.nodes
    outlinks = mycas.linksoutmcnf
    outnodes = mycas.nodesoutmcnf
    ;
    linksvar
        from = from
        to = to
        weight = wgt
        lower = min
        upper = max
    ;
    nodesvar
        node = node
        lower = min
        upper = max
    ;
    mincostflow
    ;
run;
```

Figure 4.31 shows the output for the minimum-cost network flow algorithm. It says an optimal solution was found and the objective function was 415.

Let's take a look at the results. The LINKSOUTMCNF dataset presented in Figure 4.32 shows the *mcf_flow* variable with the amount of flow to be sent throughout the network. It says how much should be flowed on each link to minimize the overall cost to flow all the goods. The variable *mcf_rc* presents the reduced cost for the optimal solution on each link.

The major difference in relation to the previous solution are the links *2–3*, *3–5*, *5–7*, and *7–8*. As node *8* demands more units, now those links are flowing the amount of 15, 10, 10, and 10, respectively.

Figure 4.33 shows the solution for the minimum-cost network flow problem considering this flexible network.

The figure with the solution flow suppresses the information about the links, like the cost to flow throughout the link, and the lower and upper bounds of possible amount to be flowed by using the link. It shows only the information about the range of the supply and demand for the nodes, and the optimal amount of flow as a solution.

Finally, suppose that node *8* requires a range of -100 and -50 units, as described in the following code. The other nodes are suppressed to simplify the explanation, but they are in the original data step that creates the nodes dataset.

```
data mycas.nodes;
   input node min max;
datalines;
8 -100 -50
;
run;
```

Even though some nodes can dispatch loads of units, with node 1 producing an infinity amount, the links are limited by upper bounds. The amount required by node 8 cannot be accomplished. On this case, proc optnetwork returns that there is no feasible solution for this problem. Figure 4.34 shows the summary results from proc optnetwork.

One way to overcome this limitation is to increase the possible number of units that can be flowed through the links *1–4*, *4–7*, and *7–8*. We can increase the upper bound limits or just make them infinity, as shown in the following code:

The OPTNETWORK Procedure

Solution Summary

Problem Type	Minimum-Cost Network Flow
Solution Status	Optimal
Objective Value	415
CPU Time	0.00
Real Time	0.00

Figure 4.31 Output by proc optnetwork running the minimum-cost network flow algorithm for a flexible network problem.

	from	to	wgt	min	max	mcf_flow	mcf_rc
1	1	4	2	5	10	10	-88
2	2	1	1	0	10	0	91
3	2	3	0	5	20	15	0
4	2	6	6	10	20	10	96
5	3	4	1	5	5	5	1
6	3	5	4	0	10	10	0
7	4	7	5	10	10	10	-6
8	5	6	2	0	20	0	96
9	5	7	7	5	15	10	0
10	6	8	8	0	10	10	-102
11	7	8	9	0	15	10	0

LINKSOUTMCNF — Table Rows: 11 — Columns: 7 of 7 — Rows 1 to 11

Figure 4.32 The minimum-cost network flow solution for the links.

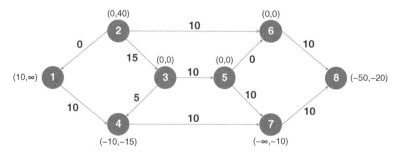

Figure 4.33 Minimum-cost network flow results for the flexible network.

The OPTNETWORK Procedure

Solution Summary

Problem Type	Minimum-Cost Network Flow
Solution Status	Infeasible
CPU Time	0.00
Real Time	0.00

Figure 4.34 Output by proc optnetwork showing that there is no feasible solution for the network flow problem considering the constraint for node 8.

The OPTNETWORK Procedure

Solution Summary

Problem Type	Minimum-Cost Network Flow
Solution Status	Optimal
Objective Value	875
CPU Time	0.00
Real Time	0.00

Figure 4.35 Output by proc optnetwork showing a feasible solution for the network flow problem considering the increase in the upper bound for some links.

```
data mycas.links;
   input from to wgt min max;
datalines;
1 4 2 5 .I
. . .
4 7 5 10 .I
. . .
7 8 9 0 .I
;
run;
```

Increasing the number of units that can be flowed throughout these links allows for a feasible solution. Proc optnetwork shows that an optimal solution can be found, and the objective function is 875. Figure 4.35 shows the summary results with the objective value.

For more complex scenarios in network flow problems, there is a SAS procedure called OPTLP that can be used. Proc OPTLP provides four methods for solving linear programs, including primal simplex algorithm, dual simplex algorithm, network simplex algorithm (similar to the mincostflow algorithm in proc optnetwork), and interior point algorithm.

4.6   Maximum Network Flow Problem

In optimization theory, maximum flows problems involve finding a feasible flow throughout a flow network that obtains the maximum possible flow rate. In graph theory, a flow network is defined as a directed graph involving a source node s and a sink node t, along several other nodes connected by multiple links. Each ink has an individual capacity, which is the maximum limit of flow that the link could allow. The feasible flow is a single source and a single sink flow network that is maximum.

Let's assume $G = (N, L)$ is a directed graph. For each link $ij \in L$, we define a nonnegative capacity u_{ij} that specifies the maximum flow that link ij can carry. We also define decision variables x_{ij} that denote the amount of flow sent across each link ij. The problem can be modeled as a linear programming problem as:

$$maxmize \sum_{ij \in \delta_i^{out}} x_{ij}$$

$$subject\ to \sum_{ij \in \delta_i^{out}} x_{ij} = \sum_{ij \in \delta_i^{in}} x_{ij} \quad i \in N \ \{s, t\}$$

$$0 \le x_{ij} \le u_{ij} \quad ij \in L$$

where δ_i^{out} represents the set of outgoing links that are connected from node i, and δ_i^{in} represents the set of incoming links that are connected to node i.

In proc optnetwork, the MAXFLOW statement invokes the algorithm that solves the maximum network flow. The input for the network flow is a standard graph input. If the input is an undirected graph, then proc optnetwork treats each link as two directed links with the same capacity. Proc optnetwork uses the Boykov-Kolmogorov algorithm to compute the maximum flow.

There are only two options for the maximum network flow algorithm in proc optnetwork. The first option is to define the source node of the network flow. The option SOURCE= specifies the node where the flow starts for the maximum network flow calculations. The option SINK= specifies the node where the flow ends for the maximum network flow calculations.

The final result for the maximal network flow algorithm is reported and saved in output tables defined by the OUTLINKS= and OUTNODES= options. The resulting optimal flow for the maximal network flow algorithm through the network is saved on the output table specified in the OUTLINKS= option. This output table includes the original information on the links dataset, which is the *from* node, the *to* node, and the upper limit to be flowed on the link, plus the optimal flow for each link, stored in the variable called *mf_flow*. If the output table for the nodes is specified in the OUTNODES= option, it stores just the list of nodes within the network.

The definition of the links in the maximal network flow algorithm is crucial. It defines the constraints for the network flow problem and how the algorithm will search for the optimal solution. The links dataset is defined using the LINKS= option. For the maximal network flow, the nodes dataset doesn't need to be defined. We can do it, but it doesn't impact the algorithm. The DIRECTION= option can be specified as directed or undirected. If the graph is directed, the links direction follows the links definition, which means, if we have a link A-B with an upper limit of 10, that would be a direct link departing from A, arriving at B, and allowing a maximal flow of 10 units. If the graph is undirected, each link is defined in both directions with the same upper limit. The previous case would turn into 2 links, from A to B with an upper limit of 10, and a second link from B to A with the same upper limit of 10 units. In the LINKSVAR statement, the variables *from=* and *to=* are used to define the link x_{ij}, and the variable *upper=* is used to define the upper limit u_{ij} for the link to flow.

Notice that differently than the minimum cost network flow problem, unlimited bounds are not allowed. The explicit upper bound of ∞ cannot be specified by using the special missing value ".I". The link must have a finite upper limit to be flowed.

4.6.1 Finding the Maximum Network Flow in a Distribution Problem

Let's consider a simple example to demonstrate the maximum network flow problem using proc optnetwork. Consider the graph presented in Figure 4.36. It shows a network with 8 nodes and 12 links. This network is represented by a directed graph, where the direction of the link matters. Each link has a maximum capacity to flow through the network. The departing node will be node A, and the arriving node will be node H. The maximum network flow algorithm searches for the optimal solution, aiming for the maximum units to be flowed in each link from node A throughout node H.

The following code describes how to create the input dataset for the maximum network flow problem. In the previous algorithm, the minimum cost network flow problem, two datasets needed to be created, one for the nodes and another one for the links. These two datasets define all the constraints and resources assigned to the problem. The maximum network flow problem is simpler. It requires only the links dataset, containing a single constraint which is the upper limit capacity to be flowed in each link.

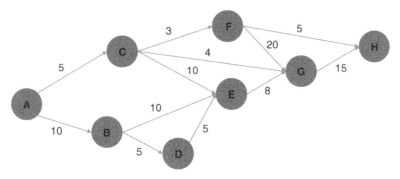

Figure 4.36 Network flow with upper limits of flow in each link.

```
data mycas.links;
    input from $ to $ max @@;
datalines;
A B 10  A C 5  B D 5  B E 10  C E 10  C F 3
C G 4  D E 5  E G 8  F G 20  F H 5  G H 15
;
run;
```

Once the input dataset is created, now we can invoke the maximum network flow algorithm using proc optnetwork. The following code describes how to do it. Notice the links definition using the LINKSVAR statements. Like all the other examples, the variables *from=* and *to=* define the link, and the variable *upper=* define the maximum capacity to be flowed in the link. The option DIRECTION= defines in this particular case that the network is a directed graph. When invoking the maximum network flow algorithm, both source and sink nodes must be defined. Here, node A is the starting point for the maximum network flow problem, and node H is the ending point. They are defined using the options SOURCE= and SINK= in the maxflow algorithm.

```
proc optnetwork
    direction = directed
    links = mycas.links
    outLinks = mycas.linksoutmnf
    ;
    linksvar
        from = from
        to = to
        upper = max
    ;
    maxflow
        source    = 'A'
        sink      = 'H'
    ;
run;
```

Figure 4.37 shows the output for the maximum network flow algorithm. It says an optimal solution was found and the objective function (the maximum units to be flowed from the starting point A throughout to the ending point H) is 13.

The following picture shows the result table for the minimum network flow algorithm. The LINKSOUTMNF dataset presented in Figure 4.38 shows the original information on the links, the *from* and *to* variables that define the link (origin and destination), the *max* variable, which defines the upper limit to be flowed on that link. The solution comes with the *mf_flow* variable. It says how much should be flowed on each link to maximize the amount of flow throughout the network departing from node A and arriving at node H.

The output table has all the links of the network flow, with the amount of flow to be flowed on each link. The output table includes even the links that are not used to send any flow. There are four links with no flowed, B-D, C-E, D-E, and F-G. All of them have a maximum capacity to flow something but they were not used to flow anything. Their *mf_flow* is zero. Five links flow less amount than their maximum capacity. The links A-B, B-E, C-G, F-H, and G-H flow 8 out of 10, 8 out of 10, 2 out of 4, 3 out of 5, and 10 out of 15, respectively. Finally, three links flow their maximum capacity. The links A-C, C-F, and E-G flow 5 out of 5, 3 out of 3, and 8 out of 8, respectively.

Figure 4.39 shows the solution for the maximum network flow problem, including the amount to flow for each link throughout the network, starting at node A and ending at node H.

Starting from the top of the network, the optimal solution defines a flow of 5 from **A** to **C**, then a split flow of 3 from **C** to **F**, and 2 from **C** to **G**. The flow of 3 from **C** to **F** continues from **F** to **H**, the destination. From the bottom part of the network, there is a flow of 8 from **A** to **B**, and then that amount of flow continues from **B** to **E**, and from **E** to **G**. Node **G** now has received 2 flows from **C** and 8 flows from **E**. Node **G** sends those 10 flows to **H**, the final destination node. With the 3 flows sent from **F**, node **H** receives the total flow of 13, the same amount initially sent by node **A**, the starting point of the network flow.

The OPTNETWORK Procedure

Problem Summary

Number of Nodes	8
Number of Links	12
Graph Direction	Directed

The OPTNETWORK Procedure

Solution Summary

Problem Type	Maximum Flow
Solution Status	Optimal
Objective Value	13
CPU Time	0.00
Real Time	0.00

Output CAS Tables

CAS Library	Name	Number of Rows	Number of Columns
CASUSER(Carlos.Pinheiro@sas.com)	LINKSOUTMNF	12	4

Figure 4.37 Output results by proc optnetwork running the maximum network flow algorithm.

« LINKSOUTMNF Table Rows: 12   Columns: 4 of 4   Rows 1 to 12

Enter expression

	from	to	max	mf_flow
1	A	B	10	8
2	A	C	5	5
3	B	D	5	0
4	B	E	10	8
5	C	E	10	0
6	C	F	3	3
7	C	G	4	2
8	D	E	5	0
9	E	G	8	8
10	F	G	20	0
11	F	H	5	3
12	G	H	15	10

Figure 4.38 The maximum network flow solution for the links.

Notice that the maximum flow depends on the entire network and the maximum capacity for all the links. The maximum amount of flow starting from node A is 15, considering 5 through the link A-C and 10 through the link A-B, the maximum flow in this scenario is 13. However, even if we increase the upper limit for these initial links, the maximum flow still relies on the rest of the flow, either in the middle of the network or at the end of the flow. For example, even

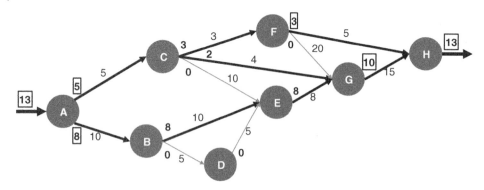

Figure 4.39 Maximum network flow results.

if we get a substantial increase in these first two links, the maximum flow still doesn't get much better as the maximum to be flowed to node H is 15. Even though node H can receive 20 units (5 from F and 15 for G), node F can receive just 3 (from C) and node G just 12 (4 from C and 8 from E). Based on that, the maximum network flow for this graph would be 15 on top.

Let's change for example the upper limit for the first two links

```
data mycas.links;
    input from $ to $ max;
datalines;
A B 100
A C 100
...
;
run;
```

Figure 4.40 shows that the objective value for the maximum network flow algorithm is now 15.

One of the great benefits from the maximum network flow algorithm is to search for the overall optimal solution. The amount to be flowed on every link is considered in order to optimize the network flow. That means, even though some links can flow more units, there is no reason to do so if the following nodes in the consecutive steps cannot accommodate that amount. Then, the optimal solution minimizes the amount of flow throughout the entire network to achieve the maximum network flow to reach out to the final destination.

For example, let's consider that new example, where the first two links have a limit capacity of 100 units. Let's assume we flow the maximum capacity at all times. Figure 4.41 presents a possible solution for this maximum network flow problem.

The process starts flowing 200 from A (100 to C and 100 to B), then from C to F there is a flow of 3, to G a flow of 4, and to E a flow of 10. Here the losses begin because of the limit capacity in the following links. Even though C receives 100, it can flow only 17 in total. The same case for B. It flows 10 to E and 5 to D. Again, B receives 100 and flows forward only 15. By now, 166 units were flowed from A to B and C for nothing as they couldn't be flowed forward (83 at C and 85 at B). From D there is a flow of 5 to E. Here there is another small waste. E receives 15 but can flow forward just 8. In addition to the 166 units that flowed before, another 7 were flowed for no reason. The last 2 steps are optimal. F has 3 units, can send 5 to H, and send those 3. G has 12 units, can send 15 H, and send those 12. In this way, the final node H receives 15. However, considering this approach there were 173 units flowed across the network for no reason. If we imagine that there is no cost to send units throughout the network, this is not a big deal. But there is no lunch free. Everything has a cost, and in real world optimization problems, sending units throughout a network flow certainly has a cost associated with it. Then, the optimal flow should account for that, minimizing the number of units sent through to achieve the maximum flow from the origin node to the destination node.

The OPTNETWORK Procedure

Solution Summary

Problem Type	Maximum Flow
Solution Status	Optimal
Objective Value	15
CPU Time	0.00
Real Time	0.00

Figure 4.40 Output results by proc optnetwork running the maximum network flow algorithm.

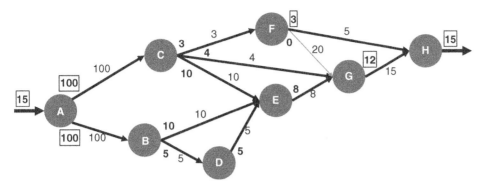

Figure 4.41 Maximum network flow results.

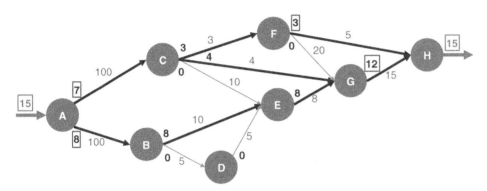

Figure 4.42 Maximum network flow results.

The optimal solution for the case we have increased the maximum capacity for the first 2 links as shown in Figure 4.42.

The optimal solution considers sending just the number of units that will be flowed forward by the consecutive nodes within the network flow. That approach would save the cost of sending units throughout the network for no reason. In order to increase the maximum network flow, it is necessary to increase the upper limit for some of the links in the middle of the network, like C-F, F-H, and E-G.

4.7 Minimum Cut

In graph theory, cut is a partition of the nodes of a graph into two disjoint subsets. The subsets are formed minimizing some particular metric. The minimum cut problem can be applied to both directed and undirected networks. If two terminal nodes are specified, they are commonly referred to as source and sink nodes. In a directed graph, the minimum cut separates the source and sink nodes by minimizing the total weight on the links that are directed from the source side of the cut to the sink side of the cut.

The minimum cut problem has many applications in areas such as network design. The minimum cut problem has two variants in terms of definitions. The first one is assigned to find the minimum set of links that if removed disconnects a particular node s from a particular node t. This variant of the problem is called the minimum s-t cut problem. The second variant is assigned to find the minimum set of links that if removed disconnects the graph. The variant is called the minimum cut problem. SAS proc optnetwork addresses both variants of the problem, for directed and undirected graphs.

A cut is a partition of the nodes N of a graph $G = (N, L)$ into two disjointed subsets S and $T = N \backslash S$. The cut S is a proper subset of N, where $S \subset N$ and $S \neq \emptyset$, N. The size of the cut with respect to S is the number of links between S and the rest of the graph $N \backslash S$. A cut set is the set of links that join a node in S to a node in T. In a regular network, we can find lots of cut sets. A minimum cut of a graph is a cut whose cut set has the smallest link metric. The link metric can be measured by distinct

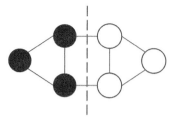

Figure 4.43 Image in pixels represented by nodes and links.

methods. For example, in unweighted graphs, the link metric can be the number of links in the cut set. On the other hand, in weighted graphs, the link metric can be the sum of the link weights in the cut set.

Another common application of the minimum cut problem nowadays is on image segmentation. Suppose we have an image created by a multitude of pixels. Imagine that we want to partition that image into two disjointed parts, or two dissimilar portions of the original image. We can turn the pixels of the image into nodes within a graph. The links will be added to connect similar pixels, or nodes within the graph. The minimum cut will represent a partition of the pixels where two portions of the image are most dissimilar. Figure 4.43 shows a representation of this graph.

The size of the cut in this image representation is 2, represented by the two links in red in the middle of the graph. These links partition the graph into two disjointed portions, the black nodes, and the white nodes.

In proc optnetwork, the MINCUT statement invokes the algorithm that solves the minimum-cut problem. The algorithm solves both variations described before. The first one, the minimum s-t cut problem, requires that the source and sink nodes be defined, and the network must be specified as a directed graph. The second variant, the minimum cut problem, runs for the entire network. The source and sink nodes cannot be defined and the network needs to be defined as an undirected graph.

There are few options for the minimum cut algorithm in proc optnetwork. They are very important in defining the problem the algorithm will search as an optimal solution. The MINCUT statement invokes an algorithm that finds the minimum link-weighted cut of an input graph. The first option is assigned to the number of cuts. The option MAXCUTS= specifies the maximum number of cuts for the algorithm to return. The minimal cut and any other cut that are found during the optimal search executed by the algorithm are returned, up to the number specified in the option. By default, the maximum number of cuts returned by the algorithm is one. The resulting minimum cuts are described in terms of nodes and links. For the nodes, the algorithm returns the partitions of the nodes within the graph. For the links, the algorithm returns the links in the cut sets. The partition of the nodes within the graph is specified in the output table by the variable *partition*. Each cut within the graph is identified in the output table by the variable *cut*. Then, for each cut, the algorithm returns an identifier for the cut and an identifier for the partition. Each node will be assigned to a value 0 or 1, defining if the node, that that particular cut, belongs to the node set S or to the node set T. The algorithm also produces an output table containing the cut sets. This table lists the links and their weight for each cut identified.

The second option refers to how the algorithm account for the weights of the links when creating the cut sets. The option MAXWEIGHT= specifies the maximum weight of the cuts to be returned by the algorithm. Only cuts whose weight is less than or equal to the number specified in the option are returned by the algorithm. The default is the largest number that can be represented by a double. That basically means that there is no limit to the total weight for the links creating the cut sets. All cut sets will be returned by the algorithm if the option is not specified.

The final result for the minimum cut algorithm in proc optnetwork is reported in the output tables specified in the options OUTCUTSETS= and OUTPARTITIONS=. The option OUTCUTSETS= specifies the output data table that contains the minimum cut sets to the minimum-cut problem. This output table contains the variable *cut* identifying the cut set, and the original variables *from*, *to*, and *weight* describing the links, which create the partitions. The option OUTPARTITIONS= specifies the output data table that contains the minimum cut partitions to the minimum-cut problem. This output table contains the variable *cut* identifying the cut, the original *node* identifier, and the *partition* identifying if the node either belongs to the node set S or to the node set T.

The options described before are mostly assigned to the second variation of the minimum cut problem, where the minimum set of links to disconnect the graph is searched. That problem is associated with undirected graphs. The first variation of the minimum cut problem is assigned to find the minimum set of links that disconnects a particular node from one partition or set S to another partition or set T. The following options are associated with that variation, called the minimum s-t cut problem.

The option SINK= specifies the sink node for minimum s-t cut calculations. If this option is specified, the option SOURCE= must be specified too. The option SOURCE= specifies the source node for minimum s-t cut calculations. Analogously, if this option is specified, the option SINK= must be specified as well. When these options are specified, the MINCUT algorithm in proc optnetwork solves the minimum s-t cut problem. The cut set in this problem intersects every path from the source node s specified in the option SOURCE= to the sink node t specified in the option SINK=. These

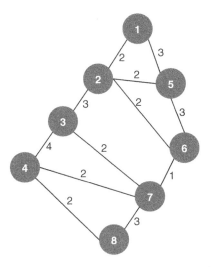

Figure 4.44 Undirected graph with weighted links.

options also imply that the graph is directed, as the cut partitions the links starting from node *s* ending at node *t*. Finally, the algorithm to solve the minimum s-t cut problem returns only one cut.

The MINCUT statement in proc optnetwork uses the Stoer-Wagner algorithm to compute the minimum cuts.

4.7.1 Finding the Minimum Cuts

Let's consider a simple example to demonstrate the minimum cut problem using proc optnetwork. Consider the graph presented in Figure 4.44. It shows a network with 8 nodes and 12 links. The network is represented by undirected links. All links within the network have weights, that will be used in searching for the minimum cuts.

The following code describes how to create the input dataset for the minimum cut problem. Notice that just the links dataset is required. The links dataset identifies the connections between all the nodes within the graph, and for each connection, it defines the respective weight, which will be accounted for when proc optnetwork calculates the minimum cuts.

```
data mycas.links;
   input from to weight @@;
datalines;
1 2 2   1 5 3   2 3 3   2 5 2   2 6 2   3 4 4
3 7 2   4 7 2   4 8 2   5 6 3   6 7 1   7 8 3
;
run;
```

Once the input dataset is defined, we can now invoke the minimum cut algorithm using proc optnetwork. The following code describes how to invoke the minimum cut statement to calculate the minimum cut sets within the graph. Notice the links definition using the LINKSVAR statement to define the links (*from* and *to* variables) and the weight for each link (*weight* variable).

```
proc optnetwork
   direction = undirected
   links = mycas.links
   ;
   linksvar
      from = from
      to = to
      weight = weight
   ;
   minCut
      outcutsets = mycas.cutsets
      outpartitions = mycas.partitions
      maxcuts = 10
   ;
run;
```

Notice that it is still possible to define the output tables using the options OUTLINKS= and OUTNODES=. However, these options here have no effect on the output results. The output table specified in the option OUTLINKS= will be exactly the same as the links dataset, containing all the links and weights identified by the variables *form*, *to*, and *weight*. The output table specified in the option OUTNODES= will have only the list of the nodes identified by the variable *node*.

The output tables with the results produced by proc optnetwork when searching for the minimum cut sets are saved through the options OUTCUSETS= and OUTPARTITIONS=.

Figure 4.45 shows the output for the minimum cut algorithm. It says an optimal solution was found and the objective function equals to 4. There is also information about the number of links and nodes in the original dataset, and the number of rows and columns in the output datasets.

The following picture shows the minimum cut results. The PARTITIONS dataset presented in Figure 4.46 shows the cuts, containing all disjoint partitions or disjoint sets of nodes. The variable *cut* indicates the cut, the variable *node* indicates which node belongs to that cut set, and the variable *partition* indicates in which partition the node belongs, either to the node set S or to the node set T.

The CUTSETS dataset presented in Figure 4.47 shows the cut sets, with the links creating the disjoint partitions within the graph. The variable *cut* identifies the cut set, and the original variables *from*, *to*, and *weight* identify the links creating the disjoint partitions.

Notice that we have seven partitions. The CUTSETS dataset identifies the links that create these partitions, and the PARTITIONS dataset identifies the nodes in each partition. Figure 4.48 shows all the seven minimum cut sets identified by proc optnetwork considering the input graph.

Also notice that, even though we have specified a maximum of 10 minimum cuts to be found (MAXCUTS = 10), proc optnetwork has found only 7 cuts, and then all these cuts were reported to the final results. Large networks are likely to have a great number of minimum cuts. A good approach is always to start small in terms of the number of cut sets to be reported in the final result and then increase progressively the maximum number of cuts in the output table.

The previous example describes the minimum cut problem based on an undirected graph. As we saw before, there are two variants for the minimum cut problem. The first one is the minimum s-t cut problem and the second one is the minimum cut problem. Let's take a look at the minimum s-t cut problem. Assume the same network we have used for the minimum cut problem. Now, for the minimum s-t cut problem, we need to define the source and sink nodes and make the graph to be directed. The graph with directed links is presented in Figure 4.49.

The OPTNETWORK Procedure

Problem Summary

Number of Nodes	8
Number of Links	12
Graph Direction	Undirected

The OPTNETWORK Procedure

Solution Summary

Problem Type	Minimum Cut
Solution Status	Optimal
Objective Value	4
CPU Time	0.00
Real Time	0.00

Output CAS Tables

CAS Library	Name	Number of Rows	Number of Columns
CASUSER(Carlos.Pinheiro@sas.com)	LINKSOUTMC	12	3
CASUSER(Carlos.Pinheiro@sas.com)	NODESOUTMC	8	1
CASUSER(Carlos.Pinheiro@sas.com)	CUTSETS	19	4
CASUSER(Carlos.Pinheiro@sas.com)	PARTITIONS	56	3

Figure 4.45 Output results by proc optnetwork running the minimum cut algorithm.

	⊞ cut	⊞ node	⊞ partition
1	1	1	1
2	1	2	1
3	1	3	0
4	1	4	0
5	1	5	1
6	1	6	1
7	1	7	0
8	1	8	0
9	2	1	1
10	2	2	1
11	2	3	1
12	2	4	1
13	2	5	1
14	2	6	1
15	2	7	1
16	2	8	0

Figure 4.46   Result table for the partitions in the minimum cut problem.

	⊞ cut	⊞ from	⊞ to	⊞ weight
1	1	2	3	3
2	1	6	7	1
3	2	4	8	2
4	2	7	8	3
5	3	1	2	2
6	3	1	5	3
7	4	3	7	2
8	4	4	7	2
9	4	4	8	2
10	4	6	7	1
11	5	3	4	4
12	5	3	7	2
13	5	6	7	1
14	6	1	2	2
15	6	2	5	2
16	6	5	6	3
17	7	2	3	3
18	7	2	6	2
19	7	5	6	3

Figure 4.47   Results table for the cut sets in the minimum cut problem.

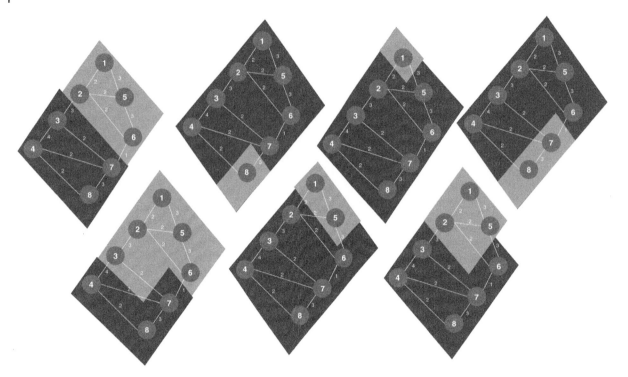

Figure 4.48 Minimum cut sets for the input graph.

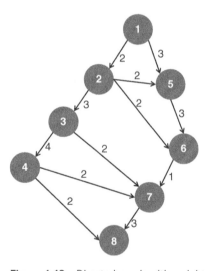

Figure 4.49 Directed graph with weighted links.

Let's also assume the node **1** as the source node and node **8** as the sink node. The following code defines these constraints and invoke the minimum cut algorithm to execute on the directed graph.

```
proc optnetwork
    direction = directed
    links = mycas.links
    ;
    linksvar
        from = from
        to = to
        weight = weight
    ;
    minCut
        outcutsets = mycas.cutsets
        outpartitions = mycas.partitions
        source = 1
        sink = 8
    ;
run;
```

The following picture shows the minimum s-t cut results. The PARTITIONS dataset presented in Figure 4.50 shows that there is one single minimum s-t cut, splitting the 8 nodes into 2 partition of 4 nodes each. The *partition* indicates as 0 or 1 in which partition each node belongs, either to the set *S* or to the set *T*.

The CUTSETS dataset presented in Figure 4.51 shows the links that create the cut sets. For this particular network, considering a directed graph and the source node as the node **1** and the sink node as the node **8**, there are just 2 links that partition the graph into 2 sets of nodes.

	⊞ cut	⊞ node	⊞ partition
1	1	1	0
2	1	2	0
3	1	3	1
4	1	4	1
5	1	5	0
6	1	6	0
7	1	7	1
8	1	8	1

≪ PARTITIONS   Table Rows: 8   Columns: 3 of 3   Rows 1 to 8   ↑ ↑ ↓ ↓   ⋮

∇ Enter expression

Figure 4.50 Result table for the partitions in the minimum s-t cut problem.

	⊞ cut	⊞ from	⊞ to	⊞ weight
1	1	2	3	3
2	1	6	7	1

≪ CUTSETS   Table Rows: 2   Columns: 4 of 4   Rows 1 to 2   ↑ ↑ ↓ ↓   ⟳ ▾   ⋮

∇ Enter expression

Figure 4.51 Results table for the cut sets in the minimum s-t cut problem.

The minimum s-t cut, and the 2 disjoint partitions are represented in Figure 4.52. The cut set are the 2 links in bold red. The cuts are the 2 groups of nodes identified in light and dark blue.

Node **1** is the source node, the *s* node of the set *S*, which comprises nodes **1**, **2**, **5**, and **6**. Node **8** is the sink node, the *t* node in the set *T*, which comprises nodes **3**, **4**, **7**, and **8**. The minimum s-t cut problem is defined by the cut set comprising the links **2-3** and **6-7**.

4.8 Minimum Spanning Tree

The minimum spanning tree problem aims to identify the minimum number of links that connect the entire network. This set of links in the minimum spanning tree keeps the same reachability of the original network, considering all original links within it. A single graph can have many different spanning trees, which may keep the same reachability of the nodes in different ways, by using a different subset of links. The minimum spanning tree problem has many applications in the design of networks. The minimum spanning tree algorithm can be applied in many varied industries like telecommunications, computers, transportation, supply chain, electrical grids, among others.

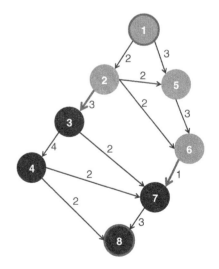

Figure 4.52 Minimum s-t cut problem.

A minimum spanning tree is a subset of the original set of weighted links within an undirected graph that connects all the nodes together with the minimum possible total link weight.

Given an undirected graph $G = (N, L)$, a spanning tree of the graph G is a tree that spans G and is a subgraph of G. That means, the spanning tree includes every node of G and all the links belonging to G. The cost of the spanning tree is the sum of the weights of all links within the tree. A regular graph can have many spanning trees. The minimum spanning tree is the spanning tree where the cost (the sum of the link weight) is minimum among all the spanning trees within the graph. Of course, it is possible that a graph has multiple minimum spanning trees with it.

When weights are assigned to the links, a minimum spanning tree is a spanning tree where the sum of link weights is less than or equal to the sum of the link weights of every other spanning tree. More generally, any undirected graph has a minimum spanning forest (considering the graph is not connected), which is a union of minimum spanning trees of its connected components.

In proc optnetwork, the MINSPANTREE statement invokes the algorithm that solves the minimum spanning tree problem. The algorithm to solve the minimum spanning tree in proc optnetwork is based on the Kruskal's algorithm. This algorithm can scale to large graphs. Kruskal's algorithm builds the spanning tree by adding links one by one into a growing spanning tree. The algorithm follows a greedy approach as in each iteration it finds the link that has the minimum weight and adds it to the growing spanning tree. A minimum spanning can't have any cycle. Then, the algorithm selects at each iteration the link with the lowest weight and checks if by adding that link to the growing spanning tree a cycle will be created. If not, it keeps searching for the next lowest link weight. If it creates a cycle, it just ignores that link and moves on searching for the next lowest link weight within the graph. In the end, all the links selected, those which avoid any cycles, will be the minimum spanning tree, and will have the cost of the sum of all link weights selected.

There is one single option for the minimum spanning tree algorithm in proc optnetwork. This option specifies the dataset with the final result of the minimum spanning tree problem. The option OUT= defines the output table containing all the resulting links within the minimum spanning tree.

The minimum spanning tree algorithm doesn't require the nodes dataset. The definition of the links within the graph is crucial for the minimum spanning tree algorithm. It defines the constraints in searching for the minimum weighted links. The links dataset is defined using the LINKS= option. As stated before, the minimum spanning tree algorithm is tailored for undirected graphs. Then, the DIRECTION= option must be specified as directed. In the LINKSVAR statement, the variables *from=* and *to=* are used to define the link x_{ij}, the variable *weight=* is used to receive the link weight. The link weights will be used in searching the optimal solution for the minimum spanning tree problem.

Notice that, like in other algorithms, it is still possible to define the output tables using the options OUTLINKS= and OUTNODES=. However, these options have no effect on the final results and only store the list of nodes and the list of links in the respective output tables.

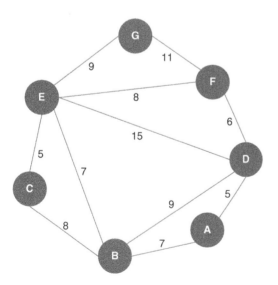

Figure 4.53 Network flow nodes and weighted links.

4.8.1 Finding the Minimum Spanning Tree

Let's consider a simple example to demonstrate the minimum spanning tree problem using proc optnetwork. Consider the graph presented in Figure 4.53. It shows a network with 7 nodes and 11 links. All nodes within the network can reach to each other, even by using multiple steps or links. For example, node **A** is directly connected to node **B**. However, node **A** is not directly connected to node **E**, but it can reach out to node **E** passing by nodes **B** and **C**. The minimum spanning tree needs to keep the same reachability, making all nodes reach each other by using a set of steps or links between them. The total link weight considering these 11 links is 90. The goal of the minimum spanning tree is to find a subset of the original set of links that keep all nodes connected with a lower total link weight.

The following code describes how to create the input dataset for the minimum spanning tree problem. The nodes dataset is not required for the minimum spanning tree algorithm. The links dataset defines the connections between the nodes, and for each connection, the weight for the link, which is used by the algorithm to calculate the minimum link weight when selecting the links to keep the reachability among the nodes.

```
data mycas.links;
    input from $ to $ weight @@;
datalines;
A B 7  A D 5  B C 8  B D 9  B E 7  C E 5
D E 15  D F 6  E F 8  E G 9  F G 11
;
run;
```

Once the links dataset is created, we can now invoke the minimum spanning tree algorithm using proc optnetwork. The following code describes how to do it. Notice the links definition using the LINKSVAR statements. The variable *from*, *to*, and *weight* receive the links definition. The option DIRECTION= defines the direction of the graph. For the minimum spanning tree problem, the network direction must be undirected.

```
proc optnetwork
    direction = undirected
    links = mycas.links
    ;
    linksvar
        from = from
        to = to
        weight = weight
    ;
    minspantree
        out = mycas.outminspantree
    ;
run;
```

Figure 4.54 shows the output for the minimum spanning tree algorithm. It says an optimal solution was found and the objective function is 39. The objective function in the minimum spanning tree problem is the total link weight for the set of links selected as minimum link weighted spanning tree.

Figure 4.55 shows the minimum spanning tree results. The OUTMINSPANTREE dataset shows the original links, which belong to the minimum link weighted spanning tree. Like in other network optimization algorithms, it is still possible to define the output tables using the options OUTLINKS= and OUTNODES=. These options however have no effect the output results. The output table specified in the option OUTLINKS= will be exactly the same as the original links dataset. The output table specified in the option OUTNODES= will have only the list of the nodes.

Notice that the minimum spanning tree has 6 links out of the original 11 links. The total link weight for these 6 links is 39. In other words, the reachability of the original network with 11 links and total link weight of 90 can be maintained by using only 6 links (54% of the original links) with a total link weight of 39 (43% of the original total link weight). In large networks the savings can be substantial when finding the minimum spanning trees.

Figure 4.56 shows the solution for the minimum spanning tree problem, highlighting in the original network the links included in the minimum spanning tree.

The optimal solution defines the links **F-D**, **D-A**, **A-B**, **B-E**, **C-E**, and **E-G** as the minimum spanning tree. These links keep the same reachability among the nodes of the original 11 links, but with a lower total link weight, 39 instead of 90. Notice that some link connections are now a little bit longer, or requiring more steps, or going throughout additional nodes. For example, node **G** could access node **F** in a straight connection, with a cost of 11. By using the minimum spanning tree set of links, node **G** needs to go through nodes **E**, **B**, **A**, and **D** to reach out to node **F**. The total cost is now 34, against the original 11. However, the minimum spanning tree requires less links in a much lower total link weight to keep the same nodes' reachability of the original network.

The OPTNETWORK Procedure

Problem Summary

Number of Nodes	7
Number of Links	11
Graph Direction	Undirected

The OPTNETWORK Procedure

Solution Summary

Problem Type	Minimum Spanning Tree
Solution Status	Optimal
Objective Value	39
CPU Time	0.00
Real Time	0.00

Output CAS Tables

CAS Library	Name	Number of Rows	Number of Columns
CASUSER(Carlos.Pinheiro@sas.com)	LINKSOUTMST	11	3
CASUSER(Carlos.Pinheiro@sas.com)	NODESOUTMST	7	1
CASUSER(Carlos.Pinheiro@sas.com)	OUTMINSPANTREE	6	3

Figure 4.54 Output results by proc optnetwork running the minimum spanning tree algorithm.

« OUTMINSPANTREE Table Rows: 6 | Columns: 3 of 3 | Rows 1 to 6

	from	to	weight
1	A	D	5
2	C	E	5
3	D	F	6
4	A	B	7
5	B	E	7
6	E	G	9

Figure 4.55 The minimum spanning tree solution.

4.9 Path

In graph theory, a path in a graph $G = (L, N)$ is an ordered sequence of links $(l_1, l_2, l_3, ..., l_{n-1})$, which connects an ordered sequence of nodes $(n_1, n_2, n_3, ..., n_n)$. In a path, all nodes n_i, where $1 \leq i \leq n$, belong to the set of nodes N. For two consecutive nodes there is a link l_{ij} that belongs to the set of links L. A simple path is a path where all nodes $n_1, ..., n_n$ are distinct. No node appears more than once in the sequence. The sequence of links is ordered in a way where the *destination* node of each link is the *origin* node of the next link. In a simple path, no node is visited more than once. A simple path has no cycles. A path between any two nodes i and j in a graph is a path that starts at i and ends at j. The starting node is called the source node, and the ending node is called the sink node. A path can be computed in both directed and undirected graphs. In undirected graphs, a duplicate link is created to represent both directions. For example, a link between nodes **A** and **B** are represented by 2 links **A-B** and **B-A**.

Paths are fundamental concepts in graph theory. They have many applications in graphs describing communications network, power distribution, transportation, and computer networks, among many others.

In proc optnetwork, the PATH statement invokes the algorithm that solves the path enumeration problem, which finds the elementary or simple paths of an input graph. Proc optnetwork finds by default all elementary paths for all pairs of nodes in the input graph. That means, the algorithm finds all paths for all possible combinations of source and sink nodes. Finding all paths possible in an input graph means not just time consuming but possibly a huge output. Even for a small network, with only a few nodes and links, the algorithm for the path enumeration can produce an output with hundreds of possible paths. An alternative method to reduce the output is to find paths in an input graph by limiting the number of source and sink nodes. The algorithm will search for all simple paths just between these sets of nodes, starting from the set of nodes defined as source nodes, and ending with the set of nodes defined as sink nodes.

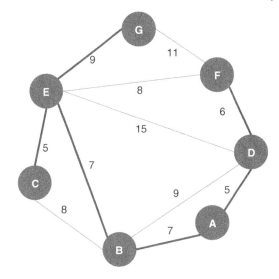

Figure 4.56 Minimum spanning tree results.

There are many options for the path enumeration algorithm in proc optnetwork. Something very important in all the input graphs we have been analyzing is the weight or the cost of each link. The link weight doesn't affect the path enumeration directly, or the way the algorithm finds all the simple paths among all the nodes, but it affects the way the algorithm evaluates the paths' total link weight. The option WEIGHT= in the LINKSVAR statement can be used to specify the weight of the links in the weighted graphs. Alternatively, there is an option AUXWEIGHT= which can also be used to specify an auxiliary weight for the links in the input graph. Eventually the link weight may be the inverse of the original weight to lead the algorithm in maximizing or minimizing an objective function, or the original link weight can be weighted in a different way to represent a particular business scenario. Finally, the direction of the link can be both directed and undirected. If we create the links dataset considering the direction of the link, the *from* node will be the source node and the *to* node will be the sink node. That link will be unique. Based on the same links dataset, we can also define an undirected graph, and then, each link will be duplicated and both directions will be evaluated by the algorithm. For example, a path A → B → C in a directed graph becomes two paths A → B → C and C → B → A in an undirected graph.

The final result for the path enumeration algorithm in proc optnetwork is reported and saved in output tables defined by the OUTPATHSLINKS= and OUTPATHSNODES= options. The output table specified in the OUTPATHSLINKS= option saves all the links comprised in all the paths, considering all possible pairs of source-sink nodes. The option OUTPATHS= has the same effect and can be used in place of the option OUTPATHSLINKS=. If there is no source-sink pair defined by the options SOURCE= and SINK=, or no source-sink pairs are defined using the option NODESSUBSET=, all nodes are used to create the combinations of source and sink nodes. This may create a huge output data table as a result, particularly for large networks. The output data table specified using the OUTPATHSNODES= option saves all the nodes comprised in all of the paths, considering again all possible pairs of source-sink nodes. Analogously, if no source-sink pair is defined, all nodes within the network are selected to create the source-sink pairs of nodes. As the output tables produced by the path enumeration algorithm can be extremely large, depending on the size of the network and/or the number of source-sink pairs of nodes defined, the time to generate these output tables can be longer than the time to compute the paths.

The output table specified in the OUTPATHSLINKS= option contains all the links in each path, including the columns to describe them. The variable *source* contains the label of the source node of the path. The variable *sink* contains the label of the sink node of the path. The variable *path* identifies a particular source-sink pair of nodes, which is the path identification. The variable *order* identifies the order of the link within the path for that particular source-sink pair of nodes, which ultimately is the order of the link in the path identified by the variable *path*. The variable used in the from= option in the

LINKSVAR statement identifies the label of the from node of the link within the path. The variable used in the to= option in the LINKSVAR statement identifies the label of the to node of the link within the path. The variable used in the weight= option in the LINKSVAR statement identifies the weight of the link within the path. Finally, the variable used in the auxweight= option in the LINKSVAR statement identifies the auxiliary weight of the link within the path. For example, if we define in the links dataset the variables *org*, *dst*, *wgt*, and *auxwgt* to define the from node, the to node, the link weight, and the auxiliary link weight, respectively, the output table will contain all these variables in addition to the output variables *source*, *sink*, *path*, and *order*.

The output table specified in the OUTPATHSNODES= option contains all the nodes within each path. This output table contains the following columns. The variable *source* identifies the label of the source node of the path. The variable *sink* identifies the label of the sink node of the path. The variable *path* identifies the source-sink pair, which ultimately defines the path identifier. The variable *order* identifies the order of the source-sink pair within the path. The variable used in the node= option in the NODESVAR statement identifies the label of the node within the path. The variable used in the weight= option in the NODESVAR statement identifies the weight of the node within the path. For example, if we define in the nodes dataset the variables *n* and *wgt* to represent the label of the node, and the weight of this node within the network, respectively, the output table will contain all these variables in addition to the original output variables *source*, *sink*, *path*, and *order*.

Therefore, we can define both input network elements to proc optnetwork, which means the links and nodes datasets. The links dataset is specified by using the option links=, and the nodes dataset is specified by using the option nodes=. The links and nodes datasets specify the input graph considered by proc optnetwork when computing the paths. The definitions of both links and nodes datasets are particularly important when the weight for both links and nodes are required to describe a particular business scenario. For example, important relations between entities within the network and important entities within the network. In banking, we can assume relations as transactions between accounts. Then, the amount transferred between these accounts is quite relevant and it will be represented by the link weight. Similarly, the account balance is important to identify relevant entities, and then it will be represented by the node weight.

By default, proc optnetwork computes all possible paths within the input graph, considering all possible combinations of source and sink nodes. The list of nodes is combined by pairs and then, all possible paths between these pairs of source and sink nodes are searched by the path enumeration algorithm. As we saw before, the final result can be exhaustive. The output tables can be extremely large. In order to prevent proc optnetwork from generating large final outcomes, we can alternatively use the options SOURCE= and SINK= to specify the source nodes, the sink nodes, or the pairs of source and sink nodes where the paths will be searched. These options work independently. For example, to fix a particular source node and find all paths departing from the fixed source node to all possible sink nodes within the input graph, we use the SOURCE= option. Analogously, to fix a particular sink node and find all paths from all possible source nodes arriving at the fixed sink node, we use the SINK= option. By using both SOURCE= and SINK= options, we fix a particular source node and a particular sink node. By doing that, proc optnetwork will search for all paths between the fixed source node to the fixed sink node, the paths departing from the fixed source node and arriving at the fixed sink node.

An alternative way is by using a subset of nodes. The options SOURCE= and SINK= define a single source node and/or a single sink node. The option NODESSUBSET= specifies a dataset containing a list of nodes and a definition if the node will be used as a source node and/or as a sink node. The NODESSUBSETVAR statement specifies a variable *node* as the node identification, a variable *source* that indicates if the node will be used as a source node, and a variable *sink* that indicates if the node will be used as a sink node. The concept of a subset of nodes within the input graph applies to the path and shortest path algorithms.

It is important to notice that the settings for the options SOURCE= and SINK override the use of the variable source and sink specified in the NODESSUBSET= option when proc optnetwork searches for the paths within an input graph.

In addition to the SOURCE= and SINK= option, there are other options that reduce the size of the final output table produced by proc optnetwork when running the path enumeration algorithm.

The option MAXLENGTH= specifies the maximum number of links contained in a path. Any path that has the number of links greater than the number specified in the option is removed from the results. The default is the largest number that can be represented by a 32-bit integer, which basically means all possible lengths. When the default is used, no paths are removed from the results. Depending on the problem, it is always a good idea to limit the length of the paths returned as a result of the path enumeration algorithm. These shorter paths may have more business meaning depending on the problem or the business scenario.

In reverse, the option MINLENGTH= specifies the minimum number of links contained in a path. Any path that has fewer links than the number specified in the option is removed from the results. By default, the value for the option MINLENGTH= equals to 1. By using the default value, no paths are removed from the results.

The option MAXLINKWEIGHT= specifies the maximum sum of link weights contained in a path. Any path that has the sum of link weights greater than the number specified in the option is removed from the results. Again, the default value is the largest number that can be represented by a double, which basically means, any possible sum of link weights. When the default is used, no paths are removed from the results. Analogously to the maximum length of the path, the maximum sum of link weights can reduce the size of the output table to more meaningful paths according to the problem or the business scenario under analysis.

The other option, MINLINKWEIGHT= specifies the minimum sum of link weights contained in a path. Any path that has the sum of link weights less than the number specified in the option is removed from the results. The default is the largest (in magnitude) negative number that can be represented by a double. That basically means any link weights in the negative value. When the default is used, no paths are removed from the results.

The option MAXNODEWEIGHT= specifies the maximum sum of node weights contained in a path. Any path that has the sum of node weights greater than the number specified in the option is removed from the results. The default is the largest number that can be represented by a double. When the default is used, no paths are removed from the results. Reducing the number of paths returned by the path enumeration based on a particular maximum node weight cannot just reduce the path range to analyze but can highlight more meaningful paths according to a particular problem or business scenario.

In reverse to the maximum node weights, the option MINNODEWEIGHT= specifies the minimum sum of node weights contained in a path. Any path that has the sum of node weights less than the number specified in the option is removed from the results. The default is the largest (in magnitude) negative number that can be represented by a double. The default value basically means any negative number, which causes no paths to be removed from the results.

Another option to reduce the number of paths to be reported in the output table is to limit the time of running. The option MAXTIME= specifies the maximum amount of time that proc optnetwork will spend searching for the possible paths. The type of time (either CPU time or real-time) is determined by the value defined in the TIMETYPE= option. The default is the largest number that can be represented by a double. Limiting the running time is also a good approach to reduce the number of paths to be analyzed according to the problem or business scenario.

4.9.1 Finding Paths

Let's consider a simple graph to demonstrate the path enumeration problem using proc optnetwork. Consider the graph presented in Figure 4.57. It shows a network with just 6 nodes and only 11 links. It is important to emphasize here that even with a very small network, the number of paths found in the path enumeration problem can be large. As the input graph grows larger, the process can be exhaustive, and the output table can be extremely large or even unfeasible.

The following code describes how to create the input dataset for the path enumeration problem. At first, only the links dataset will be defined. Later on, we will see the impact of defining both links and nodes datasets to represent the input graph. The links dataset has the nodes identification, the from and to variables, and the weight of the link. This dataset can be used to represent both directed and undirected graphs as we will see in the following examples.

```
data mycas.links;
    input from $ to $ weight @@;
datalines;
A B 1  A E 1  B C 1  C D 1  C A 6  D E 3
D F 1  E A 1  E B 1  E C 4  F E 1
;
run;
```

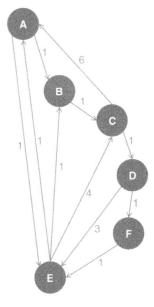

Figure 4.57 Directed input graph with weighted links.

Once the input dataset is defined, we can now invoke the path enumeration algorithm using proc optnetwork. The following code describes how to do it. Notice the links definition using the LINKSVAR statements. For the links, the variable weight receives the weight of the link, and the variables from and to receive the source and the sink nodes. As there is no SOURCE= or SINK= option being used, all possible pairs of nodes combinations will be used as sources and sinks nodes within this input graph. The input graph in this example is defined as a directed network.

```
proc optnetwork
    direction = directed
    links = mycas.links
    ;
    linksvar
        from = from
        to = to
        weight = weight
    ;
    path
        outpathslinks = mycas.outpathlinks
        outpathsnodes = mycas.outpathnodes
    ;
run;
```

Figure 4.58 shows the output for the path enumeration algorithm. It says an optimal solution was found (solution status is ok) and the number of paths found between all possible pairs of nodes is 86.

The following pictures show the path enumeration results. The OUTPATHLINKS dataset presented in Figure 4.59 shows the original variables, the *from* and *to* which defines the source and the sink nodes in the link, and the variable *weight*, which defines the weight of this link. In addition to that, the output table shows the variables *source* and *sink*, which defines from

The OPTNETWORK Procedure

Problem Summary

Number of Nodes	6
Number of Links	11
Graph Direction	Directed

The OPTNETWORK Procedure

Solution Summary

Problem Type	Path
Solution Status	OK
Number of Paths	86
CPU Time	0.00
Real Time	0.00

Output CAS Tables

CAS Library	Name	Number of Rows	Number of Columns
CASUSER(Carlos.Pinheiro@sas.com)	OUTPATHLINKS	255	7
CASUSER(Carlos.Pinheiro@sas.com)	OUTPATHNODES	341	5

Figure 4.58 Output results by proc optnetwork running the path enumeration algorithm on a directed input graph.

	source	sink	path	order	from	to	weight
1	A	B	1	1	A	B	1
2	A	E	1	1	A	E	1
3	A	C	1	1	A	B	1
4	A	C	1	2	B	C	1
5	A	B	2	1	A	E	1
6	A	B	2	2	E	B	1
7	A	C	2	1	A	E	1
8	A	C	2	2	E	C	4
9	A	D	1	1	A	B	1
10	A	D	1	2	B	C	1
11	A	D	1	3	C	D	1
12	A	C	3	1	A	E	1
13	A	C	3	2	E	B	1
14	A	C	3	3	B	C	1
15	A	D	2	1	A	E	1
16	A	D	2	2	E	C	4
17	A	D	2	3	C	D	1

OUTPATHLINKS   Table Rows: 255   Columns: 7 of 7   Rows 1 to 200

Figure 4.59 The path enumeration solution for the links.

where the path departs and to where it arrives. Finally, the output table also shows the variables *path*, which is the path identifier, and the variable *order*, which identifies the order of link within that particular path.

For example, line 1 describes a path comprising one single link. The source node is **A**, the sink node **B**, and the path has a unique link from **A** to **B**. Lines 3 and 4 describe a bit longer path. The source node is **A** again, the sink node is **C**, but this path has 2 links, from **A** to **B** and then from **B** to **C**. As we can see on the top of the picture, this output table has 255 rows, describing all 86 paths found by the path enumeration algorithm. This picture shows only the first 17 rows.

Proc optnetwork also produces an output for the nodes. The OUTPATHNODES dataset presented in Figure 4.60 shows the variables *source* to identify the source node of the path, the variable *sink* to identify the sink node of the path, the variable *path* to identify the path, the variable *order* to identify the order of the link within the path and the variable *node* to identify the node within that particular link/path.

Notice that the variable *order* describes the sequence of the links where the node is. For example, lines 1 and 2 describes the path from the source node **A** to the sink node **B**. Line 1 describes node **A** in order 1, and line 2 describes node **B** in order 2. In other words, this path goes from node **A** to node **B**. Lines 5, 6, and 7 describes a longer path. The source node is **A**, and the sink node is **C**. Line 5 describes node **A** in order 1, line 6 describes node **B** in order 2, and finally line 7 describes node **C** in order 3. This path goes from node **A** to **B** and then from **B** to **C**. This path has 2 links.

Let's take a look at the use of the options SOURCE= and SINK= to reduce the number of paths searched by the path enumeration algorithm in proc optnetwork. The next example describes the option SOURCE= to fix a particular node when searching the paths within the input graph. All paths returned by the path enumeration algorithm need to start at the node specified in the option SOURCE=. All other nodes can be the sink node. The following code show how to fix a source node when searching for the paths:

```
proc optnetwork
   direction = directed
   links = mycas.links
   ;
   path
```

	⬦ source	⬦ sink	⊞ path	⊞ order	⬦ node
1	A	B	1	1	A
2	A	B	1	2	B
3	A	E	1	1	A
4	A	E	1	2	E
5	A	C	1	1	A
6	A	C	1	2	B
7	A	C	1	3	C
8	A	B	2	1	A
9	A	B	2	2	E
10	A	B	2	3	B
11	A	C	2	1	A
12	A	C	2	2	E
13	A	C	2	3	C
14	A	D	1	1	A
15	A	D	1	2	B
16	A	D	1	3	C
17	A	D	1	4	D

OUTPATHNODES   Table Rows: 341   Columns: 5 of 5   Rows 1 to 200

Figure 4.60 The path enumeration solution for the nodes.

```
            source = 'A'
            outpathslinks = mycas.outpathlinks
            outpathsnodes = mycas.outpathnodes
    ;
run;
```

Figure 4.61 shows the output produced by proc optnetwork. Notice that only 14 paths were found when the source node is fixed as node **A**.

Figure 4.62 shows the result table for the path enumeration algorithm. Notice that the source node for all rows is node **A**. This output table shows then all paths starting at the source node **A** and ending at any possible sink node within the input graph.

The output table OUTPATHLINKS contains all 43 links comprised by the 14 paths starting at the source node **A**.

Analogously, we can define only a sink node when executing the path enumeration algorithm in proc optnetwork. The following code shows how to define a sink node to search for all paths ending at a fixed sink node.

The OPTNETWORK Procedure

Solution Summary	
Problem Type	Path
Solution Status	OK
Number of Paths	14
CPU Time	0.00
Real Time	0.00

Figure 4.61 The path enumeration solution with a fixed source node.

```
proc optnetwork
    direction = directed
    links = mycas.links
    ;
    path
        sink = 'D'
        outpathslinks = mycas.outpathlinks
        outpathsnodes = mycas.outpathnodes
    ;
run;
```

	⚠ source	⚠ sink	⊕ path	⊕ order	⚠ from	⚠ to	⊕ weight
1	A	B	1	1	A	B	1
2	A	E	1	1	A	E	1
3	A	C	1	1	A	B	1
4	A	C	1	2	B	C	1
5	A	B	2	1	A	E	1
6	A	B	2	2	E	B	1
7	A	C	2	1	A	E	1
8	A	C	2	2	E	C	4
9	A	D	1	1	A	B	1
10	A	D	1	2	B	C	1
11	A	D	1	3	C	D	1

OUTPATHLINKS   Table Rows: 43   Columns: 7 of 7   Rows 1 to 43

Figure 4.62   The path enumeration solution for the links using a fixed source node.

Figure 4.63 shows the output produced by proc optnetwork. Notice that only 11 paths were found when the sink node is fixed as node **D**.

Figure 4.64 shows the result table for the path enumeration algorithm. Notice that the sink node for all rows is node **D**. This output table then shows all paths starting at any possible node within the input graph and ending at the sink node **D**.

The output table OUTPATHLINKS contains all 34 links comprised by the 11 paths ending at the sink node **D**.

Finally let's take a look at an example where both source and sink nodes are fixed. Proc optnetwork will search for all possible paths within the input graph where the

The OPTNETWORK Procedure

Solution Summary

Problem Type	Path
Solution Status	OK
Number of Paths	11
CPU Time	0.00
Real Time	0.00

Figure 4.63   The path enumeration solution with a fixed sink node.

	⚠ source	⚠ sink	⊕ path	⊕ order	⚠ from	⚠ to	⊕ weight
1	A	D	1	1	A	B	1
2	A	D	1	2	B	C	1
3	A	D	1	3	C	D	1
4	A	D	2	1	A	E	1
5	A	D	2	2	E	C	4
6	A	D	2	3	C	D	1
7	A	D	3	1	A	E	1
8	A	D	3	2	E	B	1
9	A	D	3	3	B	C	1
10	A	D	3	4	C	D	1
11	B	D	1	1	B	C	1
12	B	D	1	2	C	D	1

OUTPATHLINKS   Table Rows: 34   Columns: 7 of 7   Rows 1 to 34

Figure 4.64   The path enumeration solution for the links using a fixed sink node.

source node is the node fixed by the option SOURCE= and the sink node is the node fixed by the option SINK=. The following code shows how to fix both source and sink nodes.

```
proc optnetwork
    direction = directed
    links = mycas.links
    ;
    path
        source = 'A'
        sink = 'D'
        outpathslinks = mycas.outpathlinks
        outpathsnodes = mycas.outpathnodes
    ;
run;
```

Figure 4.65 shows the output produced by proc optnetwork when running the path enumeration algorithm for fixed source and sink nodes. Notice that only 3 paths were found when the source node is fixed as node **A** and the sink node is fixed as node **D**.

Figure 4.66 shows the result table for the path enumeration algorithm. Notice that the source node for all rows is node **A** and the sink node for all rows is node **D**. This output table then shows all paths within the input graph starting at node **A** and ending at node **D**.

The output table OUTPATHLINKS contains all 10 links comprised by the 3 paths starting at node **A** and ending at the sink node **D**.

Figure 4.67 shows the original input graph highlighting the source node **A** and the sink node **D**.

Figure 4.68 shows the solution for the path enumeration algorithm considering the input graph with the source node fixed as node **A** and the sink node fixed as node **D**.

Notice that there are only 3 paths starting from node **A** and ending to node **D**, A → B → C → D, A → E → C → D and A → E → B → C → D.

Lastly, let's take a look at the path enumeration algorithm for an undirected graph. Also, let's use different column names for the original nodes and links datasets in order to show how proc optnetwork produces the final output tables.

The OPTNETWORK Procedure

Solution Summary	
Problem Type	Path
Solution Status	OK
Number of Paths	3
CPU Time	0.00
Real Time	0.00

Figure 4.65 The path enumeration solution for fixed source and sink nodes.

« OUTPATHLINKS   Table Rows: 10 | Columns: 7 of 7 | Rows 1 to 10

	source	sink	path	order	from	to	weight
1	A	D	1	1	A	B	1
2	A	D	1	2	B	C	1
3	A	D	1	3	C	D	1
4	A	D	2	1	A	E	1
5	A	D	2	2	E	C	4
6	A	D	2	3	C	D	1
7	A	D	3	1	A	E	1
8	A	D	3	2	E	B	1
9	A	D	3	3	B	C	1
10	A	D	3	4	C	D	1

Figure 4.66 The path enumeration solution for the links using fixed source and sink nodes.

The following code describes how to create the input datasets for the path enumeration problem. In this example we are using both links and nodes datasets. For the links dataset we have the variable *a* to represent the from node, the variable *b* to represent the to node, the variable *wgt* to represent the weight of the link, and the variable *aux* to represent the auxiliary weight for the link. As an undirected graph, the from node doesn't necessarily represent the source node in the path enumeration problem. The same concept applies to the to node. It doesn't necessarily represent the sink node. For an undirected graph, proc optnetwork duplicates the links under the hood to compute all possible paths considering both directions. For example, the following datasets create a link starting at node A and ending at node B. As an undirected graph, we can also have a link staring at node B and ending at node A. Proc optnetwork creates this second link internally in order to compute the path enumeration for an undirected input graph.

```
data mycas.links;
    input a $ b $ wgt aux @@;
datalines;
A B 1 0.5   A E 1 0.5   B C 1 0.5   C D 1 0.5   C A 6 0.17   D E 3 0.33
D F 1 0.5   E A 1 0.5   E B 1 0.5   E C 4 0.25   F E 1 0.5
;
run;
```

Here we also create a nodes dataset to show how proc optnetwork produces the final output tables. This nodes dataset has the variable *n* to define the node and the variable *wgt* to define the weight of the node.

```
data mycas.nodes;
    input n $ wgt @@;
datalines;
A 10   B 15   C 5   D 20   E 25   F 30
;
run;
```

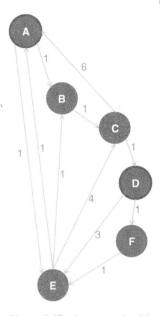

Figure 4.67 Input graph with fixed source and sink nodes.

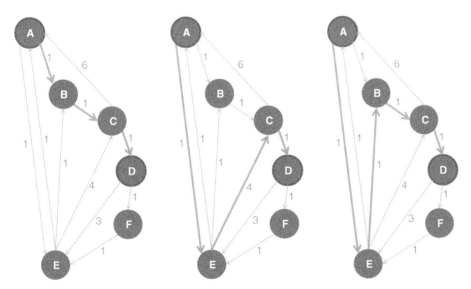

Figure 4.68 The path enumeration solution for the nodes.

The following code describes how we specify the nodes and links variables when solving the path enumeration problem with proc optnetwork. The DIRECTION= option defines the input graph as undirected. The LINKSVAR and NODESVAR statements define the links and nodes properties, respectively by using the variables *from*, *to*, *weight* and *auxweight* for the links, and the variables *node* and *weight* for the nodes.

```
proc optnetwork
    direction = undirected
    links = mycas.links
    nodes = mycas.nodes
    ;
    linksvar
        from = a
        to = b
        weight = wgt
        auxweight = aux
    ;
    nodesvar
        node = n
        weight = wgt
    ;
    path
        outpathslinks = mycas.outpathlinks
        outpathsnodes = mycas.outpathnodes
    ;
run;
```

Figure 4.69 shows the output for the path enumeration algorithm when running on an undirected input graph. An optimal solution was achieved, and the number of paths found between all possible pairs of nodes is 422. Recall that for the same number of links on a directed input graph, the path enumeration found 86 paths. The major difference is the number of internal links on the input graph. For the directed graph, the 11 links remain as 11 links respecting the from and to nodes (source and sink nodes). For the undirected graph, the 11 links are internally duplicated creating 22 links for both directions between the from and to nodes. That duplication creates more possible combinations of source and sink pairs of nodes for the input graph, and then, more paths found by the algorithm, 422 against 86. Notice that the output states 6 nodes and 11 links. These are the original 11 links in the links dataset. But as an undirected graph, proc optnetwork treats each link as having actually 2 directions.

The following pictures show the path enumeration results. The OUTPATHLINKS dataset presented in Figure 4.70 shows the original variables *a* and *b*, which defines the origin and the destination nodes for each link, the variable *wgt*, which defines the weight of the link, and the variable *aux*, which defines the auxiliary weight for the link. Additionally, the output table has the variables *source* and *sink*, which defines the origin and the destination nodes for the path. Finally, the output table also shows the variables *path*, which is the path identifier, and the variable *order*, which identifies the order of links within that particular path.

The concept is similar to the one presented before for the directed graph. A small difference should be noticed here. For example, line 2 shows a path A-C, with a unique link C-A. At line 544, we can see a path C-A with the same link C-A. The path C-A is natural, associated with the unique link C-A. But what explains the path A-C with a link C-A? This happens because of the nature of the undirected graph. Internally, proc optnetwork "duplicates" the link C-A, creating an imaginary link A-C. That makes a path A-C possible with the original link C-A (the "duplicated link A-C"). The same thing happens on lines 3 and 4. We can see the same path A-E with unique links E-A and A-E, respectively. This happens because in the links dataset we actually have both links A-E and E-A. Then, we will observe 4 paths, 2 paths A-E with unique links E-A and A-E (lines 3 and 4), and 2 paths E-A with unique links E-A and A-E again (lines 1013 and 1014).

The OUTPATHNODES dataset presented In Figure 4.71 will be similar to the one presented for the directed graph, but with many more rows in order to describe all the paths found. This output table has 1872 rows against 341 rows for the directed graph. The output table presents the variables *source* and *sink* to identify the source and sink nodes of the path,

The OPTNETWORK Procedure

Problem Summary

Number of Nodes	6
Number of Links	11
Graph Direction	Undirected

The OPTNETWORK Procedure

Solution Summary

Problem Type	Path
Solution Status	OK
Number of Paths	422
CPU Time	0.01
Real Time	0.00

Output CAS Tables

CAS Library	Name	Number of Rows	Number of Columns
CASUSER(Carlos.Pinheiro@sas.com)	OUTPATHLINKS	1450	7
CASUSER(Carlos.Pinheiro@sas.com)	OUTPATHNODES	1872	5

Figure 4.69 Output results by proc optnetwork running the path enumeration algorithm on an undirected input graph.

OUTPATHLINKS ▾   Table Rows: 1450   Columns: 8 of 8   Rows 1 to 200

	source	sink	path	order	a	b	wgt	aux
1	A	B	1	1	A	B	1	0.5
2	A	C	1	1	C	A	6	0.17
3	A	E	1	1	E	A	1	0.5
4	A	E	2	1	A	E	1	0.5
5	A	C	2	1	A	B	1	0.5
6	A	C	2	2	B	C	1	0.5
538	B	F	23	5	D	F	1	0.5
539	B	A	13	1	E	B	1	0.5
540	B	A	13	2	F	E	1	0.5
541	B	A	13	3	D	F	1	0.5
542	B	A	13	4	C	D	1	0.5
543	B	A	13	5	C	A	6	0.17
544	C	A	1	1	C	A	6	0.17
545	C	B	1	1	B	C	1	0.5
546	C	D	1	1	C	D	1	0.5
547	C	E	1	1	E	C	4	0.25
1012	D	A	19	5	A	B	1	0.5
1013	E	A	1	1	E	A	1	0.5
1014	E	A	2	1	A	E	1	0.5
1015	E	B	1	1	E	B	1	0.5

Figure 4.70 The path enumeration solution for the links.

		source		sink		path		order		n		wgt
1		A		B		1		1		A		10
2		A		B		1		2		B		15
3		A		C		1		1		A		10
4		A		C		1		2		C		5
5		A		E		1		1		A		10
6		A		E		1		2		E		25
7		A		E		2		1		A		10
8		A		E		2		2		E		25
333		B		A		1		1		B		15
334		B		A		1		2		A		10

(OUTPATHNODES ▾ Table Rows: 1872 Columns: 6 of 6 Rows 1 to 200)

Figure 4.71 The path enumeration solution for the nodes.

the variable *path* to identify the path, the variable *order* to identify the order of the link within the path, and the original variables created in the nodes dataset, *n* to identify the node and *wgt* to identify the weight of the node.

Analogously to the links output, the output table for the nodes presents the duplications appearance for the nodes. Node **A** appears in the path A-B shown in line 1 as well as in path B-A shown in line 334. The same for node **B**, it appears in path A-B showed in line 2 and in path B-A shown in line 333. This happens even though there is only the link A-B defined in the links dataset. Proc optnetwork "duplicates" that link to create an imaginary link B-A to represent the undirected graph.

4.10 Shortest Path

As we saw in the previous section, a path in an input graph is an ordered sequence of links which connects an ordered sequence of nodes. All nodes in a path are distinct. No node appears more than once in the sequence.

The shortest path problem is aimed to find the path between two nodes in an input graph where the sum of the weights of its links is minimized. There is a set of real-life problems that can be addressed by the shortest path, such as road networks, public transportation planning, traffic planning, sequence of choices in decision making, minimum delay in telecommunications, operations research, layout facility, and robotics, among many others.

Formally, the shortest path between two nodes *i* and *j* in an input graph is the path that starts at node *i* and ends at node *j* with the lowest total link weight. Like in the path enumeration problem, the starting node is referred to as the source node and the ending node is referred to as the sink node.

By default, proc optnetwork computes the shortest paths for all pairs of nodes within the network. Similarly to the path algorithm, that means to find the shortest path between all possible combinations of source nodes and sink nodes within the input graph. This outcome can result in an extremely large output table. To reduce the size of the outcome, alternatively we can select the pair of source and sink nodes to search the shortest path or select a subset of nodes to be fixed as source nodes and sink nodes.

In proc optnetwork, the SHORTESTPATH statement invokes the algorithm that solves the shortest path problem. Proc optnetwork uses multiple algorithms to find the shortest path between a pair of nodes. The algorithm used by proc optnetwork when searching for the shortest paths relies on the type of data that describes the input graph. If the graph is unweighted, which means the variable *weight* is not used to differentiate the importance or the cost of the links within the network, the algorithm that proc optnetwork uses is based on the breadth-first search. If the graph is weighted, and there is no negative weight to represent the importance or the cost of the links within the network, the algorithm that proc optnetwork uses is based on Dijkstra's algorithm. Finally, if the graph is weighted and there are positive and negative weights to represent the importance or the cost of the links within the network, the algorithm that proc optnetwork uses is based on the Bellman = Ford algorithm.

There are many options for the shortest path algorithm in proc optnetwork. As we saw before, the type of the input graph is relevant to define which algorithm proc optnetwork will use in searching for the shortest paths. The relevancy is if the graph is weighted or unweighted, and if there are positive and negative weights associated with the links within the network. For weighted graphs, proc optnetwork uses the variable *weight* defined in the links dataset to evaluate all possible paths based on the total link weight, or the total cost of the path. The option WEIGHT= in the LINKSVAR statement is then used to specify the weight of the links in weighted graphs. Alternatively, there is an option AUXWEIGHT=, which can also be used to specify an auxiliary weight for the links within the input graph. The auxiliary weight is not used in the algorithm to calculate the total cost of the path and compare it to the other possible paths when searching the shortest paths. The auxiliary weight is used to calculate a different total cost for each path, and it is reported in the output tables. The auxiliary link weight may be the inverse of the original weight to lead the algorithm in maximizing or minimizing an objective function, or the original link weight can be weighted in a different way to represent a particular business scenario. For example, if we are looking at the total cost of a path, the shortest path between a pair of nodes is the smallest total link weight between the source and sink nodes. In transportation, the shortest path would be the smallest total distance, or the smallest total time, or the smallest total cost (tolls in highways). In social network, we are primarily looking for strong relationships. The shortest path between a pair of nodes in a social scenario would be the path with the highest weight between the source node and the sink node, either based on total frequency or total time. One straightforward way to compute the shortest path in this particular case is to invert the original link weight between the nodes and assign it to the variable *weight* in the LINKSVAR statement. By doing this, the higher the original weight, the lower the inverse weight, which will contribute most to the shortest path. The original weight will be kept and assigned to the variable *auxweight* in the LINKSVAR statement. Both original and inverse weights will be included in the final output table.

Similar to the path enumeration problem, the direction for the links in the input graph for the shortest path algorithm can be both directed and undirected. In a particular links dataset, considering a directed graph, the *from* node will be the source node and the *to* node will be the sink node when the algorithm searches for the shortest path. Each link in the links dataset is unique. Based on the exact same links dataset, but defining the network as an undirected graph, each link is internally "duplicated" to account for both directions. A defined link (A,B) in the links dataset will represent the link A → B on a directed graph and two links A → B and B → A in an undirected graph.

The final result for the shortest path algorithm in proc optnetwork is reported to the output tables defined by the OUTPATHSLINKS=, OUTPATHSNODES=, OUTWEIGHTS=, and OUTSUMMARY= options. The output table specified in the OUTPATHSLINKS= option saves the links comprised in the shortest paths, considering all possible pairs of source-sink nodes. If there is no source-sink pair defined by the options SOURCE= and SINK=, or no source-sink pairs are defined using the option NODESUBSET=, all nodes are used to create the combinations of source and sink nodes. This may create a huge output data table as a result, particularly for large networks. The output data table specified using the OUTPATHSNODES= option saves the nodes comprised in the shortest paths, considering again all possible pairs of source-sink nodes. Analogously, if no source-sink pair is defined, all nodes within the network are selected to create the source-sink pairs of nodes. Similar to the path enumeration algorithm, as the output tables produced by the shortest path algorithm can be extremely large, depending on the size of the network and/or the number of source-sink pairs of nodes defined. Therefore, the time to generate the output tables can be longer than the time to search for the shortest paths.

The output table specified in the OUTPATHSLINKS= option (OUT= and OUTPATHS= options are used alternatively) contains all the links in each shortest path, including the columns to describe them. The variable *source* contains the label of the source node of the shortest path. The variable *sink* contains the label of the sink node of the shortest path. The variable *order* identifies the order of the link within the shortest path assigned to that particular source-sink pair of nodes. The variable used in the from= option in the LINKSVAR statement identifies the label of the from node of the link within the shortest path. The variable used in the to= option in the LINKSVAR statement identifies the label of the to node of the link within the shortest path. The variable used in the weight= option in the LINKSVAR statement identifies the weight of the link within the shortest path. Finally, the variable used in the auxweight= option in the LINKSVAR statement identifies the auxiliary weight of the link within the shortest path. For example, if we define in the links dataset the variables *org*, *dst*, *wgt*, and *auxwgt* to define the from node, the to node, the link weight, and the auxiliary link weight, respectively, the output table will contain all these original variables in addition to the output variables *source*, *sink*, and *order*. If we define the maximum number of ranked paths to find for each source-sink pair by using the option MAXPATHSPERPAIR=, a variable *rank* will be added to the output table to specify the rank of the link within the shortest path.

The output table specified in the OUTPATHSNODES= option contains all the nodes assigned to each shortest path. This output table contains the following columns. The variable *source* identifies the label of the source node of the shortest path. The variable *sink* identifies the label of the sink node of the shortest path. The variable *order* identifies the order of the source-sink pair within the shortest path. The variable used in the node= option in the NODESVAR statement identifies the label of the node within the shortest path. For example, if we define in the nodes dataset the variables *n* to represent the label of the node, the output table will contain that original variable in addition to the output variables *source*, *sink*, and *order*. If we define the maximum number of ranked paths to find for each source-sink pair by using the option MAXPATHSPERPAIR=, a variable *rank* will be added to the output table to specify the rank of the source-sink pair within the shortest path.

The output table specified in the OUTSUMMARY= option contains the summary statistics for the shortest paths for each source node. This output table contains the following variables. The variable *source* identifies the source node label within the shortest path. The variable *paths* identifies the number of shortest paths from the source node to the requested sink nodes within the shortest paths. The variable *path_weight_min* identifies the minimum weight of shortest paths from the source node to all the sink nodes. The variable *path_weight_max* identifies the maximum weight of shortest paths from the source node to all the sink nodes. The variable *path_weight_avg* identifies the average weight of shortest paths from the source node to all the sink nodes. The variable *path_weight_std* identifies the standard deviation of weight of shortest paths from the source node to all the sink nodes. The variable *path_weight_var* identifies the variance of weight of shortest paths from the source node to all the sink nodes. If an auxiliary weight is defined by using the option AUXWEIGHT= in the LINKSVAR statement, additional variables are added to the output table. The variables are similar to the ones added to represent the descriptive statistics for the original weights of the links. The variable *path_auxweight_min* identifies the minimum auxiliary weight of shortest paths from the source node to all sink nodes. The variable *path_auxweight_max* identifies the maximum auxiliary weight of shortest paths from the source node to all sink nodes. The variable *path_auxweight_avg* identifies the average auxiliary weight of shortest paths from the source node to all sink nodes. The variable *path_auxweight_std* identifies the standard deviation of auxiliary weight of shortest paths from the source node to all sink nodes. Finally, the variable *path_auxweight_var* identifies the variance of auxiliary weight of shortest paths from the source node to all sink nodes.

The output table specified in the OUTWEIGHTS= option contains the weights and the auxiliary weights for each shortest path found. This output table contains the following variables. The variable *source* identifies the source node label of the shortest path. The variable *sink* identifies the sink node label of the shortest path. The variable *path_weight* identifies the weight of the shortest path for the link (the pair of source and sink nodes). If the auxiliary weight is specified by using the option AUXWEIGHT= in the LINKSVAR statement, the variable *path_auxweight* is added to the output table to identify the auxiliary weight of the shortest path for the link. If the maximum number of ranked paths to find for each source-sink pair is defined by using the option MAXPATHSPERPAIR=, a variable *rank* is added to the output table to identify the rank of the source-sink pair within the shortest path.

Similar to the path enumeration algorithm, proc optnetwork computes by default all shortest paths within the input graph, considering all possible combinations of source and sink nodes. All the nodes within the input graph are combined by pairs and then, all possible shortest paths between those pairs of source and sink nodes are searched by the shortest path algorithm. As we saw before, the final output table can be very large. In order to prevent proc optnetwork from generating very large outcomes, we can alternatively use the options SOURCE= and SINK= to specify independently, all shortest paths from a particular source node, all shortest paths to a particular sink node, or all shortest paths between the pairs of a subset of source and sink nodes. These options work independently. For example, we can fix a particular source node to find all shortest paths departing from it to all possible sink nodes within the input graph. The option SOURCE= is used to fix the source node. We can fix a particular sink node to find all shortest paths to it from all possible source nodes within the input graph. We use the option SINK= to fix the sink node. We can also use both options SOURCE= and SINK= to fix a particular source-sink pair to find the shortest paths just between them.

Alternatively, we can define a subset of nodes to be used as source-sink pairs within the input graph. The option NODESSUBSET= specifies a dataset containing a list of nodes and a definition if the node will be used as a source node and/or as a sink node. The NODESSUBSETVAR statement specifies a variable *node* as the node identification, a variable *source* that indicates if the node will be used as a source node, and a variable *sink* that indicates if the node will be used as a sink node. By defining a subset nodes dataset and specifying the source and sink nodes by using the NODESSUBSETVAR statement, proc optnetwork will search for all shortest paths from the source nodes to the sink nodes defined in the subset nodes dataset, no matter the links defined in the input graph.

The options SOURCE= and SINK= override the use of the variable source and sink specified in the NODESSUBSET= option when proc optnetwork search for the shortest paths within an input graph. That means, if we define a subset of nodes by using the NODESSUBSET= option and the NODESSUBSETVAR statement, and we use the options SOURCE= and SINK=, proc optnetwork will ignore all nodes defined in the subset (NODESSUBSET=) and will search for the shortest path between the source-sink pair of nodes (SOURCE=/SINK=).

The option SEQUENCE= specifies which data variable in the subset of nodes dataset (specified by NODESSUBSET= option) defines the sequence of nodes to be visited when proc optnetwork searches for the shortest paths between the pairs of source-sink nodes. The variable identifying the sequence must be numeric. The values in the nodes subset define the order from the first node to be visited (lowest number) to the last node to be visited in the path (highest number). The resulting path contains a node subsequence that matches the specified node sequence. If we use the option SEQUENCE= to define the sequence of nodes to be visited when searching for the shortest paths, we cannot use the options SOURCE= and SINK= to define the pairs of source-sink nodes within the shortest paths. That means, if we define a variable *sequence* in the nodes subset dataset, we cannot define the variables *source* or *sink*. In other words, if we use the option SEQUENCE in the SHORTESTPATH statement, we cannot use the option SOURCE= and SINK=. If we want to specify the source and sink nodes for the shortest path, we need to include them as the first and the last nodes in the sequence.

Other options can also be used to reduce the size of the final output table produced by proc optnetwork when running the shortest path algorithm.

The options MAXABSOBJGAP= or MAXABSOLUTEOBJECTIVEGAP= specify an acceptance criterion for the sum of link weights in paths relative to the sum of link weights in the shortest path. The sum of the link weight in the paths are used to identify candidate shortest paths. The absolute objective gap is equal to the sum of the weights in the candidate path minus the sum of the weights in the shortest path. Any shortest path whose absolute objective gap is greater than number specified in the option is removed from the results. The value of number can be any nonnegative number. The default is the largest number that can be represented by a double. When the default is used, no shortest paths are removed from the results.

The options MAXRELOBJGAP= or MAXRELATIVEOBJECTIVEGAP= specify an acceptance criterion for the sum of link weights in paths relative to the sum of link weights in the shortest path. The relative objective gap is equal to the sum of the weight in the candidate path minus the sum of the weights in the shortest path divided by the sum of the sum of the weights in the shortest path. Any shortest path whose relative objective gap is greater than number specified in the option is removed from the results. The value of number can be any nonnegative number. The default is the largest number that can be represented by a double. When the default is used, no shortest paths are removed from the results.

The options MAXLINKWEIGHT= or MAXPATHWEIGHT= specify the maximum sum of link weights in a shortest path. Any shortest path whose sum of link weights is greater than the number specified in the option is removed from the results. The default is the largest number that can be represented by a double. When the default is used, no shortest paths are removed from the final results.

The options MINLINKWEIGHT= or MINPATHWEIGHT= specify the minimum sum of link weights in a shortest path. Any shortest path whose sum of link weights is less than number specified in the option is removed from the results. The default is the largest negative number that can be represented by a double. When the default is used, no shortest paths are removed from the results.

As presented before, the option MAXPATHSPERPAIR= specifies the maximum number of ranked shortest paths to find for each source-sink pair of nodes. The algorithm finds the maximum the number specified in the option of best shortest paths between each source-sink pair of nodes. The paths that are reported to the output table ordered by the variable *rank*, in ascending order of the sum of link weights for each shortest path. By default, the option MAXPATHSPERPAIR= equals to 1 and the algorithm finds one single shortest path between each source-sink pair of nodes.

4.10.1 Finding Shortest Paths

Let's consider the same input graph used in the path enumeration section to demonstrate the shortest path algorithm in proc optnetwork. The input graph has just 6 nodes and only 11 weighted links. Figure 4.72 shows the input graph with directed weighted links.

The code to create the links dataset ass presented in the previous section about the path enumeration problem. We invoke the shortest path algorithm in proc optnetwork by using the SHORTESTPATH statement. The links definition using the LINKSVAR statement assigns the from and to nodes for each link as well as the weight for the links. First, let's see how to search for the shortest paths considering all possible combinations of source-sink pairs of nodes within the input graph. The following code shows how to do it.

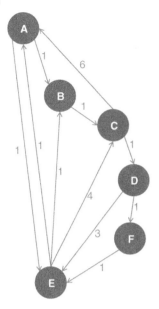

Figure 4.72 Directed input graph with weighted links.

```
proc optnetwork
    direction = directed
    links = mycas.links
    ;
    linksvar
        from = from
        to = to
        weight = weight
    ;
    shortestpath
        outsummary = mycas.shortpathsummary
        outweights = mycas.shortpathweights
        outpathslinks = mycas.shortpathlinks
        outpathsnodes = mycas.shortpathnodes
    ;
run;
```

The following pictures show the different outputs for the shortest path algorithm in proc optnetwork. Proc optnetwork results presented in Figure 4.73 show the summary information, containing the input graph information (6 nodes and 11 links), the direction of the graph (directed), the type of problem (shortest path), the solution status (OK, meaning that shortest paths were found) and the number of shortest paths found (30). Lastly, the four output tables we have specified in the proc optnetwork code, containing the results for the shortest path links, the shortest path nodes, the shortest paths weights, and the summary statistics.

The SHORTPATHLINKS dataset presented in Figure 4.74 shows the original variables, the *from* and *to* variables that define the source and sink nodes for the shortest paths, and the variable *weight* that defines the weight of the link used

The OPTNETWORK Procedure

Problem Summary

Number of Nodes	6
Number of Links	11
Graph Direction	Directed

The OPTNETWORK Procedure

Solution Summary

Problem Type	Shortest Path
Solution Status	OK
Number of Paths	30
CPU Time	0.00
Real Time	0.00

Output CAS Tables

CAS Library	Name	Number of Rows	Number of Columns
CASUSER(Carlos.Pinheiro@sas.com)	SHORTPATHLINKS	76	6
CASUSER(Carlos.Pinheiro@sas.com)	SHORTPATHNODES	106	4
CASUSER(Carlos.Pinheiro@sas.com)	SHORTPATHWEIGHTS	30	3
CASUSER(Carlos.Pinheiro@sas.com)	SHORTPATHSUMMARY	6	7

Figure 4.73 Output results by proc optnetwork running the shortest path algorithm on a directed input graph.

« SHORTPATHLINKS   Table Rows: 76 | Columns: 6 of 6 | Rows 1 to 76   ↑ ↑ ↓ ↓   ⋮

▽ | Enter expression   ⌕

	△ source	△ sink	⊞ order	△ from	△ to	⊞ weight
1	A	B	1	A	B	1
2	A	C	1	A	B	1
3	A	C	2	B	C	1
4	A	D	1	A	B	1
5	A	D	2	B	C	1
6	A	D	3	C	D	1
7	A	E	1	A	E	1
8	A	F	1	A	B	1
9	A	F	2	B	C	1
10	A	F	3	C	D	1
11	A	F	4	D	F	1
12	B	A	1	B	C	1
13	B	A	2	C	D	1
14	B	A	3	D	F	1
15	B	A	4	F	E	1
16	B	A	5	E	A	1

Figure 4.74 The shortest path solution for the links.

when the procedure searches for the minimum sum of weights for the links within the shortest path. In addition to that, the output table also shows the variables *source* and *sink*, that defines from where the shortest path departs and where it arrives. Finally, the output table shows the variable *order*, that identifies the order of link within the shortest path.

Line 1 describes a shortest path comprising one single link, from the source node **A** to the sink node **B**. Lines 2 and 3 describe a shortest path comprising two links. The source node is **A**, and the sink node is **C**. The two links comprised in this shortest path are from **A** to **B** and then from **B** to **C**. Similarly, lines 4 to 6 show a shortest path from source node **A** to sink node **D**, comprising three links, from **A** to **B**, from **B** to **C**, and finally from **C** to **D**. Notice that proc optnetwork found 30 shortest paths (in the results summary), comprising 76 links (number of rows in the output table SHORTPATHLINKS). This picture shows the first 16 rows.

The SHORTPATHNODES dataset presented in Figure 4.75 shows the variables *source* to identify the source node of the shortest path, the variable *sink* to identify the sink node of the shortest path, the variable *order* to identify the order of the link within the shortest path, and finally the variable *node* to identify the node partaking the particular shortest path.

Notice that the variable *order* identifies the sequence of the links where the node is. Shortest path **A-B**, with a single link, has nodes **A** and **B** (lines 1 and 2). Shortest path **A-C** has 2 links and 3 nodes, **A**, **B**, and **C** (lines 3, 4, and 5). Similarly, shortest path **A-D** has 3 links and 4 nodes, **A**, **B**, **C**, and **D** (lines 6–9).

The SHORTPATHWEIGHTS dataset presented in Figure 4.76 shows all shortest paths and their sum of link weights, represented by the variables *source* and *sink*, identifying the source-sink pair of node of the shortest path, and the variable *path_weight*, identifying the sum of the link weights of the shortest path.

For example, line 1 shows the shortest path **A-B** with the sum of link weights equal to 1 (the weight of the single link A-B). Lines 2 shows the shortest path **A-C** and the sum of link weights equal to 2 (1 for A-B and 1 for B-C). Line 3 shows the shortest path **A-D** and the sum of link weights equal to 3 (1 for each link within the shortest path, A-B, B-C, and C-D).

Figure 4.77 shows the input graph and highlights the links A-B, B-C, and C-D that partake shortest paths represented by lines 1, 2, and 3 on the previous output table.

The last output table, the SHORTPATHSUMMARY dataset, presented in Figure 4.78, shows the summary statistics for the shortest paths for each source node. It shows the variables *source*, identifying the source node of the shortest paths, *paths*, identifying the number of shortest paths from that particular source node, *path_weight_min*, identifying the minimum

	source	sink	order	node
1	A	B	1	A
2	A	B	2	B
3	A	C	1	A
4	A	C	2	B
5	A	C	3	C
6	A	D	1	A
7	A	D	2	B
8	A	D	3	C
9	A	D	4	D
10	A	E	1	A
11	A	E	2	E
12	A	F	1	A
13	A	F	2	B
14	A	F	3	C
15	A	F	4	D
16	A	F	5	F
17	B	A	1	B
18	B	A	2	C
19	B	A	3	D
20	B	A	4	F
21	B	A	5	E
22	B	A	6	A

SHORTPATHNODES   Table Rows: 106   Columns: 4 of 4   Rows 1 to 106

Figure 4.75 The shortest path solution for the nodes.

	source	sink	path_weight
1	A	B	1
2	A	C	2
3	A	D	3
4	A	E	1
5	A	F	4
6	B	A	5
7	B	C	1
8	B	D	2
9	B	E	4
10	B	F	3

SHORTPATHWEIGHTS   Table Rows: 30   Columns: 3 of 3   Rows 1 to 30

Figure 4.76 The sum of link weights of the shortest paths.

weight of shortest paths from that source node, *path_weight_max*, identifying the maximum weight of shortest paths from that source node, *path_weight_avg*, identifying the average weight of shortest paths from that source node, *path_weight_std*, identifying the standard deviation of weight of shortest paths from that source node, and *path_weight_var*, identifying the variance of weight of shortest paths from that source node.

Let's see how to narrow down the search for shortest paths by defining the source and sink pairs of nodes. First, let's use the option SOURCE= to fix a particular node to tell proc optnetwork to search all shortest paths departing from that particular source node. All shortest paths returned by the shortest path algorithm will start at that node specified in the option SOURCE=. The sink nodes can be any other node in the input graph. The following code shows how to do it. For this example, the output tables to provide the summary statistics and the sum of weight links are suppressed.

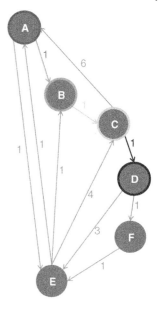

Figure 4.77 The sum of link weights of the shortest paths A-B, A-C, and C-D.

```
proc optnetwork
   direction = directed
   links = mycas.links
   ;
   linksvar
      from = from
      to = to
      weight = weight
   ;
   shortestpath
      source = 'A'
      outpathslinks = mycas.outshortpathlinks
      outpathsnodes = mycas.outshortpathnodes
   ;
run;
```

Figure 4.79 shows the output produced by proc optnetwork. Notice that only 5 shortest paths were found when the source node is fixed as node **A**. Considering the entire input graph with 6 nodes being source nodes, 30 shortest paths were found.

Figure 4.80 shows the result table for the shortest path algorithm. Notice that the source node for all rows is node **A**. This output table shows then all the shortest paths starting at the source node **A** and ending at any possible sink node within the input graph.

	source	paths	path_weight_min	path_weight_max	path_weight_avg	path_weight_std	path_weight_var
1	A	5	1	4	2.2	1.303840481	1.7
2	B	5	1	5	3	1.5811388301	2.5
3	C	5	1	4	2.8	1.303840481	1.7
4	D	5	1	4	2.6	1.1401754251	1.3
5	E	5	1	4	2.2	1.303840481	1.7
6	F	5	1	4	2.4	1.1401754251	1.3

SHORTPATHSUMMARY   Table Rows: 6   Columns: 7 of 7   Rows 1 to 6

Figure 4.78 The summary statistics for the shortest path solution.

The OPTNETWORK Procedure

Solution Summary

Problem Type	Shortest Path
Solution Status	OK
Number of Paths	5
CPU Time	0.00
Real Time	0.00

Figure 4.79 The shortest path solution with a fixed source node.

```
      from = from
      to = to
      weight = weight
   ;
  shortestpath
     sink = 'D'
     outpathslinks = mycas.outshortpathlinks
     outpathsnodes = mycas.outshortpathnodes
   ;
run;
```

The output table OUTSHORTPATHLINKS contains all 11 links comprised by the 5 shortest paths starting at the source node **A**.

Let's see now how to fix a sink node by using the option SINK=. By doing this, proc optnetwork will search for all shortest paths within the input graph that end at the sink node specified in the option SINK=. The following code shows how to define a sink node in proc optnetwork:

```
proc optnetwork
   direction = directed
   links = mycas.links
   ;
   linksvar
```

Figure 4.81 shows the output produced by proc optnetwork. Only 5 shortest paths were found when the sink node is fixed as node **D**.

Figure 4.82 shows the result table for the shortest path algorithm. As expected, the sink node for all rows is node **D**. This output table shows all shortest paths starting at any possible source node within the input graph and ending at the sink node **D**.

The output table OUTSHORTPATHLINKS shows all 13 links comprised by the 5 shortest paths ending at the sink node **D**.

Lastly, let's see how to fix both source and sink nodes when searching shortest paths and how this approach reduces dramatically the final outcomes. Proc optnetwork will search for all possible paths within the input graph where the source

⟨⟨ OUTSHORTPATHLINKS   Table Rows: 11   Columns: 6 of 6   Rows 1 to 11

	source	sink	order	from	to	weight
1	A	B	1	A	B	1
2	A	C	1	A	B	1
3	A	C	2	B	C	1
4	A	D	1	A	B	1
5	A	D	2	B	C	1
6	A	D	3	C	D	1
7	A	E	1	A	E	1
8	A	F	1	A	B	1
9	A	F	2	B	C	1
10	A	F	3	C	D	1
11	A	F	4	D	F	1

Figure 4.80 The shortest path solution for the links using a fixed source node.

node is the node fixed by the option SOURCE= and the sink node is the node fixed by the option SINK=. The following code shows how to do it:

```
proc optnetwork
    direction = directed
    links = mycas.links
    ;
    linksvar
        from = from
        to = to
        weight = weight
    ;
    shortestpath
        source = 'A'
        sink = 'D'
        outpathslinks = mycas.outshortpathlinks
        outpathsnodes = mycas.outshortpathnodes
    ;
run;
```

The OPTNETWORK Procedure

Solution Summary

Problem Type	Shortest Path
Solution Status	OK
Number of Paths	5
CPU Time	0.00
Real Time	0.00

Figure 4.81 The shortest path solution with a fixed sink node.

Figure 4.83 shows the output produced by proc optnetwork when running the shortest path algorithm for fixed source and sink nodes. As expected, by fixing both source and sink nodes, proc optnetwork returns a single shortest path, the one with the lowest sum of weight links within the input graph.

Figure 4.84 shows the result table for the shortest path algorithm. The table shows all links partaking the shortest path between the source node is **A** and the sink node is **D**.

The output table OUTSHORTPATHLINKS contains the three links comprised by the single shortest path (lowest sum of weight links) starting at source node **A** and ending at the sink node **D**.

Figure 4.85 shows the original input graph highlighting the source node **A**, and the sink node **D**, and the links partaking the shortest path between **A** and **D**.

« OUTSHORTPATHLINKS Table Rows: 13   Columns: 6 of 6   Rows 1 to 13

	source	sink	order	from	to	weight
1	A	D	1	A	B	1
2	A	D	2	B	C	1
3	A	D	3	C	D	1
4	B	D	1	B	C	1
5	B	D	2	C	D	1
6	C	D	1	C	D	1
7	E	D	1	E	B	1
8	E	D	2	B	C	1
9	E	D	3	C	D	1
10	F	D	1	F	E	1
11	F	D	2	E	B	1
12	F	D	3	B	C	1
13	F	D	4	C	D	1

Figure 4.82 The shortest path solution for the links using a fixed sink node.

The OPTNETWORK Procedure

Solution Summary

Problem Type	Shortest Path
Solution Status	OK
Number of Paths	1
CPU Time	0.01
Real Time	0.00

Figure 4.83 The shortest path solution for fixed source and sink nodes.

As we observed in the previous section, the path enumeration solution returned 3 possible paths between the source node **A** and the sink node **D**. The path highlighted as the shortest path between A and D is the one with the lowest sum of weight links. The path $A \rightarrow E \rightarrow B \rightarrow C \rightarrow D$ has a total link weight equal to 4. The path $A \rightarrow E \rightarrow C \rightarrow D$ has a total link weight equal to 6. The path $A \rightarrow B \rightarrow C \rightarrow D$ has a total link weight equal to 3.

Shortest path can also be computed for undirected graphs. In that scenario, the number of shortest paths returned by proc optnetwork may be greater than the number of shortest paths returned for a directed graph just because more links will be available to connect all nodes within the input graph. This is basically due to the bidirectional aspect of the links in the input graph. A directed link $A \rightarrow B$ turns into two possible links $A \rightarrow B$ and $B \rightarrow A$, allowing possibly more paths to be traveled.

In this example, we are also going to use different column names for the original links datasets in order to show how proc optnetwork produces the final output tables. This is also justifies the use of the LINKSVAR statement to assign the links attributes.

The input graph is the same used for the path enumeration problem. The following code creates the links dataset considering the variable *a* to represent the from node, the variable *b* to represent the to node, the variable *wgt* to represent the weight of the link, and the variable *aux* to represent the auxiliary weight for the link. As an undirected graph, the from node doesn't necessarily represent the source node in the shortest path situation. The same for the to node. It doesn't necessarily represent the sink node. For an undirected graph, proc optnetwork internally duplicates the links to compute all possible shortest paths considering both directions. For example, the link $A \rightarrow B$ in the links dataset will be represented by the links $A \rightarrow B$ and $B \rightarrow A$ in order to allow proc optnetwork to search for all possible shortest paths within the undirected input graph.

```
data mycas.links;
   input a $ b $ wgt aux @@;
datalines;
A B 1 0.5  A E 1 0.5  B C 1 0.5  C D 1 0.5  C A 6 0.17  D E 3 0.33
D F 1 0.5  E A 1 0.5  E B 1 0.5  E C 4 0.25  F E 1 0.5
;
run;
```

The following code defines the direction of the input graph as well as the links attributes. The DIRECTION= option in the following code defines the input graph as an undirected graph. The LINKSVAR statement defines the variables *from*, *to*, *weight* and *auxweight* for the links when proc optnetwork searches for the shortest paths.

```
proc optnetwork
   direction = undirected
   links = mycas.links
   ;
```

<< OUTSHORTPATHLINKS   Table Rows: 3 | Columns: 6 of 6 | Rows 1 to 3   ↑ ↑ ↓ ↓ | ↻ ▾ | ⋮

	source	sink	order	from	to	weight
1	A	D	1	A	B	1
2	A	D	2	B	C	1
3	A	D	3	C	D	1

Figure 4.84 The shortest path solution for the links using fixed source and sink nodes.

```
linksvar
    from = a
    to = b
    weight = wgt
    auxweight = aux
;
shortestpath
    outpathslinks = mycas.outshortpathlinks
    outpathsnodes = mycas.outshortpathnodes
;
run;
```

The Figure 4.86 shows the output for the shortest path algorithm when proc optnetwork runs upon an undirected input graph. An optimal solution was achieved, and the number of shortest paths found between all possible pairs of nodes is 30, the same number found upon the directed graph.

The number of shortest paths in an undirected graph can really be greater than the number of shortest paths in a directed graph as the undirected links create more possible paths between the nodes. In this particular case, the number of shortest paths were equal based on an undirected and a directed graphs because we have here a very small network. However, something interesting is the fact the on an undirected graph, we have less links to accomplish the same number of shortest paths than on a directed graph. The reason is the same. The undirected graph creates more paths between the nodes in the input graph than the directed graphs.

Figure 4.87 shows the number of links comprised by the same 30 shortest paths when running on the directed and on the undirected graphs.

The first table refers to the directed graph and the second one refers to the undirected graph. Notice that we have 76 links comprised in the 30 shortest paths on the directed graph and just 48 links comprised in the same 30 shortest paths on the undirected graph.

Figure 4.88 shows the reason for this. As the undirected links create more straight paths between the nodes, the shortest paths between any source and sink nodes may require fewer links.

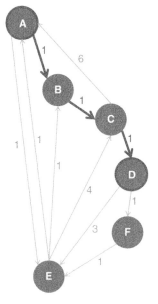

Figure 4.85 Input graph with fixed source and sink nodes.

The OPTNETWORK Procedure

Solution Summary

Problem Type	Shortest Path
Solution Status	OK
Number of Paths	30
CPU Time	0.00
Real Time	0.01

Figure 4.86 Output results by proc optnetwork running the shortest path algorithm on an undirected graph.

Output CAS Tables

CAS Library	Name	Number of Rows	Number of Columns
CASUSER(Carlos.Pinheiro@sas.com)	OUTSHORTPATHLINKS	76	7
CASUSER(Carlos.Pinheiro@sas.com)	OUTSHORTPATHNODES	106	4

Output CAS Tables

CAS Library	Name	Number of Rows	Number of Columns
CASUSER(Carlos.Pinheiro@sas.com)	OUTSHORTPATHLINKS	48	7
CASUSER(Carlos.Pinheiro@sas.com)	OUTSHORTPATHNODES	78	4

Figure 4.87 The number of links comprised in the shortest paths when proc optnetwork runs on the directed and undirected graphs.

« OUTSHORTPATHLINKS   Table Rows: 76 | Columns: 7 of 7 | Rows 1 to 76   ↑ ↑ ↓ ↓ .   ↻ ▾ | ⋮

	source	sink	order	a	b	wgt	aux
1	A	B	1	A	B	1	0.5
2	A	C	1	A	B	1	0.5
3	A	C	2	B	C	1	0.5
4	A	D	1	A	B	1	0.5
5	A	D	2	B	C	1	0.5
6	A	D	3	C	D	1	0.5
7	A	E	1	A	E	1	0.5
8	A	F	1	A	B	1	0.5
9	A	F	2	B	C	1	0.5
10	A	F	3	C	D	1	0.5
11	A	F	4	D	F	1	0.5
12	B	A	1	B	C	1	0.5
13	B	A	2	C	D	1	0.5
14	B	A	3	D	F	1	0.5
15	B	A	4	F	E	1	0.5
16	B	A	5	E	A	1	0.5

Figure 4.88 Links comprised in the shortest paths for the directed graph.

« OUTSHORTPATHLINKS   Table Rows: 48 | Columns: 7 of 7 | Rows 1 to 48   ↑ ↑ ↓ ↓   ↻ ▾ | ⋮

	source	sink	order	a	b	wgt	aux
1	A	B	1	A	B	1	0.5
2	A	C	1	A	B	1	0.5
3	A	C	2	B	C	1	0.5
4	A	D	1	A	B	1	0.5
5	A	D	2	B	C	1	0.5
6	A	D	3	C	D	1	0.5
7	A	E	1	A	E	1	0.5
8	A	F	1	A	E	1	0.5
9	A	F	2	F	E	1	0.5
10	B	A	1	A	B	1	0.5

Figure 4.89 Links comprised in the shortest paths for the undirected graph.

Observe line 1 in the first table, representing the links within the shortest paths upon the directed graph. The shortest path between the source node **A** to the sink node **B** has one single link A → B. Now, observe lines 12 through 16. The shortest path between the source node **B** to the sink node **A** has 5 links, B → C, C → D, D → F, F → E and then E → A.

Now let's take a look at the shortest paths for the undirected graph. Figure 4.89 shows the links comprised in the shortest path considering the undirected graph.

In line 1 we can see the single link A → B comprised in the shortest path between the source node **A** to the sink node **B**. In line 10, we can see the single link B → A comprised in the shortest path between the source node **B** to the sink node **A**. That is the major difference between the shortest path when running on directed and undirected graphs. The original link A → B turns into 2 links A → B (the original one comprised in the shortest path between A and B) and B → A, allowing B reaches out to A directly (not passing through C, D, E, and F). The undirected graph creates more links between the nodes allowing them to reach out to each other easier, or at least, throughout shorter paths.

As mentioned before, the shortest path computation based on directed graphs tends to return less paths than when it is runs on undirected graphs. In the previous example the outcomes were similar, with 30 shortest paths for both directed and undirected paths, even though the number of total links on the directed graph was greater than the number of links on the undirected one. The reason was the size of the network. The input graph had only 6 nodes and 11 links.

The following pictures show the same scenario based on a slightly bigger network. The network here is the famous graph based on the French classic movie Les Misérables. This network has 77 nodes and 254 links. The nodes here are the actors. The links represent when the actors play scenes together. The original network is an undirected graph, as the actors play scenes together with no specific order or direction.

Figure 4.90 shows the Les Misérables's network.

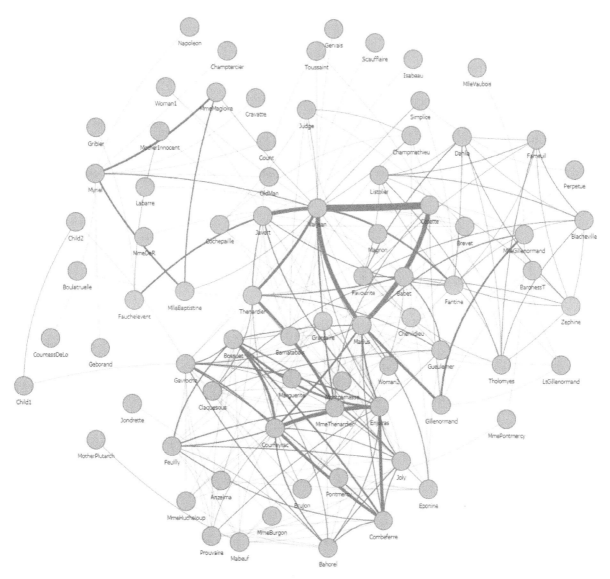

Figure 4.90 Undirected graph based on the French movie Les Misérables.

The OPTNETWORK Procedure

Problem Summary	
Number of Nodes	77
Number of Links	254
Graph Direction	Undirected

The OPTNETWORK Procedure

Solution Summary	
Problem Type	Shortest Path
Solution Status	OK
Number of Paths	5852
CPU Time	0.00
Real Time	0.00

Output CAS Tables			
CAS Library	Name	Number of Rows	Number of Columns
CASUSER(Carlos.Pinheiro@sas.com)	OUTSHORTPATHLINKS	17713	6

Figure 4.91 Links comprised in the shortest paths for Les Misérables's undirected graph.

Figure 4.91 shows the outcomes of the shortest path algorithm running upon the original undirected graph.

Based on the 254 links within the undirected graph, proc optnetwork found 5852 shortest paths, comprising 17 713 links between the actors.

Figure 4.92 shows the outcomes of the shortest path algorithm running upon the exact same links dataset but now set as a directed graph.

The OPTNETWORK Procedure

Problem Summary	
Number of Nodes	77
Number of Links	254
Graph Direction	Directed

The OPTNETWORK Procedure

Solution Summary	
Problem Type	Shortest Path
Solution Status	OK
Number of Paths	1215
CPU Time	0.00
Real Time	0.00

Output CAS Tables			
CAS Library	Name	Number of Rows	Number of Columns
CASUSER(Carlos.Pinheiro@sas.com)	OUTSHORTPATHLINKS	3362	6

Figure 4.92 Links comprised in the shortest paths for Les Misérables's directed graph.

Now, proc optnetwork found 1215 shortest paths comprising 3362 links between the actors. As the network grows in terms of size, particularly in terms of the number of links, the difference between the number of shortest paths and the links comprised in the shortest paths tend to be very different. More shortest paths and links when running on undirected graphs.

4.11 Transitive Closure

In computer science and graph theory, transitive closure can be thought of as a concept about reachability. Can one particular node A reach node B in one or more steps considering a particular network? Transitive closure can help to efficiently answer questions about reachability between nodes within an input graph. This algorithm is commonly applied to structure query languages in database management systems in order to reduce the processing time of queries. It is also applied in problems involving reachability analysis of transition networks in distributed systems, among other applications in computer science.

The transitive closure of a graph G is a graph $G^T = (N, L^T)$ where for all nodes $i, j \in N$, there is a link $(i, j) \in L^T$ if and only if there is a path from node i to node j in the input graph G. In other words, the transitive closure of an input graph is a subset of links that represent binary relations between nodes. The binary relations between the nodes represent the reachability of the original input graph, or all the nodes that can reach out to each other by using one or more links in the original input graph. Eventually, a link in the transitive closure graph G^T contains a link or relation that doesn't exist in the original input graph G, but the nodes witihn that relation are connected by multiple links within the original input graph. For example, an input graph has the links A → B, B → C, C → D, and D → E. The transitive closure graph has a binary relation A → E, even though the input graph does not contain that link A → E. This binary relation within the transitive closure graph is created because node A can reach out node E throughout the links A → B, B → C, C → D and D → E. Analogously, there is no link A → C, A → D, B → D, B → E, and C → E in the input graph, but the transitive closure graph contains the binary relations A → C, A → D, B → D, B → E, and C → E. The binary relation A → C goes thought the links A → B and B → C, the binary relation A → D goes thought the links A → B, B → C and C → D, binary relation B → D goes through the links B → C and C → D, the binary relation B → E goes through the links B → C, C → D, and D → E, and the binary relation C → E goes through the links C → D and D → E.

Figure 4.93 shows the input graph and the binary relations within the transitive closure graph.

The major benefit of the transitive closure algorithm is to answer questions about reachability, such as whether it is possible to reach node j from node i considering the input graph G. Given the transitive closure G^T of G, it is possible thenfore to simply check for the existence of the link (i, j) in G^T to answer that reachability question about how to reach node j from node i in the original input graph G. The set of binary relations describe only the nodes are reachable, for example, that node i is connected to node j no matter the number of steps or links between them. The transitive closure in that sense is another set of links.

In proc optnetwork, the TRANSITIVECLOSURE statement invokes the algorithm that solves the transitive closure problem. Proc optnetwork uses a sparse version of the Floyd-Warshall algorithm to compute the transitive closure on a given input graph.

The output table specified in the OUT= option within the TRANSITIVECLOSURE statement contains all the links that define the transitive closure for an input graph.

Proc optnetwork computes the transitive closure for both directed and undirected graphs. As the transitive closure aims to find existing paths from one node to another within the network, the original links in the input graph (the links dataset) are "duplicated" when the direction of the graph is set to undirected. By doing this, proc optnetwork creates both directions for the links connecting a particular pair of nodes. For example, if the original links dataset defines a link A → B, when proc optnetwork computes the transitive closure based on a directed graph, that original link A → B remains unique. However, when proc optnetwork computes the transitive closure based on an undirected graph, it duplicates the original link A → B into 2 links A → B and B → A, allowing node A to reach out to node B (based on the original link), and node B to directly reach out to node A (based on the duplicated link).

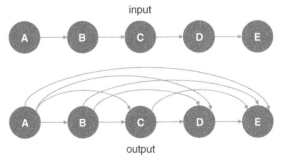

Figure 4.93 Transitive closure from a directed input graph.

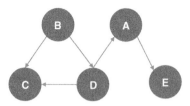

Figure 4.94 Directed input graph with unweighted links.

4.11.1 Finding the Transitive Closure

Let's consider a very simple graph to demonstrate the transitive closure problem using proc optnetwork. Consider the graph presented in Figure 4.94. It shows a tiny network with only 5 nodes and 5 directed links. The transitive closure algorithm searches in the input graph all binary relations that represent the reachability between the nodes, which means, which nodes can reach out to other nodes within the input graph.

The following code describes how to create the links dataset that represents the input graph. In this particular case, the links represent a directed network. The links dataset has only the nodes identification, which means, the *from* and *to* variables. The link weights in the transitive closure problem are irrelevant. In other words, it doesn't matter the cost or the weight of the links, the algorithm searches for the possible paths to connect the nodes within the input graph. If there is a link or a set of links, no matter the weights, that connects node i to node j, that is the matter.

```
data mycas.links;
    input from $ to $ @@;
datalines;
B C  B D  D A  D C  A E
;
```

Once the links dataset is created, we can invoke the transitive closure algorithm in proc optnetwork. The following code describes how to do it. Notice that for this particular example, we are considering the graph as directed by using the DIRECTION= option.

```
proc optnetwork
    direction = directed
    links = mycas.links
    ;
    linksvar
        from = from
        to = to
    ;
    transitiveclosure
        out = mycas.outtransclosure
    ;
run;
```

Figure 4.95 shows the output for the transitive closure algorithm. It says an optimal solution was found (solution status is ok) and the output table has 8 rows, representing all the binary relations between the possible connected nodes.

The next picture shows the transitive closure output table with the results. The OUTTRANSCLOSURE dataset presented in Figure 4.96 shows the original link variables *from* and *to* describing the binary relations between the nodes. Notice that this is not a set of links, but instead, it is the pairs of nodes that can be connected to each other by one or multiple links upon the original input graph.

As the network was set as a directed graph, the transitive closure algorithm follows the direction of the links in order to establish the possible paths between all pairs of nodes. Based on that, some nodes can reach out to other nodes, but cannot be reached by the same nodes. For example, node **B** can reach out to node **C**, but node **C** cannot reach out to node **B**. This happens because there is a directed link from node **B** to node **C**, but there is no directed link from node **C** to node **B**, or any sequence of links that connect node **C** to node **B**. Also notice that the number of steps or links between the nodes doesn't matter in the transitive closure solution, just the existing paths. For example, line 5 shows a binary relation between nodes **B** and **E**. There is no straight link from node **B** to node **E**, but node **B** can reach out to node **E** throughout the 3 links B → D, D → A, and A → E.

The OPTNETWORK Procedure

Problem Summary

Number of Nodes	5
Number of Links	5
Graph Direction	Directed

The OPTNETWORK Procedure

Solution Summary

Problem Type	Transitive Closure
Solution Status	OK
CPU Time	0.00
Real Time	0.00

Output CAS Tables

CAS Library	Name	Number of Rows	Number of Columns
CASUSER(Carlos.Pinheiro@sas.com)	OUTTRANSCLOSURE	8	2

Figure 4.95 Output results by proc optnetwork running the transitive closure algorithm on a directed input graph.

| » OUTTRANSCLOSURE | Table Rows: 8 | Columns: 2 of 2 | Rows 1 to 8 |

	from	to
1	A	E
2	B	C
3	B	D
4	B	A
5	B	E
6	D	A
7	D	C
8	D	E

Figure 4.96 The transitive closure output.

Figure 4.97 shows the path from node B to node E throughout directed links within the input graph.

In this example, nodes **C** and **E** cannot reach out to any other node in the input graph, no matter the number of links or possible paths in the network.

This scenario certainly changes if we define the network as an undirected graph. As an undirected graph, proc optnetwork will turn all unique directed links into bidirectional links. That would create more paths within the input graph, allowing more nodes to reach out other nodes and be reached.

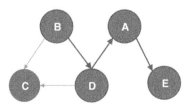

Figure 4.97 The binary relation B-E through three directed links within the input graph.

Let's use the exact same input graph. The only modification we need to do in the code is to change the option DIRECTION= to set the graph as undirected. The following code shows that modification:

```
proc optnetwork
   direction = undirected
   links = mycas.links
   ;
   linksvar
      from = from
      to = to
   ;
   transitiveclosure
      out = mycas.outtransclosure
   ;
run;
```

Figure 4.98 shows the output for the transitive closure algorithm running upon an undirected graph. Notice that we have the exact same number of nodes and links (5 for both), but the output table now has 15 rows, representing more binary relations between the nodes in the input graph.

Figure 4.99 shows the transitive closure output table with the results for the undirected graph. The OUTTRANSCLOSURE dataset shows all 15 binary relations between the nodes in the network.

As the network was now set as an undirected graph, the transitive closure algorithm follows both directions for each link defined in the links dataset. That creates more possible paths to connect the nodes within the input graph. For the directed graph, we saw that nodes C and D couldn't reach out to any other node in the network. Now, as an undirected graph, all nodes can reach out to any other node in the network.

There are two important things to notice on the results of the transitive closure for undirected graph. First, we can see that nodes can reach out to themselves. That would be auto relations, like node A can reach out to node A. That happens to all 5 nodes in the input graph. Second, as an undirected network, proc optnetwork doesn't repeat the binary relation in the output. For example, node A can reach out to all other nodes, B, C, D, and E, shown in lines 2 through 5. Node B can also reach out to all other nodes, A, B, C, D, and E. Line 6 shows the auto binary relation. Lines 7, 8, and 9 show the relations to nodes C,

The OPTNETWORK Procedure

Problem Summary

Number of Nodes	5
Number of Links	5
Graph Direction	Undirected

The OPTNETWORK Procedure

Solution Summary

Problem Type	Transitive Closure
Solution Status	OK
CPU Time	0.00
Real Time	0.00

Output CAS Tables

CAS Library	Name	Number of Rows	Number of Columns
CASUSER(Carlos.Pinheiro@sas.com)	OUTTRANSCLOSURE	15	2

Figure 4.98 Output results by proc optnetwork running the transitive closure algorithm on a input undirected graph.

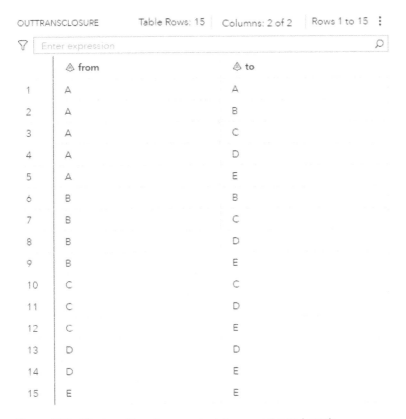

OUTTRANSCLOSURE | Table Rows: 15 | Columns: 2 of 2 | Rows 1 to 15

	from	to
1	A	A
2	A	B
3	A	C
4	A	D
5	A	E
6	B	B
7	B	C
8	B	D
9	B	E
10	C	C
11	C	D
12	C	E
13	D	D
14	D	E
15	E	E

Figure 4.99 The transitive closure output for an undirected graph.

D, and E, respectively. Line 2 actually shows the relation to node A, as A-B. As an undirected graph, the relation A-B is exact the same as the relation B-A. Therefore, there is no need to duplicate that relation as B-A in the output table.

Figure 4.100 shows the undirected graph based on the bidirectional links created by proc optnetwork, and we can see that there is a path from any node to any other node in the network.

Finally, it is important to notice that the transitive closure algorithm searches for all possible combinations of paths within the input graph in order to find binary relations between the nodes,

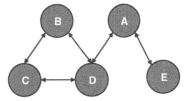

Figure 4.100 The undirected graph with bidirectional links between the nodes.

the output table can be very large if the network is large. For example, considering the network to represent the French movie Les Misérables, with only 254 links, based on an undirected graph, the transitive closure will produce an output table with 3,003 rows, representing all binary relations within the network. Figure 4.101 shows the result for the transitive closure considering the Les Misérables network.

The Les Misérables network is a very dense graph, where actors commonly play scenes together, or in terms of a graph, nodes are commonly connected. The 77 actors, or nodes, are completely connected, which means any node within that network can reach out to any other node. The combination of 77 nodes by 2 elements gives us 2926 possible binary relations. Plus, we have the 77 auto relations. That gives us the final result of 3003 rows in the output table.

4.12 Traveling Salesman Problem

The TSP aims to find the minimum-cost tour in a given input graph. The minimum cost tour is a particular elementary cycle within the input graph. A path in a graph is a sequence of nodes linked to each other. An elementary cycle is a path where the start node and the end node are the same and no node is visited more than once in the sequence. A tour is an elementary

The OPTNETWORK Procedure

Problem Summary

Number of Nodes	77
Number of Links	254
Graph Direction	Undirected

The OPTNETWORK Procedure

Solution Summary

Problem Type	Transitive Closure
Solution Status	OK
CPU Time	0.00
Real Time	0.00

Output CAS Tables

CAS Library	Name	Number of Rows	Number of Columns
CASUSER(Carlos.Pinheiro@sas.com)	OUTTRANSCLOSURE	3003	2

Figure 4.101 The transitive closure for the Les Misérables's network.

cycle where every node within the input graph is visited once and only once. The TSP algorithm aims to find the tour with the minimum total cost. The elementary cycle that visits every node is known as the Hamiltonian Cycle. The Hamiltonian path in an undirected graph is a path that visits each node exactly once. A Hamiltonian cycle is a Hamiltonian path such that there is a link from the last node to the first node of the Hamiltonian path. There is a subtle difference between the Hamiltonian Cycle and the TSP. The Hamiltonian cycle problem aims to find a tour that visits every node exactly once within the input graph (or a subgraph within a larger network). In fact, many tours can exist in the input graph that satisfies that condition. The TSP aims to find the minimum weight Hamiltonian cycle.

The TSP answer questions about optimal routes to visit a set of locations considering existing links between nodes (locations to visit) within a network. For example, given a list of locations in a city, and the distances between each pair of these locations, what is the shortest possible route that visits every selected location within the city exactly once and returns to the origin place. The origin place might be a hotel in a tourist city, and the places to visit may be a set of monuments, museums, and restaurants. The TSP searches for the optimal tour to depart from the hotel, visit all selected monuments, museums, and restaurants, and return to the hotel at the end of the tour. The optimal tour can be the total shortest distance traveled when covering all the selected places, in the shortest time, or the shortest cost. For instance, assume we have multiple transportation options to go from one place to another. We can walk, which is free, take the bus, which is cheap, take the metro, which is fast, or take the taxi, which might be more straightforward. The optimal tour can minimize the time, the distance, or the cost to travel and visit a set of locations in the city.

The TSP has applications in many areas like planning, logistics, routing, manufacturing, computer networks, astronomy, DNA sequencing, among others. TSP has an unclear origin. It was described by Willian Rowan Hamilton and Thomas Kirkman in the 1800s. The problem was mathematically formulated in the 1930s by Merrill Flood searching for a solution on school bus routing problem. The problem has been studied for decades in many fields like mathematics, computer science, chemistry, and physics, and several solutions have been theorized. The simplest solution is to try all possibilities. However, this method can be quite time consuming and exhaustive particularly in large networks. For those reasons, many solutions use heuristics, which provides probability outcomes for the possible optimal tours. The results in these cases are approximate and not always optimal. Other solutions focus more on optimization methods, which includes branch and bound, Monte Carlo, and Las Vegas algorithms to minimize the computing time while finding the optimal solutions.

Formally, the TSP finds a minimum-cost tour in a graph G that has a node set N and a link set L. When searching for the solution, each link $(i, j) \in L$ is associated with a binary variable x_{ij}, indicating whether or not the link (i, j) is part of the tour.

Additionally, each link (i, j) is also associated with another variable c_{ij} indicating the cost of the link. The total sum of the link weight is important to define the minimum cost tour in the TSP.

The TSP can be formulated as a linear programming for an undirected graph G. The mathematical formulation is as follows:

$$minimize \sum_{ij \in L} c_{ij} x_{ij}$$

$$subject\ to \sum_{ij \in \delta(i)} x_{ij} = 2 \quad i \in N \qquad (TwoMatch)$$

$$\sum_{ij \in \delta(S)} x_{ij} \geq 2 \quad S \subset N, 2 \leq |S| \leq |N| - 1 \qquad (Subtour)$$

$$x_{ij} \in \{0, 1\} \quad ij \in L$$

where for each subset S of nodes $\delta(S)$ represents the set of links (i, j), with $i \in S$ and $j \notin S$.

The first equation represents the matching constraints, which ensure that each node has degree 2 in the subgraph. The second inequality equation represents the subtour elimination constraints, which enforce connectivity.

For a directed graph G, the same formulation and solution approach are used on an expanded graph G'. Proc optnetwork constructs the expanded graph and returns the solution in terms of the original input graph.

In the context of network flow, we can think of the TSP as a routing problem. For example, each node is a location, and each link is a road connection to a pair of locations. The distance between the pairs of locations is the cost of the link. The goal of the TSP is to find the shortest possible route that connects all locations. Each location must be visited only once. Links can also be seen as pathways, metro, or bus lines, or any other mode of transportation. Then, the distance can be seen as cost or time. In that way, the TSP aims to find the cheapest or the shortest time route to visit all locations in the node set.

In proc optnetwork, the TSP statement invokes the algorithm that solves the TSP. This algorithm is based on a variant of the branch-and-cut process described in the book *The Traveling Salesman Problem: A Computational Study*, by David L. Applegate, Robert E. Bixby, Vašek Chvátal, and William J. Cook.

Proc optnetwork implements a linear programming-based branch and cut algorithm. This method uses a divide and conquer approach. It attempts to solve the original problem on general mixed linear programs by solving linear programming relaxations of a sequence of smaller subproblems. The procedure also implements advanced techniques like generating cutting planes and applying primal heuristics to improve the efficiency of the overall algorithm.

There are many options for the TSP algorithm in proc optnetwork. As usual when trying to optimize (maximize or minimize) an objective function in network flow problems, something important when defining the input graph is the weight (or the cost) of the links. The option WEIGHT= in the LINKSVAR statement is used to specify the cost of the links in the input graph and used when proc optnetwork is searching for the minimum cost tour. The direction of the link in the input graph is also crucial. Proc optnetwork can compute the TSP for both directed and undirected graphs. The results of the algorithm can be very different when searching the optimal tour in directed and undirected graphs. Undirected graphs create more paths between the nodes, allowing more possible cycles within the network. Directed graphs have fewer paths within the network as the direction of the link may continue or interrupt a particular cycle. The option DIRECTION= defines the direction of the links within the network and completely changes the way proc optnetwork computes the optimal tour. For example, if we create the links dataset considering the direction of the link, in a directed graph, the *from* node in each link will be the origin and the *to* node will be the destination. That link will be unique in the network. For the same links dataset, defined as an undirected graph, the *from* node can be origin in one link and destination in another, and the *to* node can be the destination in one link and origin in another. In the directed graph the link A $\rightarrow$ B is unique and represents A as origin and B as destination. This is a single connection from A to B. In an undirected graph, the original link A $\rightarrow$ B in the link set can represent A as origin and B as destination, as well as B as origin and A as destination. The original link A $\rightarrow$ B turns into 2 links A $\rightarrow$ B and B $\rightarrow$ A describing two possible connections, from A to B and from B to A. That creates more possible paths between the nodes, which represents more possible cycles. From more possible cycles, there are more possible optimal tours within the network.

The final result for the TSP algorithm in proc optnetwork is reported and saved in the output tables defined by the options OUT= and OUTNODES=. The output table specified in the OUT= option in the TSP statement reports the optimal tour as a sequence of links. This output table contains the following variables. The variable *tsp_order* contains the order or the sequence of the link in the tour. The variable *from* contains the node label of the origin in the link. The variable *to* contains the node label of the destination in link. The variable *weight* contains the cost of the link. The cost can be the distance, the time, or the price to travel from the origin node to the destination node. The output table specified in the OUTNODES= option in the PROC OPTNETWORK statement reports the optimal tour as a sequence of nodes. This output table contains the following variables. The variable *node* contains the node label, and the variable *tsp_order* contains the order or the sequence of the node in the optimal tour.

The option ABSOBJGAP= or ABSOLUTEOBJECTIVEGAP= specifies a stopping criterion for the algorithm when searching for the optimal tour. If the absolute difference between the best integer objective and the objective of the best remaining branch-and-bound node becomes less than the value specified in the option, the solver stops searching for more solutions. The value of number specified in the option can be any nonnegative number. The default t value is 1E–6.

The option RELOBJGAP= or RELATIVEOBJECTIVEGAP= specifies a stopping criterion for the algorithm when searching for the optimal tour. This stopping criterion is based on the best integer objective and the objective of the best remaining node. The relative objective gap is the difference between the best integer objective and the objective of the best remaining node divided by the objective of the best remaining node. When the relative objective gap becomes less than the number specified in the option, the solver stops searching for an optimal solution. The value of number can be any nonnegative number. By default, the option RELOBJGAP= is set to 0.0001.

The option TARGET= specifies a stopping criterion for minimization problems. If the best integer objective is less than or equal to the number specified in the option, the solver stops searching for an optimal solution for the TSP tour. The value of the number specified in the option can be any number. The default is the largest negative number in magnitude that can be represented by a double.

The option CUTOFF= specifies the cutoff value for any branch-and-bound nodes in a minimization problem. The main idea of the cutoff is the following. If the optimal linear programming relaxation objective is not better than the cutoff value specified, then any mixed integer programming solution of a descendant can be no better than the cutoff. Then the node in the branch-and-bond method can be fathomed and does not need to be further considered in the search. If the objective value is greater than or equal to the number specified in the option, the algorithm stops the branch-and-bound search. The default is the largest number that can be represented by a double.

The option CUTSTRATEGY= specifies the level of mixed integer linear programming cutting planes to be generated. The cutting plane method iteratively refines the feasible set of solution values or the objective function by means of linear inequalities, also called as cuts. This method is commonly used to find integer or mixed integer solutions in linear programming. The cutting plane method works by solving a non-integer linear program by using a linear relaxation of the given integer program. Specific cutting planes are always generated by proc optnetwork when searching for optimal tours in input graphs. There are four options for the cutting strategy in proc optnetwork. The option AUTOMATIC generates cutting planes on the basis of a strategy that is determined by the mixed integer linear programming solver. The option NONE disables the generation of mixed integer linear programming cutting planes (some TSP-specific cutting planes are still active for validity). The option MODERATE uses a moderate cutting strategy. Finally, the option AGGRESSIVE uses an aggressive cutting strategy. By default, proc optnetwork disables the generation of mixed integer linear programming cutting planes by using the option CUTSTRATEGY= NONE.

The option HEURISTICS= determines how frequently to apply primal heuristics during the branch-and-bound tree search. The application of primal heuristics to improve algorithm efficiency affects the maximum number of iterations that are allowed in iterative heuristics. Some computationally expensive heuristics might be disabled by the solver at less aggressive levels. There are five levels of how frequent proc optnetwork applies primal heuristics during the tree search. The option AUTOMATIC applies the default level of heuristics. The option NONE disables all primal heuristics. The option BASIC applies basic primal heuristics at low frequency. The option MODERATE applies most primal heuristics at moderate frequency. Finally, the option AGGRESSIVE applies all primal heuristics at high frequency. By default, the level of primal heuristics application in the tree search is AUTOMATIC.

The option LOGFREQ= or LOGFREQUENCY= specifies the time interval in seconds for printing information in the node log during the branch-and-bound algorithm. The node indicates the sequence number of the current node in the search tree. The number specified in the option can be any integer greater than or equal to zero. If the option is specified as 0, the node log is disabled. If the option is specified with any number greater than 0, the root node processing information is printed and, if possible, an entry is made at every number of seconds specified in the option. An entry is also made each time a better integer solution is found. By default, the option LOGFREQ= is specified at 5, which means, printing information in the node log at every five seconds.

The option MAXNODES= specifies the maximum number of branch-and-bound nodes to be processed. The default is any nonnegative integer that can be represented by a 32-bit integer. When running the mixed integer liner programming in concurrent mode, the solver stops as soon as the number specified in the option is reached on any machine. When running the mixed integer linear programming algorithm in distributed mode, the solver periodically checks the total number of nodes that are processed by all grids and stops when the number specified in the option is reached.

The option MAXSOLS= specifies a stopping criterion. When the number of solutions specified in the option is found, the procedure stops searching for new optimal solutions. The default is the largest number that can be represented by a 32-bit integer.

The option MAXTIME= specifies the maximum amount of time to spend solving the TSP. The type of time can be either CPU time or real-time. The type of time is determined by the value of the TIMETYPE= option in the proc optnetwork statement. The default is the largest number that can be represented by a double. The maximum time limits the optimization process, including the problem generation and the solution time. If the MAXTIME= option is not used, the solver does not stop based on the amount of time elapsed during the optimization process.

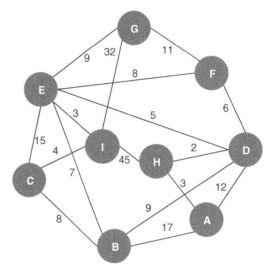

Figure 4.102 Undirected input graph with weighted links.

The option MILP= specifies the type of algorithm proc optnetwork uses to find the optimal TSP tour. Proc optnetwork might or might not use a mixed integer linear programming solver. The mixed integer linear programming solver attempts to find the overall best tour by using the branch-and-cut algorithm. This algorithm can be expensive for large-scale problems. The value TRUE in the option MILP= specifies the mixed integer linear programming solver as the algorithm used by proc optnetwork. The option FALSE determines proc optnetwork to use its initial heuristics to find a feasible, but not necessarily optimal tour, as quickly as possible. By default, proc optnetwork uses the mixed integer linear programming approach based on the branch-and-cut algorithm (option MILP = TRUE).

4.12.1 Finding the Optimal Tour

Let's consider a simple graph to demonstrate the TSP using proc optnetwork. Consider the graph presented in Figure 4.102. It shows a network with just 9 nodes and only 17 links. First, we assume this network as an undirected graph in order to compute the optimal tour to visit all nodes exactly once. Later, we assume the same network as a directed graph to see how the outcomes for the TSP can be different depending on the direction of the input graph.

The following code describes how to create the input dataset for the TSP. The links dataset has the nodes identification, the from and to variables indicating origin and destination of the possible paths within the tour, and the weight of the link, which is used by proc optnetwork to search the minimum cost tour to visit all nodes in the network. This dataset can be used to represent both directed and undirected graph as we will see in the following examples.

```
data mycas.links;
    input from $ to $ weight @@;
datalines;
G F 11   F D 6   F E 8   D A 12   D B 9   D H 2   D E 5   H A 3   A B 17
B C 8   B E 7   C E 15   I C 4   I H 45   E I 3   E G 9   I G 32
;
run;
```

With the links dataset defined, we can now use the TSP algorithm in proc optnetwork. The following code describes how to invoke the algorithm. The links definition using the LINKSVAR statement specifies the variables *from* and *to* as the origin and destination of possible paths within the network, and the variable *weight* as the cost to travel through that possible path. The input graph in this example is defined as an undirected network. In that case, proc network creates two possible directions for each. For example, first link (*G*, *F*) internally turns into two distinct links (*G*, *F*) and (*F*, *G*).

```
proc optnetwork
    direction = undirected
```

```
   links = mycas.links
   outnodes = mycas.nodesetout
   ;
   linksvar
      from = from
      to = to
      weight = weight;
   tsp
      out = mycas.tsptour
   ;
run;
```

Figure 4.103 shows the output for the TSP algorithm. It shows the size of the input graph, 9 nodes and 17 links. Later it says that an optimal solution was found (solution status is optimal), one single solution was found (number of solutions 1), and the objective value is 63. This value represents the cost of the optimal tour, the sum of link weights to visit all nodes exactly once.

The following pictures show the detailed results of the TSP generated by proc optnetwork. The NODESETOUT dataset presented in Figure 4.104 shows the variable *node* that identifies the node label and the variable *tsp_order* that identifies the order of the optimal tour. Note that the TSP represents a tour that visits all nodes within the network exactly once. The tour is a cycle where all nodes must be visited once and only once. As a cycle, there is no starting or ending point. The optimal tour can start at any node within the input graph.

For example, the sequence of the optimal tour starts with node A, then goes to node B, C, I, E, G, F, D, and then H. However, the sequence can actually start at node E, then to G and so forth. There is a sequence that needs to be followed, but no starting point.

The TSPTOUR dataset presented in Figure 4.105 shows the variables *tsp_order* that specifies the sequence of the links to be traveled in order to cover all nodes within the input graph, and the original link variables, *from* and *to* to identify origin and destination for each step of the path, and the variable *weight* to identify the cost of each of those steps.

Figure 4.106 shows the solution for the TSP algorithm considering the undirected input graph.

Following from the top of the input graph, the optimal tour departs from node G, goes to F, then D, H, A, B, C, I, E, and then returns to G to complete the cycle.

Let's now consider the same network but as a directed graph. The links dataset is created exactly as shown before. The only difference is the direction of the network defined in the DIRECTION= option in the proc optnetwork statement. Based on the links definition, the input directed graph looks like the following picture. For example, the definition of the first link in the data step creates the link $(G, F, 11)$. As a directed graph, the link G → F is unique. Proc optnetwork will not create a link F → G as it did for the undirected graph.

Figure 4.107 shows the input graph considering directed weighted links.

Based on that link set definition, we can now run the proc optnetwork to search for the optimal tour invoking the TSP algorithm.

```
proc optnetwork
   direction = directed
   links = mycas.links
   outnodes = mycas.nodesetout
   ;
   linksvar
      from = from
      to = to
      weight = weight;
   tsp
      out = mycas.tsptour
   ;
run;
```

The OPTNETWORK Procedure

Problem Summary

Number of Nodes	9
Number of Links	17
Graph Direction	Undirected

The OPTNETWORK Procedure

Solution Summary

Problem Type	Traveling Salesman Problem
Solution Status	Optimal
Number of Solutions	1
Objective Value	63
Relative Gap	0
Absolute Gap	0
Primal Infeasibility	0
Bound Infeasibility	0
Integer Infeasibility	0
Best Bound	63
Nodes	1
Iterations	11
CPU Time	0.00
Real Time	0.00

Output CAS Tables

CAS Library	Name	Number of Rows	Number of Columns
CASUSER(Carlos.Pinheiro@sas.com)	NODESETOUT	9	2
CASUSER(Carlos.Pinheiro@sas.com)	TSPTOUR	9	4

Figure 4.103 Output results by proc optnetwork running the traveling salesman algorithm based on an undirected input graph.

<< NODESETOUT   Table Rows: 9   Columns: 2 of 2   Rows 1 to 9   ↑ ↑ ↓ ↓   ⟳ ▾   ⋮

▽ | Enter expression   🔎

	node	tsp_order
1	A	1
2	B	2
3	C	3
4	D	8
5	E	5
6	F	7
7	G	6
8	H	9
9	I	4

Figure 4.104 The sequence of nodes in the traveling salesman problem output based on an undirected graph.

| ≪ TSPTOUR | Table Rows: 9 | Columns: 4 of 4 | Rows 1 to 9 | ⤒ ↑ ↓ ⤓ | ↺ ▾ | ⋮ |

▽ Enter expression 🔍

	⊞ tsp_order	⚠ from	⚠ to	⊞ weight
1	1	A	B	17
2	2	B	C	8
3	3	I	C	4
4	4	E	I	3
5	5	E	G	9
6	6	G	F	11
7	7	F	D	6
8	8	D	H	2
9	9	H	A	3

Figure 4.105 The sequence of links in the traveling salesman problem output based on an undirected graph.

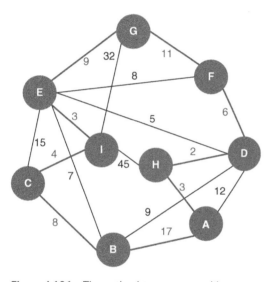

Figure 4.106 The optimal tour generated by proc optnetwork based on an undirected graph.

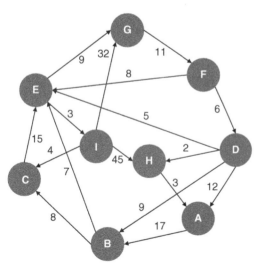

Figure 4.107 Directed input graph with weighted links.

Figure 4.108 shows the summary result produced by proc optnetwork. The output for the TSP based on the directed graph is very similar, except by the objective value. Recall that the objective value based on the undirected graph was 63. Now, based on the directed graph, the objective value is 97. A substantial difference. The reason for this high-cost tour is the absence of some of the paths that were created by the undirected graph. Now the tour has unique link directions to follow from node to node, and the final tour might end up with a higher total link weight, just because some of the paths used before are not available now.

Figure 4.109 shows the NODESETOUT dataset that identifies the node label and the order of the optimal tour. The sequence of nodes in the TSP solution now is slightly different than before, created for the undirected graph.

The optimal tour for the undirected graph was $A - B - C - I - E - G - F - D - H$. Now, for the directed input graph, the sequence is $A \rightarrow B \rightarrow C \rightarrow E \rightarrow I \rightarrow G \rightarrow F \rightarrow D \rightarrow H$.

Figure 4.110 shows the TSPTOUR dataset that identifies the sequence of the links used to visit all nodes during the optimal tour.

For the undirected graph, the tour used the links $C - I$, $I - E$, $E - G$. That path from node C to node G cost $16 (4 + 3 + 9)$. However, for the directed graph, there is no link $C \rightarrow I$ neither $I \rightarrow E$. The option for the tour is to go from C to E, costing 15, then from E to I, costing 3, and finally from I to G costing 32. The total cost for the same path is now 50, 34 more than the path for the undirected graph (16). That is exactly the difference in the objective values between the directed (97) and undirected (63) graphs.

Figure 4.111 shows the solution for the TSP algorithm considering the directed input graph.

Following from the top of the input graph, the optimal tour departs from node G, goes to F, then D, H, A, B, and C. That path so far is exactly the same as the one we saw in the undirected graph. Here resides the difference between the solutions for the directed and undirected graphs. From node C, the tour now goes to node

The OPTNETWORK Procedure

Problem Summary

Number of Nodes	9
Number of Links	17
Graph Direction	Directed

The OPTNETWORK Procedure

Solution Summary

Problem Type	Traveling Salesman Problem
Solution Status	Optimal
Number of Solutions	1
Objective Value	97
Relative Gap	0
Absolute Gap	0
Primal Infeasibility	0
Bound Infeasibility	0
Integer Infeasibility	0
Best Bound	97
Nodes	1
Iterations	15
CPU Time	0.00
Real Time	0.00

Output CAS Tables

CAS Library	Name	Number of Rows	Number of Columns
CASUSER(Carlos.Pinheiro@sas.com)	NODESETOUT	9	2
CASUSER(Carlos.Pinheiro@sas.com)	TSPTOUR	9	4

Figure 4.108 Output results by proc optnetwork running the traveling salesman problem algorithm based on a directed input graph.

E (not I), then node I (not E), and then goes back to node G. As we saw, because of this small difference, the optimal tour based on the directed graph costs 34 more than the tour for the undirected graph.

Let's change the original links dataset just a tiny bit and rerun proc optnetwork for both undirected and directed graphs. We keep the exact same set of links, but we invert the order of the nodes in the link (H, A, 3). In the following data step, we create the same links set, but with the link (A, H, 3).

```
data mycas.links;
   input from $ to $ weight @@;
datalines;
G F 11  F D 6  F E 8  D A 12  D B 9  D H 2  D E 5
A H 3
A B 17  B C 8  B E 7  C E 15  I C 4  I H 45  E I 3  E G 9  I G 32
;
run;
```

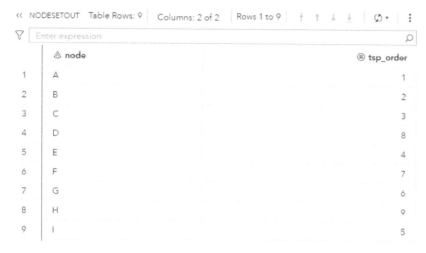

Figure 4.109 The sequence of nodes in the traveling salesman problem output based on a directed graph.

	⊕ tsp_order	⚠ from	⚠ to	⊕ weight
1	1	A	B	17
2	2	B	C	8
3	3	C	E	15
4	4	E	I	3
5	5	I	G	32
6	6	G	F	11
7	7	F	D	6
8	8	D	H	2
9	9	H	A	3

Figure 4.110 The sequence of links in the traveling salesman problem output based on a directed graph.

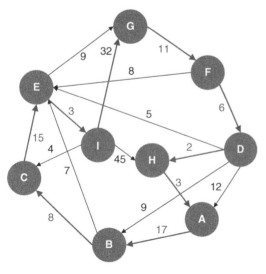

Figure 4.111 The optimal tour generated by proc optnetwork based on a directed graph.

If we run proc optnetwork based on an undirected graph, the results will be exactly the same as we found before. The objective value is 63 and the same sequence $A - B - C - I - E - G - F - D - H$ will be recommended. The reason is simple. On an undirected graph, it does not matter if we define (H, A) or (A, H) in the links dataset. Both directions will exist in the input graph when proc optnetwork searches for the optimal tour.

```
proc optnetwork
    direction = undirected
    links = mycas.links
    outnodes = mycas.nodesetout
    ;
    linksvar
        from = from
        to = to
        weight = weight;
    tsp
```

```
        out = mycas.tsptour
    ;
run;
```

Figure 4.112 shows the outcomes from proc optnetwork.

However, if we execute the TSP in proc optnetwork based on a directed graph, because of this tiny change, the solution is infeasible.

```
proc optnetwork
    direction = directed
    links = mycas.links
    outnodes = mycas.nodesetout
    ;
    linksvar
        from = from
        to = to
        weight = weight;
    tsp
        out = mycas.tsptour
    ;
run;
```

Figure 4.113 shows the new input graph based on directed links.

Notice that if we follow the original sequence based on the directed graph, we start at node G, then we go to node F, and then node D. From node D, there are 3 options, nodes H, E, and A. If we go to node H, we get stuck. There is no link to continue the path from node H. Node H is a terminal node. If we go to node E, there is no way to go back to nodes B or A. And if we return to node H, we get stuck again. If we go to node A, we can continue the path for a while, until we reach out to node H, and then we get stuck.

Figure 4.114 shows the summary outcomes from proc optnetwork when running the TSP algorithm based on the directed graph with the modification on the link A → H instead of the original link H → A.

The infeasible solution means that there is no possible sequence of links that allow us to visit all the nodes once and only once considering the available set of links in the directed input graph.

The OPTNETWORK Procedure

Solution Summary	
Problem Type	Traveling Salesman Problem
Solution Status	Optimal
Number of Solutions	1
Objective Value	63
Relative Gap	0
Absolute Gap	0
Primal Infeasibility	0
Bound Infeasibility	0
Integer Infeasibility	0
Best Bound	63
Nodes	1
Iterations	11
CPU Time	0.00
Real Time	0.00

Figure 4.112 Output results by proc optnetwork running the traveling salesman algorithm based on an undirected input graph with the link (*A*, *H*, **3**).

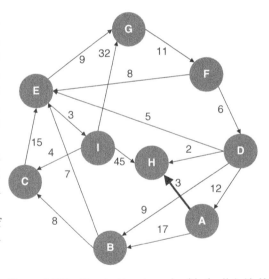

Figure 4.113 Directed input graph with the link (*A*, *H*, **3**).

4.13 Vehicle Routing Problem

The Vehicle Routing Problem (VRP) is a combinatorial optimization and integer programming problem in graph theory. The VRP aims to find a set of optimal routes for multiple vehicles to visit a specific set of locations in order to deliver goods serving a given set of customers. The VRP is one of the most challenging problems in the field of combinatorial optimization. The VRP is a generalization of the TSP covered in the previous section. If there is one single vehicle, the VRP is reduced to the TSP.

The VRP considers deterministic and heuristic methods. The VRP is an NP-hard problem (nondeterministic polynomial time problem). The size of the problems that can be optimally solved by mathematical programming or combinatorial optimization is limited. Due to the size and frequency of real-world problems, heuristics is commonly used to solve VRP in real business applications.

This algorithm addresses one of the most important problems in enterprise logistics. We can think of it as an efficient set of multiple routes for a fleet of vehicles that all start and end at the same central depot to service a given set of customers along the way, or the cycle. Customers can be warehouses, stores, schools, or any type of business associated with geographic

The OPTNETWORK Procedure

Solution Summary

Problem Type	Traveling Salesman Problem
Solution Status	Infeasible
Number of Solutions	0
Best Bound	.
Nodes	0
Iterations	0
CPU Time	0.00
Real Time	0.00

Figure 4.114 Output results by proc optnetwork running the traveling salesman algorithm based on a directed input graph with the link (**A, H, 3**).

locations. Each customer in the cycle must be visited once by only one vehicle at the lowest cost. The use of the VRP algorithm can help companies to minimize the global transportation costs based on distances traveled, or fixed costs associated with the use of vehicles and drives or minimize the numbers of vehicles needed to serve the current set of customers or reduce variation in travel time and vehicle load in logistics processes, among many other applications.

VRP in real-world applications can be complex involving factors like the number and locations of all possible stops on the routes, arrival and departure time gaps, effective loading, vehicle limitations, time windows, loading process at depots, multiple trips, inventory limitations, multiples depots where vehicles can start and end, among others. The list of possible constraints is practically endless.

In proc optnetwork, the VRP finds a set of elementary cycles in an input graph $G = (N'L)$ that start and end at a common node, visit all other nodes once, and satisfy the vehicle capacity limits at the minimum cost. Usually in the VRP, the referred common node is a depot, and the referred all other nodes are the customers. The node set $N = C \cup \{1\}$ consists of a set of customer nodes (C) and a depot node (1). A route is an elementary cycle that starts and ends at the depot node. A route is serviced by a vehicle that picks up goods at the depot node and delivers them to its assigned set of customer nodes. Each customer has a demand d_i, and each vehicle has a capacity Q. The number of routes (K) that are used by the vehicle varies depending on the vehicle fleet configuration. The number of routes is restricted to be greater than or equal to a lower limit K_l and less than or equal to an upper limit K_u.

Each link $i, j \in L$ in the VRP is associated with a binary variable x_e. Similar to the TSP, this binary variable indicates whether or not the link i, j is part of the recommended routes. Each link i, j is also associated with a continuous variable c_{ij}. This variable indicates the weight of the link, or the cost to travel throughout that link. In addition to the variables associated with the links, an integer variable K indicates the number of routes to be used by the vehicle when traveling through the customer nodes.

An integer linear programming formulation of the VRP for a directed input graph G can be formulated as follows:

$$\text{minimize} \sum_{ij \in L} c_{ij} x_{ij}$$

$$\text{subject to} \sum_{ij \in \delta_i^{out}} x_{ij} = 1 \qquad i \in C \qquad (LeaveNode)$$

$$\sum_{ij \in \delta_j^{in}} x_{ij} = 1 \qquad j \in C \qquad (EnterNode)$$

$$\sum_{ij \in \delta_1^{out}} x_{ij} = K \qquad (LeaveDepot)$$

$$\sum_{ij \in \delta_1^{in}} x_{ij} = K \qquad (EnterDepot)$$

$$\sum_{ij \in \delta(S)} x_{ij} \geq r(S) \quad S \subseteq C, S \neq \emptyset \qquad (Capacity/Subtour)$$

$$K_l \leq K \leq K_u$$

$$x_{ij} \in \{0, 1\} \qquad ij \in L$$

where for each subset S of customer nodes, $\delta(S)$ represents the set of links (i, j), with $i \in S$ and $j \notin S$. δ_i^{out} represents the set of outgoing links that are connected from node i and δ_i^{in} represents the set of incoming links that are connected to node i.

$r(S) = \left\lceil \sum_{i \in S} d_i / Q \right\rceil$ represents a lower bound on the number of vehicles needed in order to service the set of customers S.

The first equation represents the leave node. The second equation represents the enter node. Both equations together ensure that each customer node is visited once and only once during the cycle. The third equation represents the leave depot. The fourth equation represents the enter depot. Both equations together ensure that the depot node is visited K times, once for each route. The last and fifth equation represents the capacity and the subtour. These inequalities enforce connectivity throughout the cycle and ensure that the vehicle capacity restriction is observed.

In the context of network flow, we can think of the VRP as a package delivery system where each node is a customer located in some particular geographic location, that demands a certain quantity of goods, and each link is a road that connects all those geographic locations. The goods need to be sent out from the depot node (which can be a central warehouse), must be loaded into the servicing vehicles, which have a limit capacity to transport the goods, and must be delivered to all the customers in the least amount of time, or by traveling the shortest distance, or producing the lowest cost.

In proc optnetwork, the VRP statement invokes the algorithm that solves the VRP. This algorithm is based on a variant of the branch-and-cut process described by Jens Lysgaard, Adam N. Letchford, and Richard W. Eglese in the paper "A New Branch-and-Cut Algorithm for the Capacitated Vehicle Routing Problem," published in 2003 in the *Mathematical Programming Journal*. The VRP solver uses internally modified versions of two open-source packages. In order to generate primal feasible solutions, proc optnetwork uses the package VRPH. To generate cutting planes to improve the dual bound, proc optnetwork uses the package CVRPSEP.

Similar to the TSP, there are many options for the VRP algorithm in proc optnetwork. Most of the algorithms in network optimization aims to minimize an objective function, which can be assigned to reduce time, distance, or cost within paths or cycles. The time, distance, or cost are commonly assigned to the link weight. Therefore, it is important in network optimization algorithms to properly define the link weight and specify it in proc optnetwork. The option WEIGHT= in the LINKSVAR statement is used to specify the time, distance, or the cost of the links in the input graph. This link weight is used by proc optnetwork when searching for the minimum cost routes associated with the VRP. If the link weight is not specified by using the WEIGHT= option, proc optnetwork will assume all links have the weight equal to 1.

Particularly for the VRP, the nodes definition is quite important. The nodes dataset defines the customer nodes and the demand they have for delivering goods. The option NODES= in PROC OPTNETWORK statement specifies the nodes data table containing the customers and their demands. The NODESVAR statement is used to specify the demand of the customer node by using the option LOWER=. If the customer node demand is not specified by using the LOWER= option, or it is missing in the nodes dataset, proc optnetwork will assume the customer demand is equal to 0. If the depot node is included in the nodes dataset, and has a demand associated, proc optnetwork will ignore the demand for the depot node.

Another important aspect is the direction of the graph, or the direction of the link within the input graph. As we noticed in the TSP, the same link set can produce a feasible or infeasible solution depending on the direction of the input graph. Again, undirected graphs create more possible paths between the customers nodes and the depot node, allowing more possible routes within the network. Directed graphs have less paths within the network as the direction of the link may progress or interrupt a particular route. Proc optnetwork can compute the VRP for both directed and undirected graphs. The option DIRECTION= defines the direction of the links within the network. Analogous to the TSP, if we create a links dataset for the VRP considering the direction of the link, in a directed graph, the *from* node will be the origin of the path and the *to* node will be the destination of that path. The exact same links dataset defined as undirected, both *from* and the *to* nodes can work as origin and destination. A link $A \rightarrow B$ is solely $A \rightarrow B$ in a directed graph as it can be $A \rightarrow B$ and $B \rightarrow A$ in an undirected graph.

The final result for the VRP algorithm in proc optnetwork is reported and saved in the output tables defined by the options OUT= and OUTNODES=. The output table specified in the OUT= option in the VRP statement reports the optimal routes as a sequence of links. This output table contains the following variables. The variable *route* that contains the route identifier, the variable *route_order* that contains the order or the sequence of the link within the route, the variable *from* that contains the origin node label of the link within the route, the variable *to* that contains the destination node label of the link within the route, and finally the variable *weight* that contains the cost of the link within the route. The cost can be anything around distance, time, price, or whatever variable that indicates the cost to travel through that link. The output table specified in the OUTNODES= option in the PROC OPTNETWORK statement reports the optimal route as a sequence of nodes. This output table contains the following variables. The variable *node* contains the node label, and the variable *demand* that contains the customer demand, the variable *route* that contains the identification of that particular route, and finally the variable *route_order* contains the order or the sequence of the node in the optimal route.

The option ABSOBJGAP= or ABSOLUTEOBJECTIVEGAP= specifies a stopping criterion for the algorithm when searching for optimal routes. If the absolute difference between the best integer objective and the objective of the best remaining branch-and-bound node becomes less than the value specified in the option, the solver stops searching for more solutions. The value of number specified in the option can be any nonnegative number. The default value is 0.000001.

The option RELOBJGAP= or RELATIVEOBJECTIVEGAP= specifies a stopping criterion for the algorithm when searching for optimal routes. This stopping criterion is based on the best integer objective and the objective of the best remaining node. The relative objective gap is the difference between the best integer objective and the objective of the best remaining node divided by the objective of the best remaining node. When the relative objective gap becomes less than the number specified in the option, the solver stops searching for an optimal solution. The value of number can be any nonnegative number. By default, the option RELOBJGAP= is set to 0.0001.

The option TARGET= specifies a stopping criterion for minimization problems. If the best integer objective is less than or equal to the number specified in the option, the solver stops searching for optimal routes. The value of the number specified in the option can be any number. The default is the largest negative number in magnitude that can be represented by a double.

The option CUTOFF= specifies the cutoff value for any branch-and-bound nodes in a minimization problem. The main idea of the cutoff is the following. If the optimal linear programming relaxation objective is not better than the cutoff value specified, then any mixed integer programming solution of a descendant can be no better than the cutoff. Then the node in the branch-and-bond method can be fathomed and does not need to be further considered in the search. If the objective value is greater than or equal to the number specified in the option, the algorithm stops the branch-and-bound search. The default is the largest number that can be represented by a double.

The option CUTSTRATEGY= specifies the level of mixed integer linear programming cutting planes to be generated. The cutting plane method iteratively refines the feasible set of solution values or the objective function by means of linear inequalities, also called cuts. This method is commonly used to find integer or mixed integer solutions in linear programming. The cutting plane method works by solving a non-integer linear program by using a linear relaxation of the given integer program. Specific cutting planes are always generated by proc optnetwork when searching for optimal routes within input graphs. There are four options for the cutting strategy in proc optnetwork. The option AUTOMATIC generates cutting planes on the basis of a strategy that is determined by the mixed integer linear programming solver. The option NONE disables the generation of mixed integer linear programming cutting planes (some TSP-specific cutting planes are still active for validity). The option MODERATE uses a moderate cutting strategy. Finally, the option AGGRESSIVE uses an aggressive cutting strategy. By default, proc optnetwork disables the generation of mixed integer linear programming cutting planes by using the option CUTSTRATEGY= NONE.

The option HEURISTICS= determines how frequently to apply primal heuristics during the branch-and-bound tree search. The application of primal heuristics to improve algorithm efficiency affects the maximum number of iterations that are allowed in iterative heuristics. Some computationally expensive heuristics might be disabled by the solver at less aggressive levels. There are five levels of how frequent proc optnetwork applies primal heuristics during the tree search. The option AUTOMATIC applies the default level of heuristics. The option NONE disables all primal heuristics. The option BASIC applies basic primal heuristics at low frequency. The option MODERATE applies most primal heuristics at moderate frequency. Finally, the option AGGRESSIVE applies all primal heuristics at high frequency. By default, the level of primal heuristics application in the tree search is AUTOMATIC.

The option LOGFREQ= or LOGFREQUENCY= specifies the time interval in seconds for printing information in the node log during the branch-and-bound algorithm. The node indicates the sequence number of the current node in the search tree. The number specified in the option can be any integer greater than or equal to zero. If the option is specified as 0, the node log is disabled. If the option is specified with any number greater than 0, the root node processing information is printed and, if possible, an entry is made at every number of seconds specified in the option. An entry is also made each time a better integer solution is found. By default, the option LOGFREQ= is specified at 5, which means printing information in the node log at every five seconds.

The option CAPACITY= specifies the capacity of each vehicle in the VRP. The capacity is the maximum amount of goods that a particular vehicle can pick up from the depot node and deliver to the customer nodes. The value specified in the option can be any nonnegative number. The default is the largest number that can be represented by a double. That means, if the capacity is not specified, proc optnetwork assumes unlimited capacity for a vehicle to collect goods from the depot nodes and deliver to the customers nodes.

The option DEPOT= specifies the depot node for the VRP. The centralized depot node is the source of all possible goods that will be delivered to the customer nodes. The value specified in the option must be present as a node within the links dataset. This option is required for the VRP because the depot node must be specified.

The option MAXROUTES= specifies the maximum number of routes allowed to service the overall customer's demand. The value specified in the option must be an integer greater than or equal to 1. The default value is the largest number that can be represented by a 32-bit integer.

One the other hand, the option MINROUTES= specifies the minimum number of routes allowed to service the overall customers demand. The value of number must be an integer greater than or equal to 1. By default, the value of the minimum number of routes allowed equals to 1.

The option MAXNODES= specifies the maximum number of branch-and-bound nodes to be processed. The default is any nonnegative integer that can be represented by a 32-bit integer. When running the mixed integer liner programming in concurrent mode, the solver stops as soon as the number specified in the option is reached on any machine. When running the mixed integer linear programming algorithm in distributed mode, the solver periodically checks the total number of nodes that are processed by all grids and stops when the number specified in the option is reached.

The option MAXSOLS= specifies a stopping criterion. When the number of solutions specified in the option is found, the procedure stops searching for new optimal solutions. The default is the largest number that can be represented by a 32-bit integer.

The option MAXTIME= specifies the maximum amount of time to spend solving the TSP. The type of time can be either CPU time or real-time, and is determined by the value of the TIMETYPE= option in the proc optnetwork statement. The default is the largest number that can be represented by a double. The maximum time limits the optimization process, including the problem generation and the solution time. If the MAXTIME= option is not used, the solver does not stop based on the amount of time elapsed during the optimization process.

The option MILP= specifies the type of algorithm proc optnetwork uses to find the optimal TSP tour. Proc optnetwork might or might not use a mixed integer linear programming solvert. The mixed integer linear programming solver attempts to find the overall best tour by using the branch-and-cut algorithm. This algorithm can be very expensive for large-scale problems. The value TRUE in the option MILP= specifies the mixed integer linear programming solver as the algorithm used by proc optnetwork. The option FALSE determines proc optnetwork to use its initial heuristics to find a feasible, but not necessarily optimal tour, as quickly as possible. By default, proc optnetwork uses the mixed integer linear programming approach based on the branch-and-cut algorithm (option MILP = TRUE).

4.13.1 Finding the Optimal Vehicle Routes for a Delivery Problem

Let's use a network flow graph to demonstrate the VRP using proc optnetwork. The graph presented in Figure 4.115 shows a network with 8 nodes and only 17 links. From the set of nodes, node X is the depot node. All the other 7 nodes, A, B, C, D, E, F, and G are the stores, or the customers. Each store or customer node also has a specific demand. Each link has a cost associated with flow the goods from the depot node to the store nodes. Initially, we assume this network as an undirected graph, which means, each link in the network has two directions, connecting nodes to outflow and inflow goods.

The following code describes how to create the input datasets for the VRP. The links dataset has the nodes identification, the from and to variables indicating origin and destination for the possible routes between depot and stores, as well as between stores. The cost for each link is also defined in the links dataset by weight variable. The nodes dataset has the stores or customer identification plus the demand of each node.

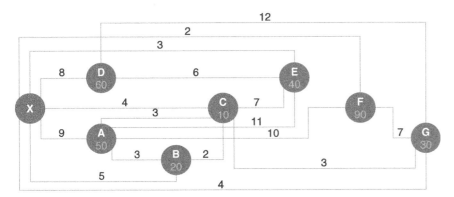

Figure 4.115 Undirected input graph with weighted links and nodes demands.

```
data mycas.links;
   input from $ to $ weight @@;
datalines;
X A 9  X B 5  X C 4  X D 8  X E 3  X F 2  X G 4  A B 3  A C 3
A E 11  A F 10  B C 2  C E 7  C G 3  D E 6  D G 12  F G 7
;
run;

data mycas.nodes;
   input node $ demand @@;
datalines;
A 50  B 20  C 10  D 60  E 40  F 90  G 30
;
run;
```

Once the links and nodes datasets are created, defining possible routes with costs, and stores with demands, we can invoke the VRP algorithm in proc optnetwork to solve the VRP. The following code describes how to do that. The links definition using the LINKSVAR statement specifies the variables *from* and *to* as the origin and destination for all possible routes within the network. The variable *weight* specifies the cost to flow goods throughout those possible routes. The input graph in this example is defined as an undirected network. In that case, proc network creates two possible directions for each link, allowing outflow and inflow through the same link. For example, the first original link (X, A) turns into two distinct links (X, A) and (A, X) when proc optnetwork is solving the VRP. The nodes definition using the NODESVAR statement specifies the store node identification and the demand for each node, or how many product deliveries each store requires. The output table defined in the OUTNODES= option saves for each node in each route it will be served and in what order. The output table defined in the OUT= option within the VRP statement saves the routes and the order of the links that will be used in each route to deliver the goods to the store nodes. The DEPOT= option in the VRP statement defines the depot node in the VRP. The CAPACITY= option in the VRP statement defines the capacity of the vehicle when delivering the goods throughout the store nodes.

```
proc optnetwork
   direction = undirected
   links = mycas.links
   nodes = mycas.nodes
   outnodes = mycas.nodesout;
   linksvar
      from = from
      to = to
      weight = weight
   ;
   nodesvar
      node = node
      lower = demand
   ;
   vrp
      depot = X
      capacity = 100
      out = mycas.routes;
run;
```

Figure 4.116 shows the output for the VRP algorithm. It shows the size of the input graph, 8 nodes and 17 links. It also states that an optimal solution was found (solution status is optimal), one single solution was found (number of solutions 1) and the objective value is 44, which represents the routing cost to deliver the demanded goods from the stores from the depot node.

The OPTNETWORK Procedure

Problem Summary

Number of Nodes	8
Number of Links	17
Graph Direction	Undirected

The OPTNETWORK Procedure

Solution Summary

Problem Type	Vehicle Routing Problem
Solution Status	Optimal
Number of Solutions	1
Objective Value	44
Relative Gap	0
Absolute Gap	0
Primal Infeasibility	0
Bound Infeasibility	0
Integer Infeasibility	0
Best Bound	44
Nodes	1
Iterations	18
CPU Time	0.01
Real Time	0.01

Figure 4.116 Output results by proc optnetwork running the vehicle routing problem algorithm based on an undirected input graph.

« NODESOUT   Table Rows: 8   Columns: 4 of 4   Rows 1 to 8

	node	demand	route	route_order
1	A	50	1	3
2	B	20	1	2
3	C	10	1	4
4	D	60	2	2
5	E	40	2	3
6	F	90	3	2
7	G	30	4	2
8	X	.	.	1

Figure 4.117 The set of store nodes and in which route and order they will be served in the vehicle routing problem solution based on an undirected graph.

The following pictures show the detailed results of the VRP generated by proc optnetwork. The NODESETOUT dataset presented in Figure 4.117 shows the variable *node* that identifies the node label, the variable *demand*, which specifies the demand for each store node, the variable *route*, which specifies what route will serve that store node, and the variable *route_order*, which specifies in what order of that route the store node will be served.

In this example, the store nodes A, B, and C will be served by route 1. The vehicle loads at the depot node X and departs to the store node B, delivering the 20 goods demanded. The order of the route is 2, as the first step (*route_order* 1) is at the depot node X. The vehicle then goes from the store node B to the store node A (*route_order* 3), delivering the 50 goods demanded, and finally serves the store node C (*route_order* 4) delivering the 10 goods demanded. After that, the vehicle returns to the depot node X to reload. Notice that the first route considers a total amount of goods as 80, 20 for node B, 50 for node A, and 10 for node C. As the capacity of the vehicle is 100, the vehicle cannot take any more goods to deliver to other store nodes. That is the reason it returns to the depot node X to reload and continues the delivering. It is important to notice here that the first loading process at the depot node X does not need to be the maximum capacity of the vehicle, 100. Based on the optimal routing, the vehicle will deliver only 80 goods throughout the 3 store nodes B, A, and C. The vehicle can therefore load only the 80 units required. The second route serves nodes D and E. At the depot node X, the vehicle reloads with 100 units. From the depot node X the vehicle goes to the store node D (*route_order* 2), delivering the 60 goods demanded. From store node D, the vehicle proceeds to the store node E (*route_order* 3) and deliver the 40 goods demanded. As these 2 store nodes demand a total of 100 units, the maximum capacity of the vehicle, the routing stops and the vehicle returns to the depot node X to reload. Now the vehicle reloads 90 units. From the depot node X, the vehicle goes to the store node F and delivers the 90 goods demanded. Due to the limit capacity of the vehicle, there are no more store nodes that can be served, and the route 3 has only one single stop (*route_order* 2 only). The vehicle goes once again to the depot node X to reload. Now it gets only 30 units, which is the demand of the last store node. The vehicle departs from the depot node X and goes to the store node G, delivering the 30 goods demanded. As node G is the last store node to be served, the route 4 has one single stop (*route_order* 2 only). As all store nodes are properly served, the vehicle can return to the depot node X, finalizing the routing.

The ROUTES dataset presented in Figure 4.118 shows the variable *route* that specifies the route taken by the vehicle to serve the store nodes, the variable *route_order* that specifies the order of the route when the vehicle deliver the goods to the store nodes, the variables *from* and *to* that represent the link taken by the vehicle at each step of the route, and finally the variable *weight* that represents the cost for taking that link.

Similar to the previous output, the ROUTES dataset shows the routes and the order of the links taken by the vehicle in order to serve all the store nodes within the network. Route 1 comprises 4 links. From the depot node X to the store node B, then from the store node B to the store node A (notice here that proc optnetwork uses the original link definition (A, B) from the links dataset, even though the direction for this link in the route is from B to A), from the store node A to the store node C, and finally from the store node C to the depot node X. Similarly, proc optnetwork uses the original link (X, C) from the links dataset, even though the direction of this link in the route is from C to X). The total cost for this route is 15, considering all the links (X, B, 5), (A, B, 3), (A, C, 3), and (X, C, 4). The route 2 comprises 3 links, from the depot node X to the store node

	route	route_order	from	to	weight
1	1	1	X	B	5
2	1	2	A	B	3
3	1	3	A	C	3
4	1	4	X	C	4
5	2	1	X	D	8
6	2	2	D	E	6
7	2	3	X	E	3
8	3	1	X	F	2
9	3	2	X	F	2
10	4	1	X	G	4
11	4	2	X	G	4

Figure 4.118 The order of the links for each route in the vehicle routing problem based on an undirected graph.

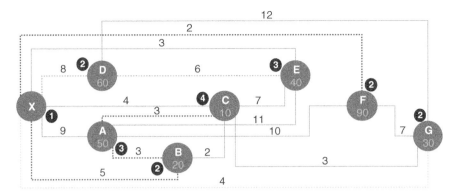

Figure 4.119 The vehicle routing problem solution generated by proc optnetwork based on an undirected graph.

D, then from the store node D to the store node E, and finally from the store node E back to the depot node X. The total cost of this route is 17, considering the links $(X, D, 8)$, $(D, E, 6,)$ and $(X, E, 3)$. Route 3 has 2 links, from the depot node X to the store node F, and then from the store node F back to the depot node X. The total cost of this route is 4, with the links $(X, F, 2)$ and $(X, F, 2)$. Finally, route 4 has also just 2 links, from depot node X to store node G and then back to depot node X. The total cost of this route is 8, with the links $(X, G, 4)$ and $(X, G, 4)$. The overall cost for the VRP is 44, as stated before, considering all 4 routes with individual costs at 15, 17, 4, and 8, respectively.

Figure 4.119 shows a possible solution for the VRP considering the input graph.

When we look at the network flow diagram, we can probably visualize that the first route would be something like, from the depot node X to the store node A, then from the store node A to the store node B, from the store node B to the store node C and finally from the store node C back to the depot node X. The total cost of this route would be 9 $(X, A, 9)$ plus 3 $(A, B, 3)$ plus 2 $(B, C, 2)$ plus 4 $(C, X, 4)$, which sums up to 16. Proc optnetwork searches for the minimum cost for all routes within the VRP. The optimal route involving nodes X, A, B and C is indeed $X \rightarrow B \rightarrow A \rightarrow C \rightarrow X$, with a total cost of 15 (instead of 16), as 5 $(X, B, 5)$ plus 3 $(B, A, 3)$ plus 3 $(A, C, 3)$ plus 4 $(C, X, 4)$.

The VRP solution provided by proc optnetwork for this small example found four distinct routings to deliver goods to all store nodes and meet the supply demands. This solution can be accomplished by using 1 single vehicle, going around three times in a sequence. For example, load at depot node X, deliver goods to the customer nodes A, B, and C, return to the depot node X to reload, deliver goods to the customer nodes D and E, return to the depot node X to reload, deliver goods to the customer node F, return to the depot node X to reload, deliver goods to the last customer node G, and finally return to the depot node X. This solution can also be accomplished by using four different vehicles. The first vehicle takes the routing for customer nodes A, B, and C, the second vehicle takes the routing for customer nodes D and E, the third vehicle takes the routing for customer node F, and the fourth vehicle takes the routing for customer node G. Of course, this solution can also be accomplished by using two or three vehicles, splitting the routings among them, and supplying the customers' demands by using the found routings in parallel.

Key information in the VRP though is the capacity of the vehicle. It changes the solution completely found by proc optnetwork. To lower the capacity, requires more trips the vehicle needs to perform to supply all the customer nodes demands. Depending on how low the maximum capacity of the vehicle is, the solution for the VRP can be infeasible, as the vehicle cannot deliver the units demanded by a particular customer node. Of course, the vehicle could make multiple trips to the same customer node until the demand is supplied, but this option is not considered by proc optnetwork. On the other hand, the higher the maximum capacity of the vehicle, the fewer trips the vehicle needs to perform to deliver all goods and fulfill the customer nodes demands. For example, an unlimited capacity for an imaginary vehicle would allow one single trip to deliver all goods to all customer nodes demands.

Let's take a look at how these options work in practice when using proc optnetwork to solve the VRP. First, we'll check on the first lower boundary extreme. The following code defines the maximum capacity for the vehicle as 89:

```
proc optnetwork
   direction = undirected
   links = mycas.links
   nodes = mycas.nodes
   outnodes = mycas.nodesout;
```

```
linksvar
   from = from
   to = to
   weight = weight
;
nodesvar
   node = node
   lower = demand
;
vrp
   depot = X
   capacity = 89
   out = mycas.routes;
run;
```

As the store node F has a demand of 90, the vehicle defined with a maximum capacity of 89 cannot deliver enough units at once to the store node F and the solution is infeasible. Figure 4.120 shows the summary result reported by proc optnetwork with the infeasible solution.

Now let's go to the other extreme and make the maximum capacity of the vehicle much higher. The following code increases the maximum capacity for the vehicle to 500:

```
proc optnetwork
   direction = directed
   links = mycas.links
   nodes = mycas.nodes
   outNodes = mycas.nodesout;
   nodesVar
      node = node
```

The OPTNETWORK Procedure

Problem Summary

Number of Nodes	8
Number of Links	17
Graph Direction	Undirected

The OPTNETWORK Procedure

Solution Summary

Problem Type	Vehicle Routing Problem
Solution Status	Infeasible
Number of Solutions	0
Best Bound	.
Nodes	0
Iterations	0
CPU Time	0.00
Real Time	0.00

Figure 4.120 The infeasible solution for the vehicle routing problem when using a low maximum capacity for the vehicle.

```
      lower = demand
  ;
  linksvar
      from = from
      to = to
      weight = weight
  ;
  vrp
      depot = 'X'
      capacity = 500
      out = mycas.routes;
run;
```

A vehicle with a maximum capacity of 500 units makes an optimal solution for the VRP possible. The outputs from proc optnetwork are shown in the following pictures. Two optimal solutions were found by proc optnetwork, and the total cost of the optimal routing is 39. Figure 4.121 shows the summary result reported by proc optnetwork with the optimal solution.

The NODESETOUT dataset presented in Figure 4.122 shows the depot node X and all store nodes, including the chosen routes and the orders within the routes.

The OPTNETWORK Procedure

Problem Summary

Number of Nodes	8
Number of Links	17
Graph Direction	Undirected

The OPTNETWORK Procedure

Solution Summary

Problem Type	Vehicle Routing Problem
Solution Status	Optimal
Number of Solutions	2
Objective Value	39
Relative Gap	0
Absolute Gap	0
Primal Infeasibility	0
Bound Infeasibility	0
Integer Infeasibility	0
Best Bound	39
Nodes	1
Iterations	17
CPU Time	0.01
Real Time	0.01

Figure 4.121 The feasible solution for the vehicle routing problem when using a high maximum capacity for the vehicle.

« NODESOUT   Table Rows: 8 | Columns: 4 of 4 | Rows 1 to 8   ↑ ↑ ↓ ↓   ↺ ▾   ⋮

▽ Enter expression   ⌕

	△ node	⊞ demand	⊞ route	⊞ route_order
1	A	50	1	3
2	B	20	1	2
3	C	10	1	4
4	D	60	1	6
5	E	40	1	7
6	F	90	2	2
7	G	30	1	5
8	X	.	.	1

Figure 4.122 The set of nodes and the demands.

« ROUTES   Table Rows: 9 | Columns: 5 of 5 | Rows 1 to 9   ↑ ↑ ↓ ↓   ↺ ▾   ⋮

▽ Enter expression   ⌕

	⊞ route	⊞ route_order	△ from	△ to	⊞ weight
1	1	1	X	B	5
2	1	2	A	B	3
3	1	3	A	C	3
4	1	4	C	G	3
5	1	5	D	G	12
6	1	6	D	E	6
7	1	7	X	E	3
8	2	1	X	F	2
9	2	2	X	F	2

Figure 4.123 The routes and the sequence of nodes.

The total demand considering all 7 store nodes are 300 units. Notice that even though the vehicle maximum capacity is 500 units, there are still 2 routes to supply the goods to all the store nodes. The ROUTES dataset presented in Figure 4.123 shows the routes.

The constraint here though is not the maximum capacity of the vehicle but the existing links within the network and its weights. In the optimal route, the vehicle departs from depot node X and covers the store nodes B, A, C, G, D, and E in one single route, returning to the depot node X with a total cost of 35. The vehicle then departs from the depot node X and goes to the store node E and returns to the depot node X in a single trip with a total cost of 4. The overall cost of the VRP is 39. The vehicle could take a route to go over all the store nodes without stopping at the depot node X to resume the delivering process. For example, the vehicle can go from X to B, A, C, E, D, G, F, and then back to X. The total cost for this route is 45 ($5 + 3 + 3 + 7 + 6 + 12 + 7 + 2$). The minimum cost to cover all the store nodes is indeed by returning at some point to the depot node X and resuming the route. If we consider that the paths from the store node E to the depot node X, from the depot node X to the store node F, and then back to the depot node X are just existing links, we would have just one single route. However, if we consider that there is a cost associated with carrying over the goods in the vehicle, we can make the overall route in 2 steps. Instead of making the vehicle depart from the depot node X with all 300 units, we can make it depart with only 210 units to deliver the goods to B, A, C, G, D, and E. As the vehicle needs to pass by the depot node X to minimize the cost of the trip, it can stop by and reload with the additional 90 units to serve the store node F. That is the reason proc optnetwork considers the overall vehicle routing with 2 routes, as the vehicle needs to go back to the depot node X to proceed

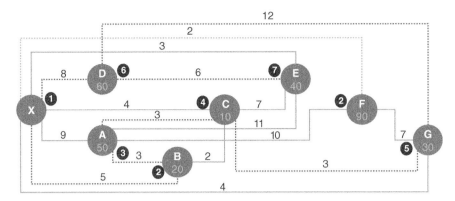

Figure 4.124 The vehicle routing problem solution with a high vehicle maximum capacity.

in the minimum cost path. If there was a link between E and F with a low-cost weight, for example, 2, the minimum cost path would likely be X, B, A, C, G, F, E, D, X, with total cost at 37. The vehicle wouldn't need to pass by the depot node X at any time but to start and end the routing.

Figure 4.124 shows the solution considering the maximum capacity for the vehicle as 500 units.

Now, let's consider the network as a directed graph. A directed graph completely changes the existing links within the network and of course, changes the possible solutions found by proc optnetwork in the VRP.

The exact same network as a direct graph is shown in Figure 4.125.

Using the same links and nodes dataset definitions, we have a directed graph with 17 links and 8 nodes. The following code invokes the VRP algorithm to run on a directed graph.

```
proc optnetwork
   direction = directed
   links = mycas.links
   nodes = mycas.nodes
   outnodes = mycas.nodesout;
   linksvar
      from = from
      to = to
      weight = weight
   ;
   nodesvar
      node = node
      lower = demand
   ;
   vrp
      depot = 'X'
      capacity = 500
      out = mycas.routes;
run;
```

The output from the proc optnetwork shows that there is no feasible solution for this problem, considering there is no sufficient existing links to allow the vehicle to depart from the depot node X and cover all the store nodes in this particular directed network. Figure 4.126 shows the summary result reported by proc optnetwork with the infeasible solution for the directed graph.

Based on the links definition, there are no links returning to the depot node X. That makes the solution infeasible. Then, let's add some links from the store nodes to the depot node X. The following code adds another 3 links to the network, creating possible paths from the store nodes E, F and G to the depot node X.

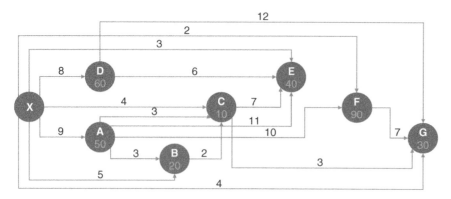

Figure 4.125 The vehicle routing problem based on a directed graph.

The OPTNETWORK Procedure

Problem Summary

Number of Nodes	8
Number of Links	17
Graph Direction	Directed

The OPTNETWORK Procedure

Solution Summary

Problem Type	Vehicle Routing Problem
Solution Status	Infeasible
Number of Solutions	0
Best Bound	.
Nodes	0
Iterations	0
CPU Time	0.00
Real Time	0.00

Figure 4.126 The solution for the vehicle routing problem with a high vehicle maximum capacity.

```
data mycas.links;
   input from $ to $ weight @@;
datalines;
X A 9   X B 5   X C 4   X D 8   X E 3   X F 2   X G 4   A B 3   A C 3
A E 11  A F 10  B C 2   C E 7   C G 3   D E 6   D G 12  F G 7
G X 8   F X 12  E X 15
;
run;
```

The new directed graph is presented in Figure 4.127.

There are now possible paths to return the vehicle to the depot node X, from the store nodes E, F, and G. We can rerun the exact same code shown previously to invoke the VRP algorithm in proc optnetwork to search for an optimal solution based on this new directed graph.

The outputs from proc optnetwork are shown in the following pictures. Figure 4.128 shows that one optimal solution was found by proc optnetwork and the total cost of the optimal routing is 68.

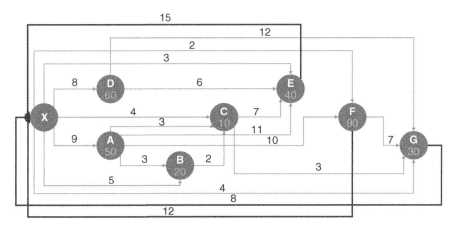

Figure 4.127 The vehicle routing problem based on a directed graph with additional links to the depot node.

The OPTNETWORK Procedure

Problem Summary

Number of Nodes	8
Number of Links	20
Graph Direction	Directed

The OPTNETWORK Procedure

Solution Summary

Problem Type	Vehicle Routing Problem
Solution Status	Optimal
Number of Solutions	1
Objective Value	68
Relative Gap	0
Absolute Gap	0
Primal Infeasibility	0
Bound Infeasibility	0
Integer Infeasibility	0
Best Bound	68
Nodes	1
Iterations	16
CPU Time	0.01
Real Time	0.01

Figure 4.128 The feasible solution for the vehicle routing problem based on a directed graph.

The NODESETOUT dataset presented in Figure 4.129 shows the depot node X and all the store nodes, including routes and the sequence of nodes within the routes.

There are three routes considered by the optimal solution. The first route serves the store nodes A, B, C, and G, supplying 50, 20, 10, and 30 units of demand, respectively. The second route serves the store nodes D and E, supplying 60 and 40 units, respectively. Finally, the third route serves the store node F, supplying 90 units. The same concept applies here. It could be

« NODESOUT   Table Rows: 8 | Columns: 4 of 4 | Rows 1 to 8 | ↑ ↑ ↓ ↓ | ↺▾ | ⋮

⊽ Enter expression   ☌

	⚠ node	⊕ demand	⊕ route	⊕ route_order
1	A	50	1	2
2	B	20	1	3
3	C	10	1	4
4	D	60	2	2
5	E	40	2	3
6	F	90	3	2
7	G	30	1	5
8	X	.	.	1

Figure 4.129   The set of nodes, the routes, and the sequence of nodes within the routes.

« ROUTES   Table Rows: 10 | Columns: 5 of 5 | Rows 1 to 10 | ↑ ↑ ↓ ↓ | ↺▾ | ⋮

⊽ Enter expression   ☌

	⊕ route	⊕ route_order	⚠ from	⚠ to	⊕ weight
1	1	1	X	A	9
2	1	2	A	B	3
3	1	3	B	C	2
4	1	4	C	G	3
5	1	5	G	X	8
6	2	1	X	D	8
7	2	2	D	E	6
8	2	3	E	X	15
9	3	1	X	F	2
10	3	2	F	X	12

Figure 4.130   The routes and the sequence of nodes.

one single vehicle covering all the store nodes by doing three trips, or it could be three vehicles covering all the store nodes by doing one single trip each.

The ROUTES dataset presented in Figure 4.130 shows the routes and the sequence of links performed by the vehicle when delivering the goods to all the store nodes. For a directed graph, this output dataset is particularly important for the VRP solution because it shows the real links traveled by the vehicle, considering the actual direction of the links. Remember that for the undirected graph, even though we know that the vehicle goes from node B to node A in the optimal routing solution, the link presented in the output dataset is the link $(A, B, 3)$, because this is the original link defined in the links dataset.

The vehicle loads 110 units in the depot node X. It goes to the store node A and delivers 50 units at the cost of 9 $(X, A, 9)$. Then the vehicle goes to the store node B and delivers 20 units at the cost of 3 $(A, B, 3)$. It progresses on the routing and goes to the store node C and delivers 10 units, at the cost of 2 $(B, C, 2)$. Finally, the vehicle goes to the store node D and delivers the last 30 units, at the cost of 3 $(C, G, 3)$. The vehicle then returns to the depot node X at the cost of 8 $(G, X, 8)$. The total cost for this route is 25. On the second trip, the vehicle loads 100 units at the depot node X and departs for the store node D, delivering 60 units at the cost of 8. $(X, D, 8)$. Then the vehicle goes to the store node E and delivers 40 units at the cost of 6 $(D, E, 6)$. The vehicle then returns to the depot node X at the cost of 15 $(E, X, 15)$. The total cost for this route is 29. Finally, the vehicle loads 90 units and goes straight to the store node F, delivering those goods at the cost of 2 $(X, F, 2)$. The vehicle then returns to the depot node X at the cost of 12 $(F, X, 12)$. This route has a total cost of 14. The overall cost for the optimal solution is 68 $(25 + 29 + 14)$.

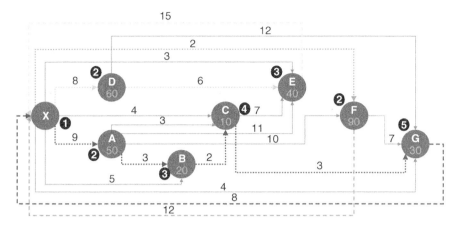

Figure 4.131 The vehicle routing problem solution based on a directed graph.

Figure 4.131 shows the optimal solution considering the directed graph.

Once again, the same concepts of multiple vehicles and multiple trips applies to the directed graph. Most of the time, real-world problems are based on directed graphs, as physical roads (streets, avenues, highways, etc.) not always having two-way directions. Many real-world problems in the VRP are much more complex than was described here. Those problems can include time window constraints, where the goods need to be delivered in a certain time frame. Deliveries out of the time window can incur penalties. Groceries delivered by the major supermarket chains represent real-life scenarios like this. Other problems can involve pickup and delivery constraints. Routes need to be dynamically optimized allowing drivers to pick up and deliver goods. Food delivery companies represent real-life scenarios like this. Capacity constraint is also common in real-world problems. Multiple vehicles have different capacities, and these differences must be considered when optimizing the vehicles routings. Sometimes vehicles also have constraints on what they can transport, like heavy equipment, frozen food, delicate goods, etc.

As we can see, there are many variants and specializations of the VRP to address real-world business problems. And to make matters worse, most of the life-scenarios happen with a few combinations of the specializations described previously. The VRP includes manual solving and optimization solvers, sometimes working together with heuristics approaches to narrow down the number of possible solutions based on the size of the problem (depot/store/customer nodes, links/paths/routes, resources, constraints, etc.). The VRP is a field in optimization that continuously evolve, with many applications in different industries and businesses.

4.14 Topological Sort

A topological sort or topological ordering of a directed graph is a linear ordering of its nodes, where for every directed link (i, j) from node i to node j, node i comes before node j in that ordering. A classic example is when tasks are considered to be represented by a graph. Each node within the graph represents a task, and the links between the nodes represent some constraints associated with the order of the tasks. For example, in a brewing process, simplistically, the first step is the mashing process, the second step is the boiling process, and finally, the third step is the fermentation process. In a graph representation, mashing can be represented by node i, boiling can be represented by node j, and fermentation can be represented by node k. The links between the nodes can represent the sequence in which the tasks need to be performed. The mashing process needs to be completed before the boiling process starts. A directed link from node i to node j can represent that ordering. The fermentation process can only start after the boiling process is completed. Then, a directed link from node j to node k can represent that mandatory sequence. Figure 4.132 illustrates that sequence.

A simple directed graph to represent the tasks associated with the nodes i, j, and k, and the ordering between them is shown in Figure 4.133.

In this example, the topological sort is just a valid sequence for the tasks. Formally, a topological sort is a graph traversal in which each node n is visited only after all its dependencies are visited, or all other previous nodes are visited. A topological

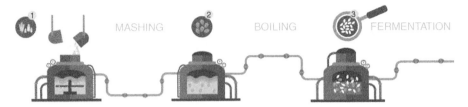

Figure 4.132 The first steps of the brewing process (USA Hops).

Figure 4.133 The directed graph representing task and their ordering.

sort cannot happen if the directed graph has cycles. This property is commonly referred to as a directed acyclic graph, or simply DAG. Any DAG has at least one topological sort.

The most common application of topological sort is to schedule a sequence of jobs or tasks based on their dependencies. The tasks are represented by nodes and the dependencies are represented by links. The directed link connecting the nodes defines the dependency or ordering of the tasks. The task in the origin node must be completed before the task in the destination node can be started. The topological sort gives the order in which tasks must be performed.

In proc optnetwork, the TOPOLOGICALSORT (or TOPSORT) statement invokes the algorithm that calculates a topological ordering of the nodes of a directed acyclic input graph.

The final result for the topological sort algorithm in proc optnetwork is reported and saved in an output table defined by the OUTNODES= option. The output table contains the variable *node* that identifies the node in the topological sort, and the variable *top_order* that identifies the order of the node in the topological sort.

4.14.1 Finding the Topological Sort in a Directed Graph

Let's consider the graph presented in Figure 4.134 to demonstrate the topological sort problem using proc optnetwork. The directed graph describes a network with just 8 nodes and only 10 links.

The following code describes how to create the input dataset for the topological sort scenario. A problem involving a topological sort requires only the links dataset. The links dataset represents both the tasks as origin and destination nodes, and the dependencies. The origin nodes represent the tasks that need to be performed and completed before the tasks represented by the destination nodes are executed. The links don't need to have weights as only the sequence of the nodes is taken into consideration when calculating the ordering of the topological sort.

```
data mycas.links;
    input from $ to $ @@;
datalines;
A B   A C   A D   C B   E B   E F   D F   D H   B G   G H
;
run;
```

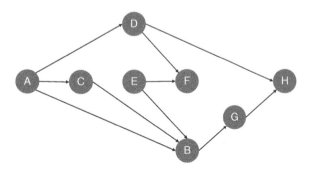

Figure 4.134 Input graph with directed links.

The OPTNETWORK Procedure

Problem Summary

Number of Nodes	8
Number of Links	10
Graph Direction	Directed

The OPTNETWORK Procedure

Solution Summary

Problem Type	Topological Sort
Solution Status	OK
CPU Time	0.00
Real Time	0.00

Output CAS Tables

CAS Library	Name	Number of Rows	Number of Columns
CASUSER(Carlos.Pinheiro@sas.com)	NODESETOUT	8	2

Figure 4.135 Output results by proc optnetwork running the topological sort algorithm.

Once the links dataset is created, we can now invoke the topological sort algorithm using proc optnetwork. The following code describes how to do it. The links are defined in proc optnetwork by using the LINKSVAR statement. No link weights are needed. Therefore, just the *from* and *to* variables are used to define origin and destination nodes in terms of dependencies. The network needs to be defined as a directed graph.

```
proc optnetwork
   direction = directed
   links = mycas.links
   outnodes = mycas.nodesetout
   ;
   linksvar
      from = from
      to = to
   ;
   topologicalsort
   ;
run;
```

Figure 4.135 shows the output for the topological sort algorithm. It says that a solution was found, as the solution status is ok.

Figure 4.136 presents the results for the topological sort problem. The OUTNODES dataset shows the variable *node* to identify the task, and the variable *top_order* to identify the order of the task in the topological sort.

Based on the solution provided by proc optnetwork, the task represented by the node E needs to be performed first. Then, the task represented by the node A can be executed. Following the task A, the task D can be performed. Then, the task F, the task C, the task B, then task G, and finally the task H. Figure 4.137 shows graphically the final solution in terms of the tasks ordering.

Notice that we can perform task A, and from there, execute tasks B, C, and D. Once task D is completed, we could perform task F. However, task F can only be executed when both tasks D and E are completed. Then, even if we execute task A, we still need to execute task E to run task F. For all other tasks, including G and H, task A is the predecessor. But again, there is

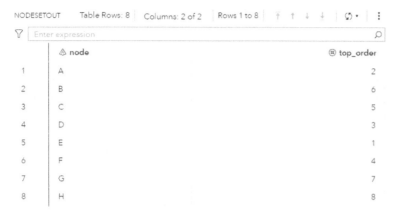

Figure 4.136   The topological sort solution.

Figure 4.137   Solution for the topological sort.

no benefit to execute task A first, allowing all other tasks to be performed in a sequence, if task E is not completed. This is the reason that task E is the very first task in the topological sort.

We can also think about the topological sort in terms of the terminal nodes, and then go backwards in the dependencies. In this example, we have 2 terminal nodes, tasks F and H. Let's take the first terminal node F. In order to perform task F, we need to complete tasks E and D. Task E has no dependencies. In order to perform task D, we need to complete task A. Task A has no dependencies. That ordering could be E → A → D → H. Let's take the second terminal node H. In order to perform task H, we need to complete tasks D and G. The order for task D is already defined, A → D. In order to perform task G, we need to complete task B. In order to perform task B, we need to complete tasks A, C, and E. Task E has no dependencies. In order to perform task C, we need to complete task A. That ordering could be E → A → C → B → G → H. Combining both paths, we have E → A → D → F → C → B → G → H.

4.15   Summary

Network optimization is a fundamental tool to explore, understand, and search for optimal solutions in network flow problems. In several industries, many real-world problems can benefit from network optimization algorithms.

There is an endless number of problems that can be represented by a network or a graph. For all those problems, most of the solutions pass by the algorithms described in this network optimization chapter. These solutions are under the field of combinatorial optimization. These solutions are in the form of algorithms.

Network optimization is a branch of mathematics focusing on the study of finite or countable objects, like existence, enumeration, structure, and so forth.

As we can see, there are indeed an endless number of business problems that can be solved by using network optimization algorithms.

This chapter presented many different concepts about network optimization, and how these concepts can be applied to solve real-world problems.

Cliques have multiple applications in industries like bioinformatics, social sciences, electrical engineering, and chemistry, among many others. Cliques can be used in chemistry to find chemicals that match a target structure. The analysis of transportation systems and traffic routes can also benefit from the concept of cliques. Even analytical models in banking, insurance, and money laundering can use outcomes provided by clique analysis.

Cycles can also be used in chemistry, as well as in biology, scheduling, and even in surface reconstruction. Social sciences, computer networks, and logistics are common candidates for cycle enumeration use when solving real-world problems. Like clique enumeration, cycles can also be used in transportation systems and as part of a broader set of analytical models to find anomaly transactions in banking, insurance, and mostly money laundering.

Linear assignment can be used to find the optimal match for available agents and tasks that need to be performed. It can be used to minimize the time to complete the set of tasks by using the agents available. Or it can be used to maximize the profit by producing the right products based on the cost of their parts.

Minimum-cost network flow can be used to find the best flow to deliver good from one or multiple points to the rest of the network. Commonly used in supply chain, the minimum-cost network flow problem is aimed to find the cheapest possible way of sending a certain amount of flow through a network. This problem is very useful for a set of real-life situations involving networks with associated costs and flows to be sent through. The minimum-cost network flow algorithm is commonly applied in telecommunications networks, energy networks, and computer networks.

Maximum network flow can be used in airline scheduling or image segmentation. Maximum flow problems involve finding a feasible flow throughout a flow network that obtains the maximum possible flow rate.

Minimum cut can also be used in image segmentation, as well as in telecommunications and computer networks. The minimum cut problem has many applications in areas such as network design.

Minimum spanning tree can be used for clustering networks and to find strengths and weaknesses in electrical grids and computer and communications networks. The minimum spanning tree problem aims to identify the minimum number of links that connect the entire network. The minimum spanning tree problem has many applications in the design of networks. The minimum spanning tree algorithm can be applied in many different industries like telecommunications, computers, transportation, supply chain, and electrical grids, among many others.

Path can be used to find possible interconnectivity in any type of network. Paths are fundamental concepts in graph theory and have many applications in graphs describing communications network, power distribution, transportation, and computer networks, among many others.

Shortest path can be used to find the minimum cost to traverse any type of network considering the existing paths. There is a set of real-life problems that can be addressed by the shortest path, such as road networks, public transportation planning, traffic planning, sequence of choices in decision making, minimum delay in telecommunications, operations research, and layout facility, robotics, among many others.

Transitive closure can be used to find reachability in any type of network and can be thought of as a concept about reachability. Transitive closure is commonly applied to structure query languages in database management systems in order to reduce the query processing time. It is also applied in problems involving reachability analysis of transition networks in distributed systems, among other applications in computer science.

TSP can be used to find optimal tours, and has applications in many areas like planning, logistics, routing, manufacturing, computer networks, astronomy, and DNA sequencing, among others.

The VRP can be used to find the best dispatch for goods from depots to customers. The VRP in real-world applications can be highly complex involving factors like the number and locations of all possible stops on the routes, arrival and departure time gaps, effective loading, vehicle limitations, time windows, loading processes at depots, multiple trips, inventory limitations, multiples depots where vehicles can start and end, among others. The list of possible constraints is practically endless.

Finally, topological sort can be used to find the sequence to execute tasks based on their dependencies. The most common application of topological sort is to schedule a sequence of jobs or tasks based on their dependencies. It can also be applied to detect deadlocks in operation systems, sentence orderings, critical path analysis, course schedule problems, and in manufacturing workflows, data serialization, and context-free grammar.

5

Real-World Applications in Network Science

5.1 Introduction

This chapter presents some use cases where network analysis and network optimization algorithms were used to solve a particular problem, or at least were used as part of an overall solution. The first two cases were created to demonstrate some of the network optimization capabilities, providing practical examples on using network algorithms to understand a particular scenario. The first case explores multiple network algorithms to understand the public transportation system in Paris, France, its topology, reachability, and specific characteristics of the network created by tram, train, metro, and bus lines. On top of that, the case presents a comparison on the Traveling Salesman (TSP) Problem in Paris by using the multimodal public transportation and by just walking. A tourist wants to visit a set of places and can do it by walking or by using the transportation lines available. The case offers an overall perspective about distance and time traveled.

The second case covers the vehicle routing problem (VRP). Similar to the first case on the (TSP), the main idea is to demonstrate some crucial features of this important network optimization algorithm. The city chosen for this demonstration is Asheville, North Carolina, U.S. The city is known to have one of the highest ratios of breweries per capita in the country. The case demonstrates how a brewery can deliver beer kegs demanded by its customers (bars and restaurants), considering the available existing routes, one or multiple vehicles, and different vehicle capacities.

The third case is a real use case developed at the beginning of the COVID-19 pandemic in early 2020. In this case, network analysis and network optimization algorithms play a crucial part of the overall solution to identify new outbreaks. Network algorithms were used to understand population movements along geographic locations, create correlation factors to virus spread, and ultimately feed supervised machine learning models to predict the likelihood of COVID-19 outbreaks in specific regions.

Partially similar to the third case, the fourth one describes the overall urban mobility in a metropolitan area and how to use that outcome to improve distinct aspects of society such as identifying the vectors of spread diseases, better planning transportation networks, and strategically defining public surveillance tasks, among many others. Network analysis and network optimization algorithms play an important role in this case study in providing most of the outcomes used in further advanced statistical analysis.

The remaining cases focus more on the network analysis algorithms, looking into multiple types of subgraph detection, centrality metrics calculation, and ultimately correlation analysis between those outcomes from some specific business events like churn, default, fraud, or consumption and the customer's profile in terms of social structures. The fifth case for instance focuses on the analysis of exaggeration in auto insurance. Based on the insurance transactions, a network is built associating all actors within the claims, like policy holders, drivers, witnesses, etc., and then connecting all similar actors throughout the claims within a particular period of time to build up the final social structure. Upon the final network, a series of subgroup algorithms are executed along a set of network centrality metrics. Finally, an outlier analysis on the different types of subgroups as well as on the centralities are performed in order to highlight suspicious activities on the original claims.

The sixth case describes the influential factors in business events like churn and product adoption within a communications company. Based on a correlation analysis over time between business events and the subscribers' profile in terms of network metrics, an influential factor is calculated to estimate the likelihood of customers affecting their peers on events such as churn and product adoption afterwards.

The seventh and last case study describers a methodology to detect fraud in telecommunications based on a combination of multiple analytical approaches. The first step is to create the network based on the mobile data. The second stage is to

Network Science: Analysis and Optimization Algorithms for Real-World Applications, First Edition. Carlos Andre Reis Pinheiro.
© 2023 John Wiley & Sons, Inc. Published 2023 by John Wiley & Sons, Inc.

detect the communities within the overall network. Then a set of network centralities is computed for each node considering their respective community. The average value for all these metrics is calculated and an outlier analysis is performed on them, highlighting the unusual communities in terms of relationships and connections between its members.

5.2 An Optimal Tour Considering a Multimodal Transportation System – The Traveling Salesman Problem Example in Paris

A good public transportation system is crucial in developing smart cities, particularly in great metropolitan areas. It helps to effectively flow the population in commuting and allow tourists to easily move around the city. Public transportation agencies around the globe open and share their data for application development and research. All these data can be used to explore the network topology in order to optimize tasks in terms of public transportation offerings. Network optimization algorithms can be applied to better understand the urban mobility, particularly based on a multimodal public transportation network.

Various algorithms can be used to evaluate and understand a public transportation network. The Minimum Spanning Tree algorithm can reveal the most critical routes to be kept in order to maintain the same level of accessibility in the public transportation network. It basically identifies which stations of trains, metros, trams, and buses need to be kept in order to maintain the same reachability of the original network, considering all the stations available. The Minimum-Cost Network Flow can describe an optimal way to flow the population throughout the city, allowing better plans for commuting based on the available routes and its capacities, considering all possible types of transportation. The Path algorithm can reveal all possible routes between any pair of locations. This is particularly important in a network considering a multimodal transportation system. It gives the transportation authorities a set of possible alternate routes in case of unexpected events. The Shortest Path can reveal an optimal route between any two locations in the city, allowing security agencies to establish faster emergency routes in case of special situations. The Transitive Closure algorithm identifies which pairs of locations are joined by a particular route, helping public transportation agencies to account for the reachability in the city.

Few algorithms are used to describe the overall topology of the network created by the public transportation system in Paris. However, the algorithm emphasized here is the TSP. It searches for the minimum-cost tour within the network. In my case, the minimum cost is based on the distance traveled, considering a set of locations to be visited, based on all possible types of transportations available in the network. Particularly, the cost is the walking distance, which means we will try to minimize the walking distance in order to visit all the places we want.

Open public transportation data allows both companies and individuals to create applications that can help residents, tourists, and even government agencies in planning and deploying the best possible public services. In this case, we are going to use the open data provided by some transportation agencies in Paris (RAPT – Régie Autonome des Transports Parisiens – and SNCF – Société Nacionale des Chemins de fer Français).

In order to create the public transportation systems in terms of a graph, with stations as nodes, and the multiple types of transport as links (between the nodes or stations), the first step is to evaluate and collect the appropriate open data provided by these agencies and to create the transportation network. This particular data contains information about 18 metro lines, 8 tram lines, and 2 major train lines. It comprises information about all lines, stations, timetables, and coordinates, among many others. This data was used to create the transportation network, identifying all possible stops and the sequence of steps performed using the public transportation system while traveling throughout the city.

In addition to some network optimization algorithms to describe the overall public transportation topology, we want to particularly show the TSP algorithm. We selected a set of 42 places in Paris, a hotel in Les Halles, that works as the starting and ending point, and 41 places of interest, including the most popular tourist places in Paris, and some popular cafes and restaurants. That is probably the hardest part in this problem, to pick only 41 cafes in Paris! In this case study, we are going to execute mainly two different optimal tours. The first one is just by walking. It is a beautiful city, so nothing is better than just walking through the city of lights. The second tour considers the multimodal public transportation system. This tour actually does not help us to enjoy the wonderful view, but it surely helps us enjoy longer visits to the cafes, restaurants, and monuments.

The first step is to set up the 42 places based on x (latitude) and y (longitude) coordinates. The following code shows how to create the input data table with all places to visit.

```
data places;
   length name $20;
   infile datalines delimiter=",";
   input name $ x y;
datalines;
Novotel,48.860886,2.346407
Tour Eiffel,48.858093,2.294694
Louvre,48.860819,2.33614
Jardin des Tuileries,48.86336,2.327042
Trocadero,48.861157,2.289276
Arc de Triomphe,48.873748,2.295059
Jardin du Luxembourg,48.846658,2.336451
Fontaine Saint Michel,48.853218,2.343757
Notre-Dame,48.852906,2.350114
Le Marais,48.860085,2.360859
Les Halles,48.862371,2.344731
Sacre-Coeur,48.88678,2.343011
Musee dOrsay,48.859852,2.326634
Opera,48.87053,2.332621
Pompidou,48.860554,2.352507
Tour Montparnasse,48.842077,2.321967
Moulin Rouge,48.884124,2.332304
Pantheon,48.846128,2.346117
Hotel des Invalides,48.856463,2.312762
Madeleine,48.869853,2.32481
Quartier Latin,48.848663,2.342126
Bastille,48.853156,2.369158
Republique,48.867877,2.363756
Canal Saint-Martin,48.870834,2.365655
Place des Vosges,48.855567,2.365558
Luigi Pepone,48.841696,2.308398
Josselin,48.841711,2.325384
The Financier,48.842607,2.323681
Berthillon,48.851721,2.35672
The Frog & Rosbif,48.864309,2.350315
Moonshiner,48.855677,2.371183
Cafe de lIndustrie,48.855655,2.371812
Chez Camille,48.84856,2.378099
Beau Regard,48.854614,2.333307
Maison Sauvage,48.853654,2.338045
Les Negociants,48.837129,2.351927
Les Cailloux,48.827689,2.34934
Cafe Hugo,48.855913,2.36669
La Chaumiere,48.852816,2.353542
Cafe Gaite,48.84049,2.323984
Au Trappiste,48.858295,2.347485
;
run;
```

The following code shows how to use SAS data steps to create a HTML file in order to show all the selected places on a map. The map is actually created based on an open-source package called Leaflet. The output file can be opened in any browser to show the places in a geographic approach. All the following maps are created based on the same approach, using multiple features of LeafLet.

```
data _null_;
   set places end=eof;
   file arq;
   length linha $1024.;
   k+1;
   if k=1 then
      do;
         put '<!DOCTYPE html>';
         put '<html>';
         put '<head>';
         put '<title>SAS Network Optimization</title>';
         put '<meta charset="utf-8"/>';
         put '<meta name="viewport" content="width=device-width, initial...';
         put '<link rel="stylesheet" href="https://unpkg.com/leaflet@1.5...';
         put '<script src="https://unpkg.com/leaflet@1.5.1/dist/leaflet....';
         put '<style>body{padding:0;margin:0;}html,body,#mapid{height:10...';
         put '</head>';
         put '<body>';
         put '<div id="mapid"></div>';
         put '<script>';
         put 'var mymap=L.map("mapid").setView([48.856358, 2.351632],14);';
         put 'L.tileLayer("https://api.tiles.mapbox.com/v4/{id}/{z}/{x}/...';
      end;
   linha='L.marker(['||x||','||y||']).addTo(mymap).bindTooltip("'||name...
   if name = 'Novotel' then
      do;
         linha='L.marker(['||x||','||y||']).addTo(mymap).bindTooltip("'|...
         linha='L.circle(['||x||','||y||'],{radius:75,color:"'||'blue'||...
      end;
   put linha;
   if eof then
      do;
         put '</script>';
         put '<body>';
         put '</html>';
      end;
run;
```

Figure 5.1 shows the map with the places to be visited by the optimal tour.

Considering the walking tour first, any possible connection between two places is a link between them. Based on that, there are 1722 possible steps (or links) to be considered when searching for the optimal tour. It is important to notice that here we are not considering existing possible routes. The distance between a pair of places, as well as the route between them, is just the Euclidian distance between the two points on the map. The next case study will consider the existing routes. The reason to do that in this first case is that we want to compare the walking tour with the multimodal public transportation tour. The transportation network does not follow the existing routes (streets, roads, highways, etc.), but it considers almost a straight line between the stations, similar to the Euclidian distance. To make a fair comparison between walking or taking the public transportation, we consider all distances as Euclidian distances instead of routing distances.

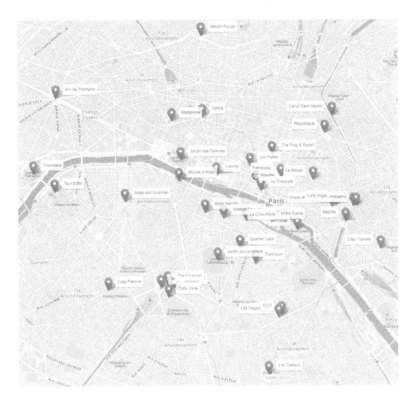

Figure 5.1 Places to be visited in the optimal tour in Paris.

The following code shows how to create the possible links between all pairs of places.

```
proc sql;
   create table placeslinktmp as
      select http://a.name as org, a.x as xorg, a.y as yorg, http://b.name as dst,
             b.x as xdst, b.y as ydst
         from places as a, places as b;
quit;
```

Figure 5.2 shows the possible links within the network.

Definitely the number of possible links makes the job quite difficult. The TSP is the network optimization algorithm that searches for the optimal sequence of places to visit in order to minimize a particular objective function. The goal may be to minimize the total time, the total distance, or any type of costs associated with the tour. The cost here is the total distance traveled.

As defined previously, the distance between the places is computed based on the Euclidian distance, rather than the existing road distance. The following code shows how to compute the Euclidian distance for all links.

```
data mycas.placesdist;
   set placeslink;
   distance=geodist(xdst,ydst,xorg,yorg,'M');
   output;
run;
```

Once the links are created, considering origin and destination places, and the distance between these places, we can compute the optimal tour to visit all places by minimizing the overall distance traveled.

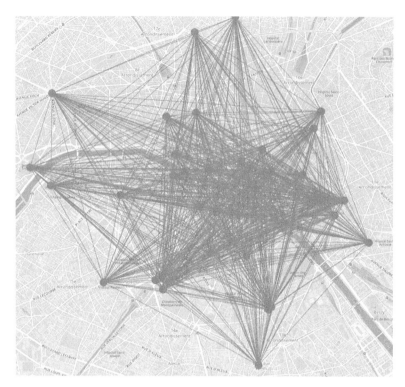

Figure 5.2 Links considered in the network when searching for the optimal tour.

```
proc optnetwork
   direction = directed
   links = mycas.placesdist
   out_nodes = mycas.placesnodes;
   linksvar
      from = org
      to = dst
      weight = distance;
   tsp
      cutstrategy = none
      heuristics = none
      milp = true
      out = mycas.placesTSP;
run;
```

Figure 5.3 shows the optimal tour, or the sequence of places to be visited in order to minimize the overall distance traveled.

The best tour to visit those 41 locations, departing from and returning to the hotel, requires 19.4 miles of walking. This walking tour would take around six hours and 12 minutes.

A feasible approach to reduce the walking distance is to use the public transportation system. In this case study, this transportation network considers train, tram, and metro. It could also consider the buses, but it is not the case here. The open data provided by the RATP is used to create the public transportation network, considering 27 lines (16 metro lines, 9 tram lines out of the existing 11, and 2 train lines out of the existing 5), and 518 stations.

The code to access the open data and import into SAS is suppressed here. We can basically use two approaches to fetch the open data from RATP. The first one is downloading the available datasets and importing them into SAS. The second one is by using the available APIs to fetch the data automatically. SAS has procedures to perform this task without too much work. Once the data representing the transportation network is loaded, we can create the map showing all possible routes within the city, as shown in Figure 5.4.

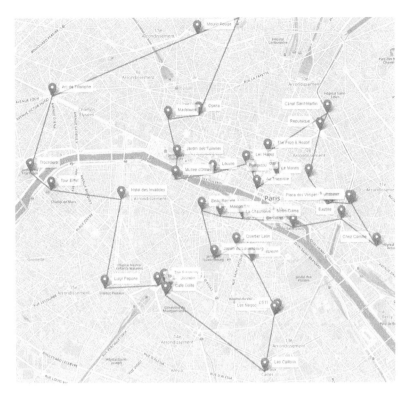

Figure 5.3 Links considered in the network when searching for the optimal tour.

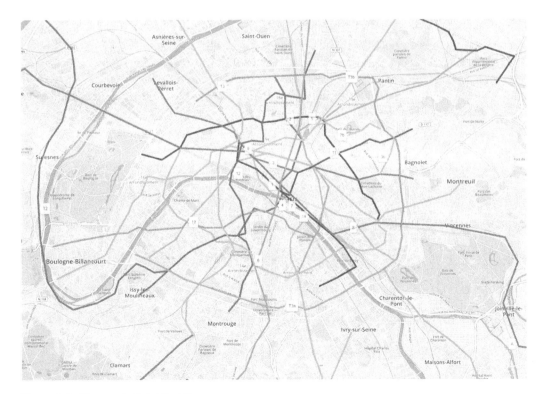

Figure 5.4 Transportation network.

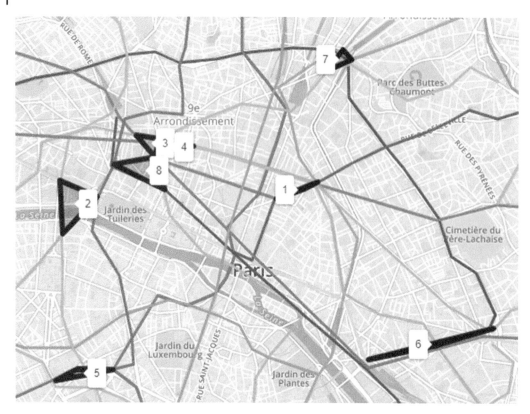

Figure 5.5 Cliques within the transportation network.

Before we move to the optimal tour considering the multimodal transportation systems, let us take a look at some other network optimization algorithms in order to describe the network topology we are about to use.

An option to find alternative stations in case of outage in the transportation lines is to identify stations that can be reached from each other, as a complete graph. The Clique algorithm can identify this subgraph within the network. To demonstrate the outcomes, the Clique algorithm was executed with a constraint of a minimum of three stations in the subgraphs, and eight distinct cycles were found. These subgraphs represent groups of stations completely connected to each other, creating a complete graph. Figure 5.5 shows the cliques within the transportation network.

The following code shows how to search for the cliques within the transportation network.

```
proc optnetwork
   direction = undirected
   links = mycas.metroparislinks;
   linksvar
      from = org
      to = dst
      weight = dist;
   clique
      maxcliques = all
      minsize = 3
      out = mycas.cliquemetroparis;
run;
```

The Cycle algorithm with a constraint of a minimum of 5 and a maximum of 10 stations found 18 cycles, which means groups of stations that creates a sequence of lines starting from one specific point and ending at that same point. Figure 5.6 shows the cycles within the transportation network.

The following code shows how to search for the cycles within the transportation network.

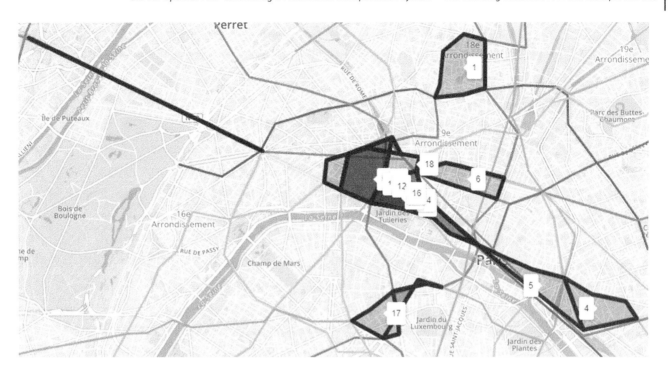

Figure 5.6 Cycles within the transportation network.

```
proc optnetwork
   direction = directed
   links = mycas.metroparislinks;
   linksvar
      from = org
      to = dst
      weight = dist;
   cycle
      algorithm = build
      maxcycles = all
      minlength = 5
      maxlength = 10
      out = mycas.cyclemetroparis;
run;
```

The Minimum Spanning Tree algorithm found that 510 stations can keep the same reachability from all of the 620 existing stations in the transportation network. The minimum spanning tree actually finds the minimum set of links to keep the same reachability of the original network. The stations associated with these links are the ones referred to previously. It is not a huge savings, but this difference tends to increase as the network gets more complex (for example by adding buses). Figure 5.7 shows all necessary pairs of stations, or existing links, to keep the same reachability of the original public transportation network. The thicker lines in the figure represent the links that must be kept maintaining the same reachability of the original network.

The following code shows how to search for the minimum spanning tree within the transportation network.

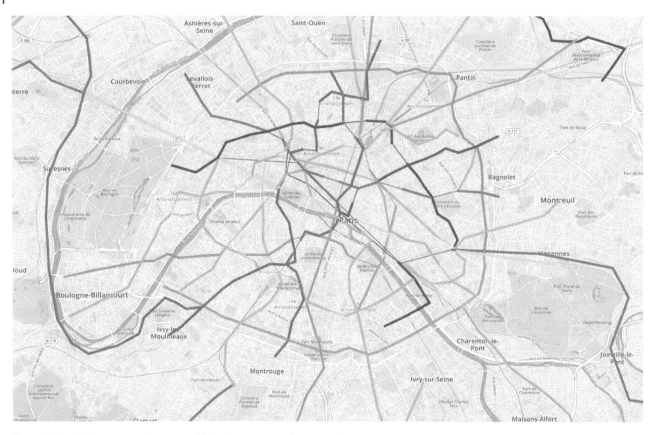

Figure 5.7 Minimum spanning tree within the transportation network.

```
proc optnetwork
   direction = directed
   links = mycas.metroparislinks;
   linksvar
      from = org
      to = dst
      weight = dist;
   minspantree
      out = mycas.minspantreemetroparis;
run;
```

The transportation network in Paris is a dense network. It holds a great number of possible routes to connect a pair of locations. For example, considering two specific locations like Volontaires and Nation, there are 7875 possible routes to connect them, even considering a constraint of 20 stations as a maximum length for the route. All possible paths between Volontaires and Nation can be found by running the Path algorithm and they are presented in Figure 5.8. In the figure we cannot see all of these multiple lines representing all the paths as most of steps use at least partially the same lines between stations. There are multiple interchange stations in the transportation network, and one single different step between stations can represent a whole new path. For that reason, there are a huge number of possible paths between a pair of locations like Volontaires and Nation.

The following code shows how to search for the paths between two locations within the transportation network.

```
proc optnetwork
   direction = directed
   links = mycas.metroparislinks;
   linksvar
      from = org
      to = dst
```

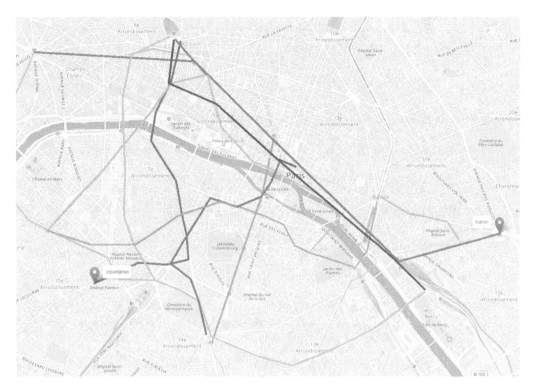

Figure 5.8 Paths within the transportation network between two locations.

```
    weight = dist;
  path
     source = Volontaires
     sink = Nation
     maxlength = 20
     outpathslinks = mycas.pathmetroparis;
run;
```

As there are many possible paths between any pair of locations, it is good to know which one is the shortest. The Shortest Path algorithm reveals important information about the best route between any pair of locations, which improves the urban mobility within the city. For example, between Volontaires and Nation, there are paths varying from 11 steps to 20 (set as a constraint in the Path algorithms), ranging from 4.8 to 11.7 miles. The shortest path considers 16 steps summarizing 4.8 miles. Figure 5.9 shows the shortest path between Volontaires and Nation. Notice that the shortest path considers multiple lines in order to minimize the distance traveled. Perhaps changing lines that much increases the overall time to travel from one location to another. If the goal is to minimize the total time of travel, information about departs and arrivals would need to be added.

The following code shows how to search for the shortest paths between two locations within the transportation network.

```
proc optnetwork
   direction = directed
   links = mycas.metroparislinks;
   linksvar
      from = org
      to = dst
      weight = dist;
   shortestpath
      source = Volontaires
      sink = Nation
      outpaths = mycas.shortpathmetroparis;
run;
```

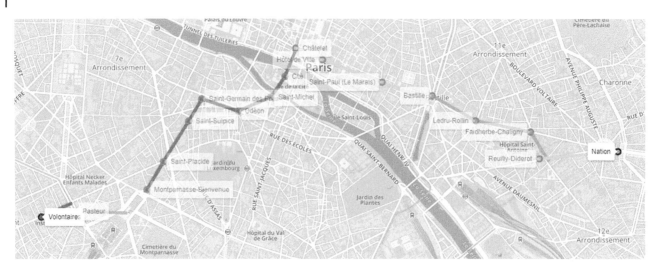

Figure 5.9 The shortest path between two locations within the transportation network.

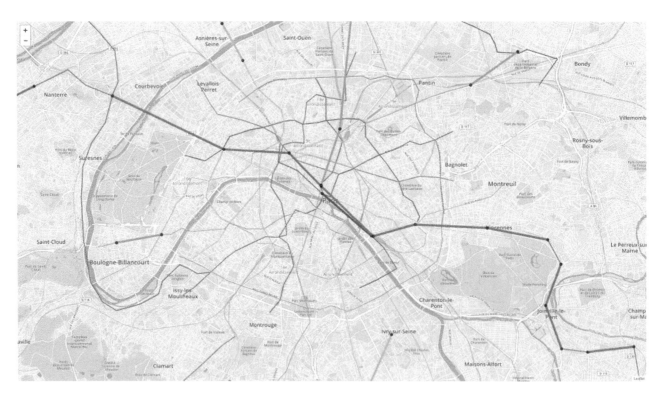

Figure 5.10 The Pattern Match within the transportation network.

Finally, the Pattern Match algorithm searches for a pattern of interest within the network. For instance, one possible pattern is the distance between any two consecutive stations in the same line. Out of the 620 steps between stations in the same line, there are 44 steps over one mile of distance, as shown in Figure 5.10.

The following code shows how to create the query graph and then execute the pattern matching within the transportation network.

```
data mycas.pmmetroparislinks;
   set mycas.metroparislinks;
   if dist ge 1 then
      weight = 1;
   else
      weight = 0;
run;
proc sql;
   select count(*) from mycas.pmmetroparislinks where weight eq 1;
quit;
data mycas.linksquery;
   input from to weight;
datalines;
1 2 1
;
run;
proc network
   direction = undirected
   links = mycas.pmmetroparislinks
   linksquery = mycas.linksquery;
   linksvar
      from = org
      to = dst
      weight = weight;
   linksqueryvar
      vars = (weight);
   patternmatch
      outmatchlinks = mycas.matchlinksmetroparis;
run;
```

Now we have a better understanding of the transportation network topology, and we can enhance the optimal tour considering the multimodal transportation system, by adding the public lines options to the walking tour. A simple constraint on this new optimal tour is that the distance we need to walk from two points of interest and their respective closest stations should be less than the distance between those two points. For example, if we want to go from place A to place B. If the total distance of walking from the origin place A to the closest station to A plus from the closest station to B to the destination place B is greater than the distance to just walk from A to B, there is no reason to take the public transportation. We go by walking. If that distance is less than the distance from A to B, then we take the public transportation. Remember we are minimizing the walking distance, not the overall distance neither time.

In order to account for this small constraint, we need to calculate the closest station to each point of interest we want to visit. What stations serves those places we want to go, and which one is the closest. For each possible step in our best tour, we need to verify if it is better to walk or to take the public transportation.

The following code shows how to find the closest stations for each place. We basically need to calculate the distance between places and places, places and stations, and stations and stations. Then we order the links and pick the pair with the shortest distance. Figure 5.11 shows the locations and their closest stations on the map.

```
proc sql;
    create table placesstations as
        select http://a.name as place, a.x as xp, a.y as yp, b.node as station,
             b.lat as xs, b.lon as ys
        from places as a, metronodes as b;
quit;
```

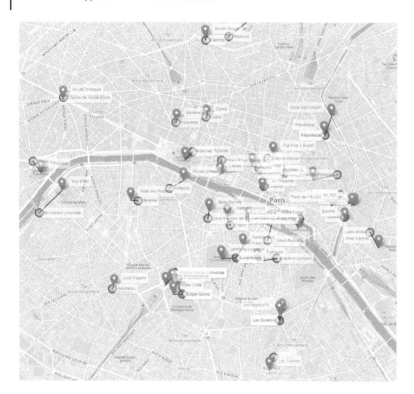

Figure 5.11 The closest stations to the locations to be visited.

```
data placesstationsdist;
   set placesstations;
   distance=geodist(xs,ys,xp,yp,'M');
   output;
run;
proc sort data=placesstationsdist;
   by place distance;
run;
data stationplace;
   length ap $20.;
   ap=place;
   set placesstationsdist;
   if ap ne place then
      do;
         drop ap;
         output;
      end;
run;
```

Once we calculate the distances between all the places we want to visit and those places to the closest stations, we can compare all possible steps in our paths in terms of shortest distances to see if we take the public transportation or if we just walk. We then execute the TSP algorithm once again to search for the optimal tour considering both walking and the public transportation options. At this point, we know the optimal sequence of places to visit in order to minimize the walking distance, and we also know when we need to walk and when we need to take the public transportation.

Figure 5.12 shows the final optimal tour considering the multimodal system.

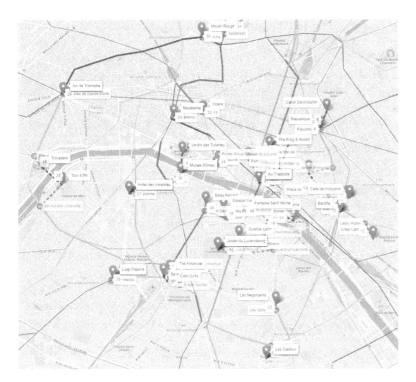

Figure 5.12 The final optimal tour considering the transportation network.

The solid lines in different colors (according to the transportation line) represent the shortest path between a pair locations (place-station or station-place) by taking the public transportation. The dashed lines represent the walking between a pair of locations (place-station, station-place, or place-place).

For example, in step 21 of our tour, we stop at Les Cailloux, an Italian restaurant in the Butte aux Cailles district, an art deco architectural heritage neighborhood. From there we walk to the Covisart station, and we take line 6 until Edgar Quinet. From there we walk to Josselin and we grab a crepe (one of the best in the area). Then we walk to Café Gaité and we enjoy a beer, just people watching. The concept of a shot of coffee lasting for hours on the outside tables works for the beer, too. It is a very nice area with lots of bars and restaurants. Then we walk to the Tour Montparnasse. Probably the best view of Paris, because from there you can see the Tour Eiffel, and from the Tour Eiffel you cannot see the Tour Eiffel. We do not need to buy the ticket (18€) for the Observation Deck. We can go to the restaurant Ciel de Paris at the 56th floor and enjoy a coffee or a glass of wine. We may lose one or two floors of height, but we definitely save 13€ or 9€ depending on what we pick. From there we walk to the Finan^{ci}er, a nice pub. From there we walk to the Montparnasse station, and we take line 6 again. We drop off at the Pasteur station and switch to line 12. We drop off at the Volontaires station and we walk to the Pizzeria Luigi Pepone (the best pizza in Paris). And from there our tour continues. We are still in step 26, and we have 14 more steps to go.

The walking tour took us 19.4 miles and around six hours and 12 minutes. The multimodal tour takes us 27.6 miles (more than the first one) but we will walk just 2.8 miles. The entire tour will take about two hours and 30 minutes. Here is the dilemma. By stopping at so many restaurants and bars, we should walk more. But by walking more, we could not stop at so many bars and restaurants. Perhaps there is an optimization algorithm that helps us to solve this complex problem.

5.3 An Optimal Beer Kegs Distribution – The Vehicle Routing Problem Example in Asheville

The VRP algorithm aims to find optimal routes for one or multiple vehicles visiting a set of locations and delivering a specific amount of goods demanded by these locations. Problems related to distribution of goods, normally between warehouses and customers or stores, are generally considered a VRP. The VRP was first proposed by Dantzig and Ramser in the paper "The Truck Dispatching Problem," published in 1959 by Informs in volume 6 of the Management Science. The paper describes the search for the optimal routing of a fleet of gasoline deliveries between a bulk terminal and a large number of service stations supplied by the terminal. The shortest routes between any two points in the system are given and a demand for one or several products is specified for a number of stations within the distribution systems. The problem aims to find a way to assist stations to trucks in such a manner that station demands are satisfied and total milage covered by the fleet is a minimum.

The VRP is a generalization of the TSP. As in the TSP, the solution searches for the shortest route that passes through each point once. A major difference is that there is a demand from each point that needs to be supplied by the vehicle departing from the depot. Assuming that each pair of points is joined by a link, the total number of different routes passing through n points is $\frac{n!}{2}$. Even for a small network, or a reduced number of points to visit, the total number of possible routes can be extremely large. The example in this case study is perfect to describe that. This exercise comprises only 22 points ($n = 22$). Assuming that a link exists between any pair of locations set up in this case, the total number of different possible routes is 562 000 363 888 803 840 000.

VRPs are extremely expensive, and they are categorized as NP-hard because real-world problems involve complex constraints like time windows, time-dependent travel times, multiple depots to originate the supply, multiple vehicles, and different capacities for distinct types of vehicles, among others. When looking at the TSP, trying to minimize a particular objective function like distance, time, or cost, the VRP variation sounds much more complex. The objective function for a VRP can be quite different mostly depending on the particular application, its constraints, and its resources. Some of the common objective functions in VRP may consist of minimizing the global transportation cost based on the overall distance traveled by the fleet, minimizing the number of vehicles to supply all customersnn' demands, minimizing the time of the overall travels considering the fleet, the depots, the customers, and the loading and unloading times, minimizing penalties for time-windows restrictions in the delivery process, or even maximizing a particular objective function like the profit of the overall trips where it is not mandatory to visit all customers.

There are lots of VRP variants to accommodate multiple constraints and resources associated with the .

- The Vehicle Routing Problem with Profits (VRPPs) as described before.
- The Capacitated Vehicle Routing Problem (CVRP) where the vehicles have limit capacity to carry the goods to be delivered.
- The Vehicle routing Problem with Time Window (VRPTW) where the locations to be served by the vehicles have limited time windows to be visited.
- The Vehicle Routing Problem with Heterogenous Fleets (HFVRP) where different vehicles have different capacities to carry the goods.
- The Time Dependent Vehicle Routing Problem (TDVRP) where customers are assigned to vehicles, which are assigned to routes, and the total time of the overall routes needed to be minimized.
- The Multi Depot Vehicle Routing Problem (MDVRP), where there are multiple depots from which vehicles can start and end the routes.
- The Open Vehicle Routing Problem (OVRP), where the vehicles are not required to return to the depot.
- The Vehicle Routing Problem with Pickup and Delivery (VRPPD), where goods need to be moved from some locations (pickup) to others locations (delivery), among others.

The example here is remarkably simple. We can consider one or more vehicles, but all vehicles will have the same capacity. The depot has a demand equal to zero. Each customer location is serviced by only one vehicle. Each customer demand is indivisible. Each vehicle cannot exceed its maximum capacity. Each vehicle starts and ends its route at the depot. There is one single depot to supply goods for all customers. Finally, customers demand, distances between customers and depot, and delivery costs are known.

In order to make this case more realistic, perhaps more pleasant, let us consider a brewery that needs to deliver its beer kegs to different bars and restaurants throughout multiple locations. Asheville, North Carolina, is a place quite famous for beer. The city is famous for other things, of course. It has a beautiful art scene and architecture, the colorful autumn, the historic Biltmore estate, the Blue Ridge Parkway, and the River Arts District, among many others. But Asheville has more breweries per capita than any city in the US, with 28.1 breweries per 100 000 residents.

Let us select some real locations. These places are the customers in the case study. For the depot, let us select a brewery. This particular problem has then a total of 22 places, the depot, and 21 restaurants and bars. Each customer has its own demands, and we are starting off with one single pickup truck with a limited capacity.

The following code creates the list of places, with the coordinates, and the demand. It also creates the macro variables so we can be more flexible along the way, changing the number of trucks available and the capacity of the trucks.

```
%let depot = 'Zillicoach Beer';
%let trucks = 4;
%let capacity = 30;
data places;
   length place $20;
   infile datalines delimiter=",";
   input place $ lat long demand;
datalines;
Zillicoach Beer,35.61727324024679,-82.57620125854477,12
Barleys Taproom, 35.593508941040184,-82.55049904390695,6
Foggy Mountain,35.594499248395294,-82.55286640671683,10
Jack of the Wood,35.5944656344917,-82.55554641291447,4
Asheville Club,35.595301953719876,-82.55427884441883,7
The Bier Graden,35.59616807405638,-82.55487446973056,10
Bold Rock,35.596168519758706,-82.5532906435109,9
Packs Tavern,35.59563366969523,-82.54867278235423,12
Bottle Riot,35.586701815340874,-82.5664137278939,5
Hillman Beer,35.5657625849887,-82.53578181164393,6
Westville Pub,35.5797705582317,-82.59669352112562,8
District 42,35.59575560112859,-82.55142220123702,10
Workshop Lounge,35.59381883030113,-82.54921206099571,4
TreeRock Social,35.57164142260938,-82.54272668032107,6
The Whale,35.57875963366179,-82.58401015401363,2
Avenue M,35.62784935175343,-82.54935140167011,4
Pillar Rooftop,35.59828820775747,-82.5436137644324,3
The Bull and Beggar,35.58724435501913,-82.564799389471,8
Jargon,35.5789624127538,-82.5903739015448,1
The Admiral,35.57900392043434,-82.57730888246586,1
Vivian,35.58378161962331,-82.56201676356083,1
Corner Kitchen,35.56835364052998,-82.53558091251179,1
;
run;
```

Here we are using the same open-source package demonstrated in the previous case study to show the outcomes (places and routes) in a map. The code to create the HTML file showing the customersn' locations and their demands is quite similar to the code presented before, and it will be suppressed here. This code generates the map shown in Figure 5.13. The map presents the depot (brewery) and the 21 places (bars and restaurants) demanding different amounts of beer kegs. Notice in parentheses the number of beer kegs required by each customer.

Most of the customers are located in the downtown area, as shown in Figure 5.14.

The next step is to define the possible links between each pair of locations and compute the distance between them. Here, in order to simplify the problem, we are calculating the Euclidian distance between a pair of places, not the existing road distances. The VRP algorithm will take into account that Euclidian distance when searching for the optimal routing, trying to minimize the overall (Euclidian) distance for vehicle travel.

The following code shows how to create the links, compute the Euclidian distance, and create the customers' nodes without the depot.

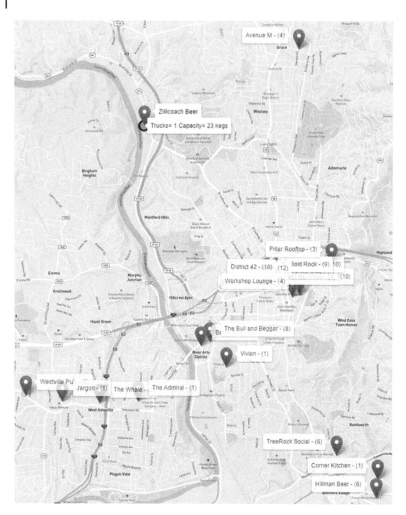

Figure 5.13 The brewery and the customers with their demands of beer kegs.

Figure 5.14 Customers located in downtown.

```
proc sql;
   create table placeslink as
      select a.place as org, a.lat as latorg, a.long as longorg,
  b.place as dst, b.lat as latdst, b.long as longdst
         from places as a, places as b
```

```
               where a.place<b.place;
quit;
data mycas.links;
   set placeslink;
   distance=geodist(latdst,longdst,latorg,longorg,'M');
   output;
run;
data mycas.nodes;
   set places(where=(place ne &depot));
run;
```

Once the proper input data is set, we can use proc optnetwork to invoke the VRP algorithm. The following code shows how to do that.

```
proc optnetwork
   direction = undirected
   links = mycas.links
   nodes = mycas.nodes
   outnodes = mycas.nodesout;
   linksvar
      from = org
      to = dst
      weight = distance
   ;
   nodesvar
      node = place
      lower = demand
      vars=(lat long)
   ;
   vrp
      depot = &depot
      capacity = &capacity
      out = mycas.routes;
run;
```

Figure 5.15 shows the results produced by proc optnetwork. It presents the size of the graph, the problem type, and details about the solution.

The output table ROUTES shown in Figure 5.16 presents the multiple trips the vehicle needs to perform in order to properly service all customers with their beer keg demands.

The VRP algorithm searches for the optimal number of trips the vehicle needs to perform, and the sequence of places to visit for each trip in order to minimize the overall travel distance, and properly supply the customers' demands. With one single vehicle, with a capacity of 23 beer kegs, this truck needs to perform six trips, going back and forth from the depot to the customers. Depending on the customersnnn' demands, and the distances between them, some trips can be really short, like route 2 with one single visit to Avenue M, while some trips can be substantially longer, like routes 4 and 6 with six customer visits. Figure 5.17 shows the outcomes of the VRP algorithm showing the existing routing plans in a map.

Let us take a step-by-step look at the overall routes. ho first route departs from the brewery and serves three downtown restaurants. The truck delivers exactly 23 beer kegs and returns to the brewery to reload. Figure 5.18 shows the first route.

The second route in the output table should be actually, at least logistically, the last trip the truck would make, delivering the remaining beer kegs from all customers' demands. This route has one single customer to visit, supplying only four beer kegs. To keep it simple, let us keep following the order of the routes produced by proc optnetwork, even though this route would be the last trip. The truck goes from the brewery to Avenue M and delivers four beer kegs. Figure 5.19 shows the second route.

The OPTNETWORK Procedure

Problem Summary	
Number of Nodes	22
Number of Links	231
Graph Direction	Undirected

The OPTNETWORK Procedure

Solution Summary	
Problem Type	Vehicle Routing Problem
Solution Status	Optimal within Relative Gap
Number of Solutions	6
Objective Value	33.438438429
Relative Gap	0.0000895192
Absolute Gap	0.0029931157
Primal Infeasibility	7.105427E-15
Bound Infeasibility	2.242651E-14
Integer Infeasibility	2.242651E-14
Best Bound	33.435445313
Nodes	709
Iterations	18148
CPU Time	1.77
Real Time	1.76

Output CAS Tables			
CAS Library	Name	Number of Rows	Number of Columns
CASUSER(Carlos.Pinheiro@sas.com)	NODESOUT	22	6
CASUSER(Carlos.Pinheiro@sas.com)	ROUTES	27	5

Figure 5.15 Output results from the VRP algorithm in proc optnetwork.

For the third route, the truck departs from the brewery and serves another three customers, delivering exactly 23 beer kegs, 4, 9, and 10, which is the maximum capacity of the vehicle. Figure 5.20 shows the third route.

Notice that the initial route is similar to the first one, but now the truck is serving three different customers.

Once again, the truck returns to the brewery to reload and departs to the fourth route. In the fourth route, shown in Figure 5.21, the truck serves six customers, delivering 23 beer kegs, 5, 1, 6, 6, 1, and 4.

Notice that the truck serves customers down to the south. It stops for two customers on the way to the south, continues on the routing serving another three customers down there, and before returning to the brewery to reload, it stops in the downtown to serve the last customer with the last four beer kegs.

« ROUTES   Table rows: 27 | Columns: 5 of 5 | Rows 1 to 27   ↑ ↑ ↓ ↓ | ⟳ ▾ | ⋮

	⊕ route	⊕ route_order	△ org	△ dst	⊕ distance
1	1	1	Asheville Club	Zillicoach Beer	1.9539486125
2	1	2	Asheville Club	Foggy Mountain	0.0968932622
3	1	3	Barleys Taproom	Foggy Mountain	0.1497745251
4	1	4	Barleys Taproom	Zillicoach Beer	2.1859454511
5	2	1	Avenue M	Zillicoach Beer	1.6780892392
6	2	2	Avenue M	Zillicoach Beer	1.6780892392
7	3	1	Jack of the Wood	Zillicoach Beer	1.9557214799
8	3	2	Bold Rock	Jack of the Wood	0.172966819
9	3	3	Bold Rock	The Bier Graden	0.0891842381
10	3	4	The Bier Graden	Zillicoach Beer	1.8865069828
11	4	1	Bottle Riot	Zillicoach Beer	2.1785155025
12	4	2	Bottle Riot	Vivian	0.3191378093
13	4	3	TreeRock Social	Vivian	1.3714641047
14	4	4	Hillman Beer	TreeRock Social	0.5632932474
15	4	5	Corner Kitchen	Hillman Beer	0.1789900119
16	4	6	Corner Kitchen	Workshop Lounge	1.9161338368
17	4	7	Workshop Lounge	Zillicoach Beer	2.2189564835
18	5	1	District 42	Zillicoach Beer	2.0364268237
19	5	2	District 42	Packs Tavern	0.1550469639
20	5	3	Packs Tavern	Zillicoach Beer	2.1512632062
21	6	1	Pillar Rooftop	Zillicoach Beer	2.2537353348
22	6	2	Pillar Rooftop	The Bull and Beggar	1.4152582126
23	6	3	The Admiral	The Bull and Beggar	0.9050376421
24	6	4	The Admiral	The Whale	0.377800963
25	6	5	Jargon	The Whale	0.358688182
26	6	6	Jargon	Westville Pub	0.3602623347
27	6	7	Westville Pub	Zillicoach Beer	2.8313079208

Figure 5.16 Output table with each route and its sequence.

The fifth route is a short one, serving only two customers in downtown. The truck departs from the brewery but now not with full capacity as this route supplies only 22 beer kegs, 10 for the first customer, and 12 for the second. Figure 5.22 shows the fifth route.

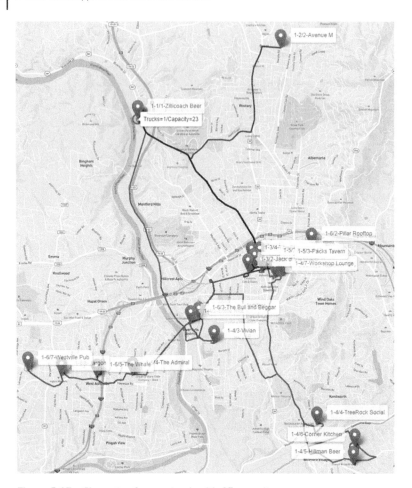

Figure 5.17 Six routes for one truck with 23 capacity.

As most of the bars and restaurants are located in the downtown, and the vehicle has a limit capacity, the truck needs to make several trips there. Notice in all the figures all different routes in the downtown marked in different colors and thicknesses. The thicker dark line represents the fifth route serving the customers in this particular trip.

The last trip serves customers in the west part of the town. However, there is still one customer in downtown to be served. Again, the truck serves six customers delivering 23 beer kegs. Figure 5.23 shows the last route for 1 route with 23 beer kegs of capacity.

Notice that the truck departs from the brewery and goes straight to downtown to deliver the first three beer kegs to the Pillar Rooftop. Then, it goes to the west part of the town to continue delivering the kegs to the other restaurants.

As a tour, after delivering all the beer kegs and supplying all the customers' demands, the truck needs to return to the brewery.

As we can observe, with only one single vehicle, there are many trips to the downtown, where most of the customers are located. Eventually the brewery will realize that in order to optimize the delivery, it will need more trucks, or a truck with a bigger capacity. Let us see how these options work for the brewery.

Originally, the VRP algorithm in proc optnetwork computes the VRP considering only one single vehicle with a fixed capacity.

For this case study, to simplify the delivery process, we are just splitting the original routes by the available vehicles. Most of the work here is done while creating the map to identify which truck makes a particular set of trips. The following code shows how to present on the map the multiple trips performed by the available trucks.

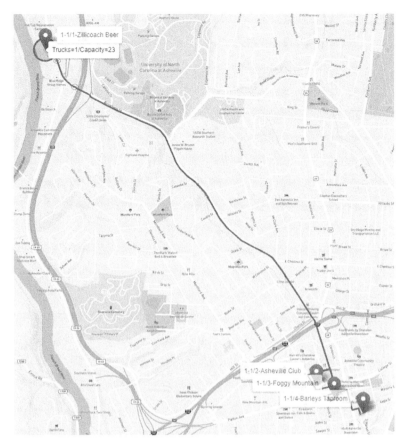

Figure 5.18 Route 1 considering one truck with 23 capacity.

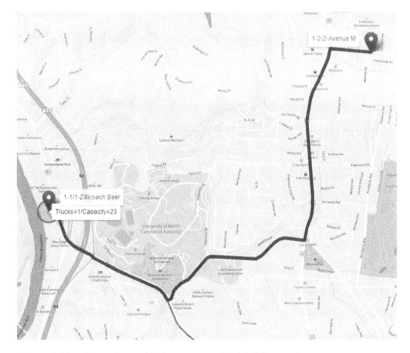

Figure 5.19 Route 2 considering one truck with 23 capacity.

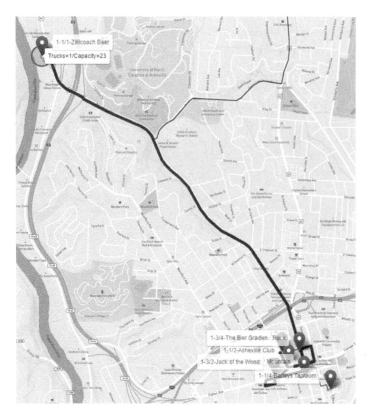

Figure 5.20 Route 3 considering one truck with 23 capacity.

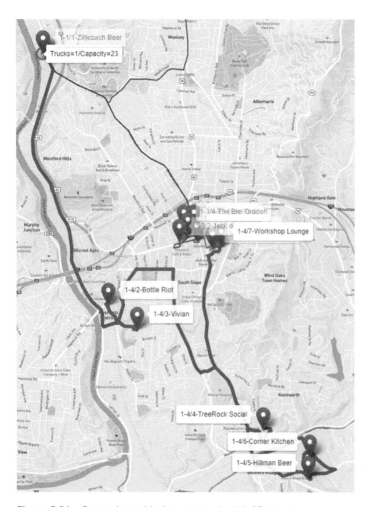

Figure 5.21 Route 4 considering one truck with 23 capacity.

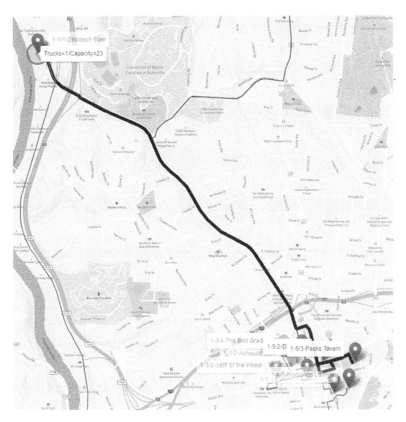

Figure 5.22 Route 5 considering one truck with 23 capacity.

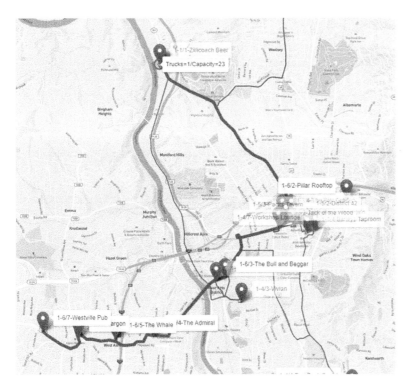

Figure 5.23 Route 6 considering one truck with 23 capacity.

```
proc sql;
   create table routing as
select a.place, a.demand, case when a.route=. then 1 else a.route
        end as route, a.route_order, b.lat, b.long
          from mycas.nodesout as a
            inner join places as b
              on a.place=b.place
            order by route, route_order;
quit;
proc sql noprint;
   select place, lat, long into :d, :latd, :longd from routing where
      demand=.;
quit;
proc sql noprint;
   select max(route)/&trucks+0.1 into :t from routing;
quit;
filename arq "&dm/routing.htm";
data _null_;
   array color{20} $ color1-color20 ('red','blue','darkred','darkblue',…
   length linha $1024.;
   length linhaR $32767.;
   ar=route;
   set routing end=eof;
   retain linhaR;
   file arq;
   k+1;
   tn=int(route/&t)+1;
   if k=1 then
      do;
          put '<!DOCTYPE html>';
          put '<html>';
          put '<head>';
          put '<title>SAS Network Optimization</title>';
          put '<meta charset="utf-8"/>';
          put '<meta name="viewport" content="width=device-width, ...
          put '<link rel="stylesheet" href="https://unpkg.com/leaf...
          put '<script src="https://unpkg.com/leaflet@1.7.1/dist/l...
          put '<link rel="stylesheet" href="https://unpkg.com/leaf...
          put '<script src="https://unpkg.com/leaflet-routing-mach...
          put '<style>body{padding:0;margin:0;}html,body,#mapid{he...
          put '</head>';
          put '<body>';
          put '<div id="mapid"></div>';
          put '<script>';
          put 'var mymap=L.map("mapid").setView([35.59560349262985...
          put 'L.tileLayer("https://api.mapbox.com/styles/v1/{id}/...
          linha='L.circle(['||"&latd"||','||"&longd"||'],{radius:8...
          put linha;
          linhaR='L.Routing.control({waypoints:[L.latLng('||"&latd...
      end;
   if route>ar then
      do;
```

```
            if k > 1 then
                do;
                    linhaR=catt(linhaR,'],routeWhileDragging:fal...
                    put linhaR;
                    linhaR='L.Routing.control({waypoints:[L.latL...
                end;
        end;
    else
        linhaR=catt(linhaR,',L.latLng('||lat||','||long||')');
    linha='L.marker(['||lat||','||long||']).addTo(mymap).bindToolt"'("'|...
    put linha;
    if eof then
        do;

    linhaR=catt(linhaR,'],routeWhileDragging:false,showAlternatives:fals...
            put linhaR;
            put '</script>';
            put '</body>';
            put '</html>';
        end;
run;
```

On the top of the map, where the brewery is located, we can see the number of trucks (2) and the capacity of the trucks (23) assigned to the depot. Now, each truck is represented by a different line size. First routes have thicker lines. Last routes have thinner lines so we can see the different trips performed by the distinct trucks, sometimes overlapping.

The first truck makes the routes 1, 2, and 3. Route 1 (thicker red) visits three customers in downtown, identified by the labels 1-1/2-Asheville Club, 1-1/3-Foggy Mountain, and 1-1/4-Barleys Taproom. This truck returns to the brewery, reloads, and visits in route 2 (thicker dark red) one single customer in the north, identified by the label 1-2/2-Avenue M. The truck returns to the depot, reloads and visits three other customers in route 3 (marron), once again in downtown, identified by the labels 1-3/2-Jack of the Wood, 1-3-3-Bold Rock, and 1-3/4-The Bier Garden. It returns to the depot and stops the trips. In parallel, truck 2 goes west and down to the south in route 4 (brown), visiting six customers, labels 2-4/2, 2-4/3, 2-4/4, 2-4/5, 2-4/6, and 2-4/7. It returns to the depot, reloads, and goes for route 5 (violet) to visit two customers in downtown, identified by the labels 2-5/2 and 2-5/3. It returns to the brewery, reloads, and goes for the route 6 (purple) to visit the final six customers, 1 to the east, label 2-6/2, 1 to the west, label 2-6/4, and 4 to the far western part of the city, labels 2-6/5, 2-6/5, 2-6/6, and 2-6/7. The trucks return to the brewery and all routes are done in much less time. Figure 5.24 shows the routes for both trucks.

Finally, let us see what happens when we increase the number of trucks to 4 and the capacity of the trucks to 30 beer kegs. Now there are only four routes, and each truck will perform one trip to visit the customers.

The first truck takes route 1 (red) to downtown and the north part of the city visiting five customers, labeled 1-1/2 to 1-1/6. The second truck takes route 2 (blue) visiting two customers in downtown, labeled 2-2/2 and 2-2/3, three customer in the south, labeled 2-2/4 to 2-2/6, and one customer in the west, labeled 2-2/7 before returning to the depot. Truck 3 takes route 3 (dark red) visiting three customers in downtown, labeled 3-3/2 to 3-3/4. Finally truck 4 takes route 4 (dark blue) visiting seven customers in the west and far west part of the city, labeled 4-4/2 to 4-4/8. Figure 5.25 shows the routes for all four trucks.

In this scenario, the downtown is served by trucks 1, 2, and 3, the north of the city is served by truck 1, while the south by truck 2. Finally, the west part of the city is served by truck 4.

Increasing the number of trucks and the capacity of the trucks certainly reduces the time to supply all customers' demands. For example, in this particular case study, not considering the time to load and unload the beer kegs, just considering the routing time, the first scenario with one single truck, the brewery would take 122 minutes to cover all six trips and supply the demand of the 21 customers. The second scenario, with two trucks with the same capacity, the brewery would take 100 minutes to cover all six trips, three for each truck, 18% of reduction in the overall time. Finally, the last

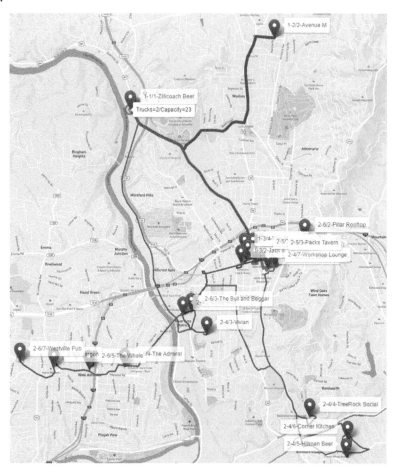

Figure 5.24 Routes considering two truck with 23 capacity.

scenario with four trucks and a little higher capacity, the brewery would take only 60 minutes to cover all four trips, 1 for each truck, to serve all 21 customers around the city, with 51% of reduction in the overall time.

The previous two case studies were created to advertise the SAS Viya Network Analytics features, particularly in the new TSP and VRP algorithms. For that reason, all the code used to create the cases were added to the case study description. Also, both case studies are pretty simple and straightforward, using only the TSP and VRP algorithms, with minor data preparation and post analysis procedures. The following case studies were real implementation in customers, considering multiple analytical procedures combined, such as data preparation and feature extraction, machine learning models, and statistical analysis, among others. The codes assigned to these cases are substantially bigger and much more complex than the ones created for the previous cases. For all these reasons, the code created for the following case studies are suppressed.

5.4 Network Analysis and Supervised Machine Learning Models to Predict COVID-19 Outbreaks

Network analytics is the study of connected data. Every industry, in every domain, has information that can be analyzed in terms of linked data in a network perspective. Network analytics can be applied to understand the viral effects in some traditional business events, such as churn and product adoption in telecommunications, service consumption in retail, fraud in insurance, and money laundering in banking. In this case study, we are applying network analytics to correlate population movements to the spread of the coronavirus. At this stage, we attempt to identify specific areas to target for social

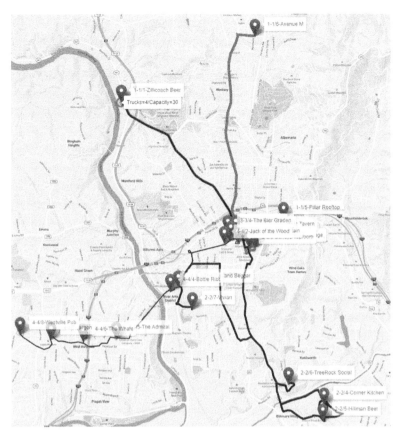

Figure 5.25 Routes considering four truck with 30 capacity.

containment policies, either to better define shelter in place measures or gradually opening locations for the new normal. Notice that this solution was developed at the beginning of the COVID-19 pandemic, around April and May of 2020. At some point, identifying geographic locations assigned to outbreaks were almost useless as all locations around the globe were affected by the virus.

Using mobile data, penetration of cell phones, companies market share, and population, we can infer the physical amount of movements over time between geographic areas. Based on this data, we use network algorithms to define relevant key performance indicators (KPIs) by geographic area to better understand the pattern of the spread of the virus according to the flow of people across locations.

These KPIs drive the creation of a set of interactive visualization dashboards and reports using visual analytics – which enable the investigation of mobility behavior and how key locations affect the spread of the virus across geographic areas over time.

Using network analytics and the KPIs, we can understand the network topology and how this topology is correlated to the spread of the virus. For example, one KPI can identify key locations that, according to the flow of people, contribute most to the velocity of the spread of the virus.

Another KPI can identify locations that serve as gatekeepers, locations that do not necessarily have a high number of positive cases but serve as bridges that spread the virus to other locations by flowing a substantial number of people across geographic regions. Another important KPI helps in understanding clusters of locations that have an elevated level of interconnectivity with respect to the mobility flow, and how these interconnected flows impact the spread of the virus among even distant geographic areas.

Several dashboards were created (in SAS Visual Analytics) in order to provide an interactive view, over time, of the mobility data and the health information, combined with the network KPIs, or the network metrics computed based on the mobility flows over time.

Figure 5.26 shows how population movements affects the COVID-19 spread over the weeks in particular geographic locations. On the map, the blue circles indicate key locations identified by centrality metrics computed by the network

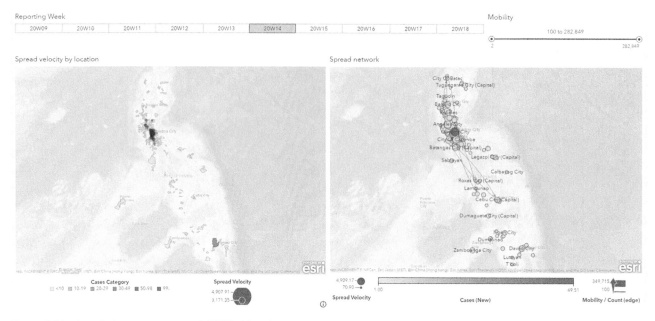

Figure 5.26 Population movement and COVID-19 outbreaks over the weeks.

algorithms. These locations play a key role in flowing people across regions. We can see the shades of red representing the number of positive cases over time. Notice that the key locations are also hot spots for the virus, presenting a substantial number of cases. These key locations are central to the flow of people in and out of geographic areas, even if those areas are distant from each other.

Notice that all those hot spots on the map are connected to each other by the flow of people. In other words, a substantial number of people flowing in and out between locations can affect the spread of the virus, even across a wide geographic region. The mobility behavior tells us how people travel between locations, and the population movement index basically tells us that a great volume of people flowing in and out increases the likelihood of the virus also flowing in and out between locations.

Here, we correlate the movement behavior and the spread of the virus over time. On the left-side of the map, the areas in shades of red represent locations with positive cases and the blue bubbles represent the spread velocity KPI. In addition, note that most of the red areas are correlated to key locations highlighted by the KPI, on the right-side of the map you can see the in and out flows between all these locations, which in fact drives the creation of this KPI. You can easily see how the flows between even distant locations can possibly also flow the virus across widespread geographic areas.

On the right side, the shades of red indicate how important these locations are in spreading the virus. These locations play an important role in connecting geographic regions by flowing people in and out over time. The right-side map shows how all those hot spots on the left-side map are connected to each other by the flow of people. The side-by-side maps show how movements between locations affect the spread of the virus.

As we commonly do when applying network analysis for business events, particularly in marketing, we performed community detection to understand the mobility behavior 'groups' locations and the flow of people traveling between them. And it's no surprise that most of the communities group together locations, which are geographically close to each other. That means people tend to travel to near locations. Of course, there are people that probably need to commute long distances. But most people try to somehow stay close to work, school, or any important community to them. If they must travel constantly to the same place, it makes sense to live as close as possible to that place. Therefore, based on the in and out flows of people traveling across geographic locations, most communities comprise locations in close proximity. In terms of virus spread, this information can be quite relevant. As one location turns out to be a hot spot, all other locations in the same community might be eventually at a higher risk, as the number of people flowing between locations inside communities are greater than between locations outside communities. Figure 5.27 shows the communities identified by the network analysis algorithm based on the in and out flows of people across the geographic locations.

Community

☐ 1 ■ 10 ■ 11 ■ 12 ■ 13 ■ 14 ■ 15 ☐ 16 ☐ 17 ■ 18 ☐ 19 ■ 2 ☐ 20 ■ 21 ■ 22 ■ 23 ■ 24 ☐ 3 ☐ 4 ☐ 5 ■ 6 ■ 7 ☐ 8 ■ 9

Figure 5.27 Communities based on the population movements.

Core decomposition is a way of clustering locations based on similar levels of interconnectivity. Here, interconnectivity means mobility. Core locations do not necessarily show a correlation to geographic proximity but instead, it shows a correlation to interconnectivity, or how locations are close to each other in terms of the same level of movements between them.

One of the most important outcomes from core is the high correlation to the wider spread of the virus. Locations in the most cohesive core do correlate over time to locations where new positive cases arise over time. By identifying cores, social containment policies can be more proactive in identifying groups of locations that should be quarantined together – rather than simply relying on geographic proximity to hotspots.

Locations within the most cohesive core are not necessarily geographically close, but hold between them a high level of interconnectivity, which means they consistently flow people in and out between them. Then they spread the virus wider. This explains the spread of the virus over time throughout locations geographically distant from each other, but close in terms of interconnectivity. Figure 5.28 shows the most cohesive core in the network and how the locations within that core are correlated to the spread of the virus across multiple geographic regions.

A combination of network metrics, or network centralities, creates important KPIs to describe the network topology, which explain the mobility behavior and then how the virus spreads throughout geographic locations over time.

Considering a specific timeframe, we can see the number of positive cases rising in some locations by the darker shades of red on the map. At the same time, we can see the flow of people between some of those hot spots. We see a great amount of people flowing between those areas, spreading the virus across different regions even if they are geographically distant from each other.

As time goes by, we notice the increase of the dark shades of red going farther from the initial hot spots, but also, we notice the flow of people between those locations. Again, the great volume of people moving from one location to another explains the spread of the virus throughout distant geographic regions. Even when you start getting even farther from the initial hot spots, we still see a substantial flow of people between locations involved in the spread of the virus. The mobility behavior, or

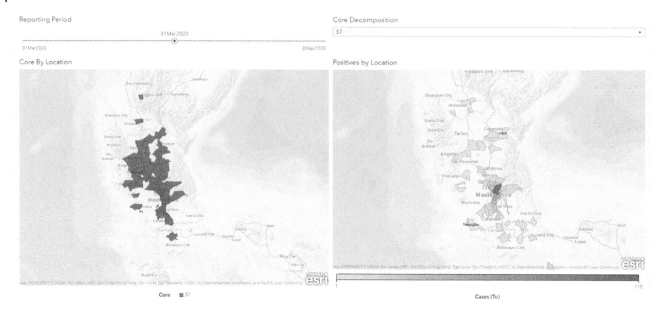

Figure 5.28 Key cities highlighted by the network metrics and the hot spots for COVID-19.

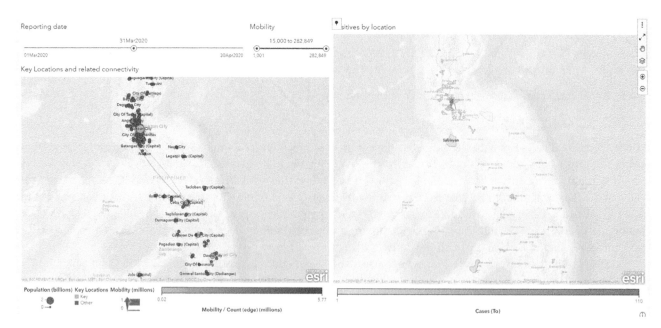

Figure 5.29 Key locations based on the network centrality measures.

the flow of people between locations, explains the spread across the most distant regions in the country. Figure 5.29 shows the key locations based on the centrality measures and the spread of the virus across the country.

A series of networks were created based on the population movements between geographic regions and the number of positive cases in these regions over time. For each network metric and each geolocation, we compute the correlation coefficient of the change in the network metric to the change in the number of positive cases. These coefficients provide evidence supporting the hypothesis that the way the network evolves over time in terms of mobility is directly correlated to the way the number of positive cases changes across regions.

The correlation between the geographic locations and the number of positive cases over time can also be used to create a risk level factor. Each geolocation has a coefficient of correlation between the combined network metrics and the number of positive cases at the destination and origin locations. These coefficients were used to categorize the risk of virus spreading. The risk level is binned into five groups. The first bin has about 1% of the locations; it is considered substantial risk. All locations in this bin have the highest risk for the number of positive cases to increase over time because of their incoming connections. The second bin has about 3–4% of the locations and is considered medium-high risk. The third bin has about 5% of the locations and is considered medium risk. The fourth bin has about 40% of the locations and is considered medium-low risk. Finally, the fifth bin has about 50% of the locations and is considered low risk. Figure 5.30 shows all groups on the map. The risk level varies from light shades of green for minimal risk to dark shades of red for high risk.

Based on the clear correlation between the computed network metrics and the virus spread over time, we decided to use these network measures as features to supervise machine learning models. In addition to the original network centralities and topologies measures, a new set of features were computed to describe the network evolution and the number of positive cases. Most of the derived variables are based on ratios of network metrics for each geographic location over time to determine how this changing topology affects the virus spread.

A set of supervised machine learning models were trained on a weekly basis to classify each geographic location in terms of how the number of positive cases would behave, increasing, decreasing, or remaining stable. Figure 5.31 shows the ROC and Lift curves for all models developed.

Local authorities can use the predictive probability reported by the supervised machine learning models to determine the level of risk associated with each geographic location in the country. This information can help them to better define effective social containment measures. The higher the probability, the more likely the location will face an increase in the number of positive cases. This risk might suggest stricter social containment. Figure 5.32 shows the results of the gradient boosting model in classifying the locations that are predicted to face an increase in the number of positive cases the following

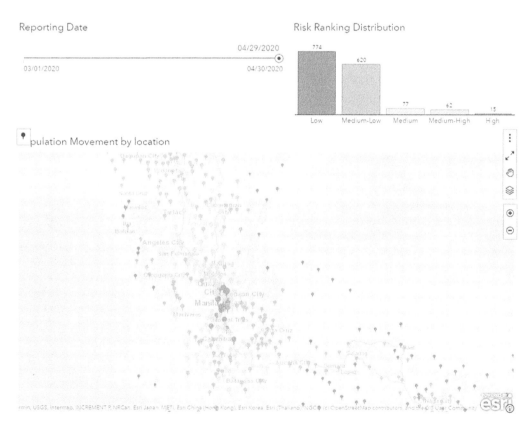

Figure 5.30 Final links associated to the optimal tour.

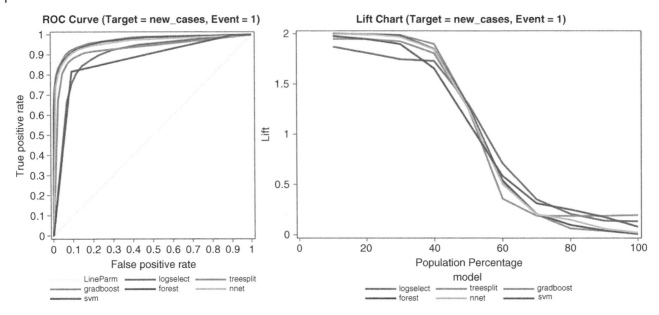

Figure 5.31 Supervised machine learning models performance.

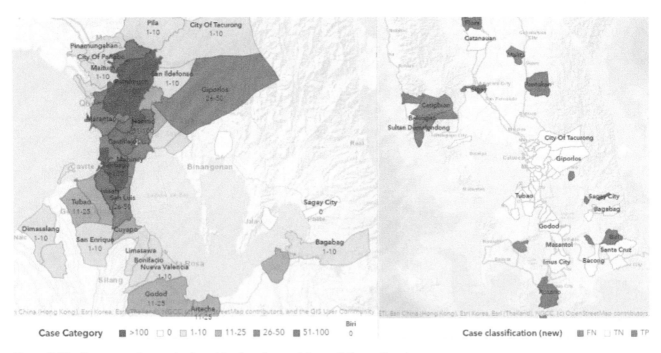

Figure 5.32 Outcomes of supervised machine learning models predicting outbreaks.

week. The model's performance on average is about 92–98%, with an overall sensitivity around 90%. The overall accuracy measures the true positive rates as well as the true negative, false positive, and false negative rates. The sensitivity measures only the true positive rates, which can be more meaningful for local authorities when determining what locations would be targeted to social containment measures. The left side on the map shows the predicted locations. The shades of red determined the posterior probability of having an increase in the number of positive cases. The right side on the map shows how the model performed on the following week, with red showing the true-positives (locations predicted and observed to increase cases), purple showing the false-negatives (locations not predicted and observed to increase cases), and white showing the true-negatives (locations predicted and observed to not increase cases).

One of the last steps in the mobility behavior analysis is the calculation of inferred cases, or how the cases supposedly travel from one location to another over time. The mobility data collected from the carrier give us the information about how the population moves around geographic regions over time. The health data collected from the local authorities gives us the information about the number of cases in each location on a specific day. As the population moves between geographic regions within the country, and we assume that these moves affect the spread of the virus, we decided to infer the possible impact of the positives from one location to another. This information basically created a network of cases in a spatiotemporal perspective. Based on the inferred cases traveling from one location to another, we can observe what locations are supposedly flowing cases out, and what locations are supposedly receiving cases from other locations. The number of cases flowing between locations over time is an extrapolation of the number of people flowing in and out between locations and the number of positive cases in both origin and destination locations. Figure 5.33 shows the network of cases presenting possible flows of positive cases between geographic locations over time.

A spatiotemporal analysis on the mobility behavior can reveal valuable information about the virus spread throughout geographic regions over time. This type of analysis enables health authorities to understand the impact of the population movements on the spread of the virus and then foresee possible outbreaks in specific locations. The correlation between mobility behavior and virus spread allows local authorities to identify particular locations to be put in social containment together, as well as the level of severity required. Outbreak prediction provides sensitive information on the pattern of the virus spread, supporting better decisions when implementing shelter-in-place policies, planning public transportation services, or allocating medical resources to locations more likely to face an increase in positive cases. The analysis of mobility behavior can be used for any type of infectious disease, evaluating how population movements over time can affect virus spread in different geographic locations.

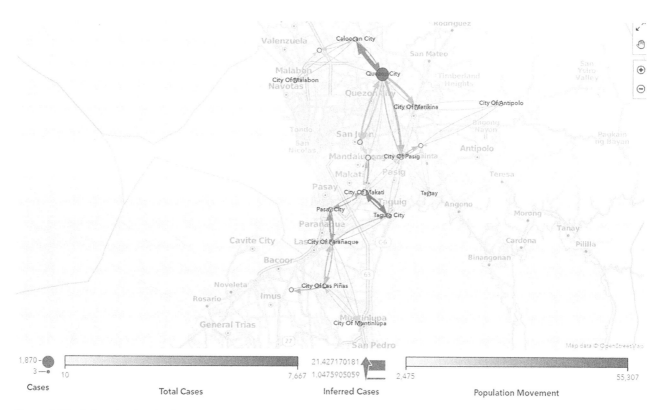

Figure 5.33 Inferred network of cases.

5.5 Urban Mobility in Metropolitan Cities

This case looks into urban mobility analysis, which has received reasonable attention over the past years. Urban mobility analysis reveals valuable insights about frequent paths, displacements, and overall motion behavior, supporting disciplines such as public transportation planning, traffic forecasting, route planning, communication networks optimization, and even spread diseases understanding. This study was conducted using mobile data from multiple mobile carriers in Rio de Janeiro, Brazil.

Mobile carriers in Brazil have an unbalanced distribution of prepaid and postpaid cell phones, with the majority (around 80%) of prepaid, which does not allow the proper customers address identification. Valuable information to identify frequent paths in urban mobility is the home and workplace addresses for the subscribers. For that, a presumed domicile and workplace addresses for each subscriber were created based on the most frequent cell visited during certain period of time. For example, overnight as home address and business hours as work address. Home and work addresses allow the definition of an Origin–Destination (OD) matrix and crucial analysis on the frequent commuting paths.

A trajectory matrix creates the foundation for the urban mobility analysis, including the frequent paths, cycles, average distances traveled, and network flows, as well as the development of supervised models to estimate number of trips between locations. A straightforward advantage of the trajectory matrix is to reveal the overall subscribers' movements within urban areas. Figure 5.34 shows the most common movements performed in the city of Rio de Janeiro.

Two distinct types of urban mobility behavior can be analyzed based on the mobile data. The first one is related to the commuting paths based on the presumed domiciles and workplaces, which provides a good understanding of the population behavior on the home-work travels. The second approach accounts for all trips considering the trajectory matrix. Commuting planning is a big challenge for great metropolitan areas, particularly considering strategies to better plan public transportation resources and network routes. Figure 5.35 shows the distribution of the presumed domiciles and workplaces throughout the city. Notice that some locations present a substantial difference in volume of domiciles and workplaces, which affects the commuting traffic throughout the city. For example, downtown has the highest volume of workplaces and a low volume of domiciles. Some places on the other hand show a well-balanced distribution on both, like the first neighborhood in the chart, presenting almost the same number of domiciles and workplaces. As an industrial zone, workers tend to search home near around.

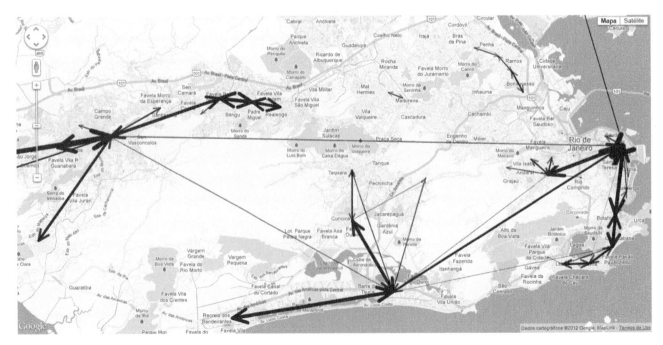

Figure 5.34 Frequent trajectories in the city of Rio de Janeiro.

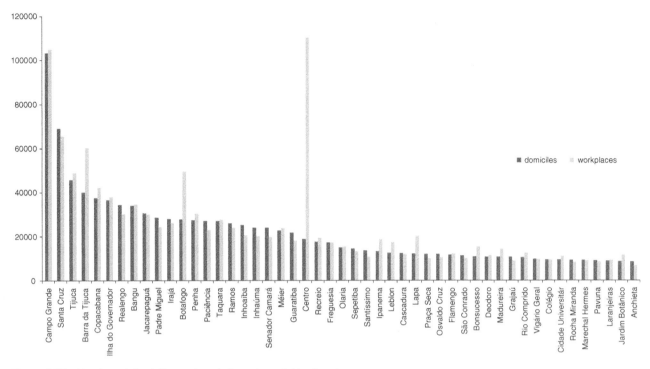

Figure 5.35 Number of domiciles and workplaces by neighborhoods.

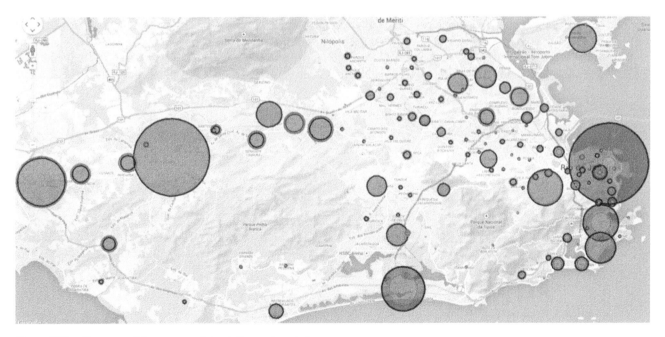

Figure 5.36 Geo map of the presumed domiciles and workplaces.

Figure 5.36 shows graphically the distribution of domiciles and workplaces throughout the city. The diameter of each circle indicates the population living or working on the locations, with light (red) circles representing the presumed domiciles and the dark (blue) circles representing the presumed workplaces. When the number of presumed domiciles and workplaces are similar, the circles overlap, and less traffic or commuting is expected. When the sizes of the circles are substantially different, it means significant commuting traffic is expected in the area.

The network created by the mobile data is defined by nodes and links. Here, the nodes represent the cell towers, and the links represent the subscribers moving from one cell tower to another. The nodes (cell towers) can be aggregated by different levels of geographic locations, like neighborhoods or towns. The links (subscribers' movements) can also be aggregated in the same way, like, subscribers' movements between cell towers, between neighborhoods, or between towns.

The most frequent commuting paths are identified based on the presumed domiciles and workplaces and are presented in Figure 5.37. The thicker the lines, the higher the traffic between the locations. These paths include commuting travels between various places within the western zone (characterized by domiciles and workplaces on the left side of the map) and within the eastern zone (mostly characterized by workplaces on the far right side of the map). The paths indicate that there are a reasonable number of people living near their workplaces because these displacements are concentrated in specific geographic areas (short movements). According to this study, 60% of movements are intra-neighborhood, and 40% of them are inter-neighborhood. Neighborhoods in Rio are sometimes quite large, around the size of town in U.S. or small cities in Europe. However, there are also significant commuting paths between different areas represented by long and high-volume movements. On the map, we can observe a thicker line in the west part between Santa Cruz and Campo Grande, two huge neighborhoods on the left side of the map. There are also several high volume commuting paths between Santa Cruz and Campo Grande to Barra da Tijuca, a prominent neighborhood in the south zone, on the bottom of the map, and between those three neighborhoods to Centro (downtown), in the east zone of the city, on the far-right side of the map. From Centro, the highest commuting traffic to Copacabana, the famous beach, and Tijuca, a popular neighborhood. One interesting commuting path is between downtown and the airport's neighborhood, Ilha do Governador. Notice a thicker line between downtown and the airport crossing the bay. Here we have the Euclidian distances between each pair of locations, not the real routing.

Figure 5.38 shows the number of subscribers' displacements in the city of Rio de Janeiro during the week. This distribution is based on six months of subscribers' movement aggregated by the day of the week and one-hour time slots. We can observe that subscribers' mobility behavior is approximately similar for all days of the week. For each day, regardless of weekday or weekend, the mobility activity is proportionally equal, varying similar during each day, and presenting two peaks during the 24 time slots. The first peak is around time slot 12, and the second peak is recorded at time slot 18. Both peaks occur daily. The difference in the mobility behavior on weekdays over weekends relies on the total number of movements and the peaks. The first peak of mobility activity is always during lunch time, both on weekdays and weekends. The second peak of mobility activity is always early evening. This period includes mostly commuting and leisure. At the beginning of the week – i.e. Mondays and Tuesdays – and at the end – i.e. Fridays, Saturdays, and Sundays, the first peak is higher than the second one, which means higher traffic at 12 than at 18. However, there is no difference on Wednesdays, where the traffic is virtually the same. On

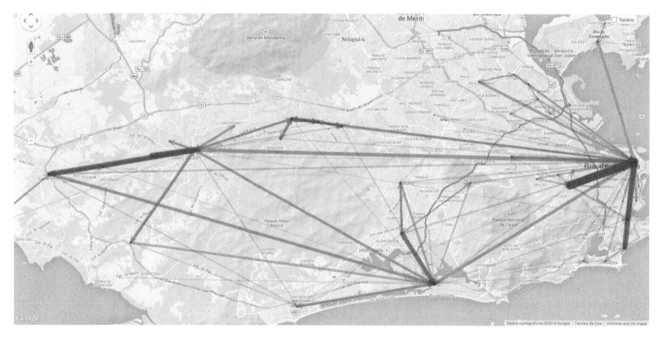

Figure 5.37 Frequent commuting paths in the city of Rio de Janeiro.

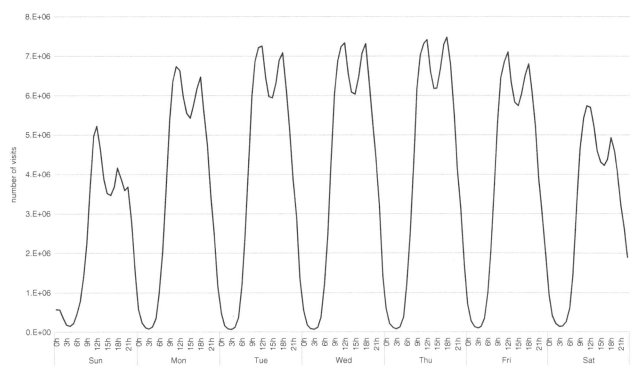

Figure 5.38 Traffic volume by day.

Thursdays, the highest peak is at 18, where the traffic is higher than at 12. A likely reason for this difference in mobility behavior is the happy hour, very popular on Thursdays. Nightlife in Rio is famous for locals and tourists. The city offers a variety of daily events, from early evening till late night. But a culture habit in the city is the happy hour, when people leave their jobs and go to bars, cafes, pubs, and restaurants to socialize (the formal excuse is to avoid the traffic jam). As in many great cosmopolitan cities around the globe, many people go for a happy hour after work. Even though all nights are busy in Rio, it is noticeable that on Thursdays it is the most frequented day for happy hour. Wednesdays are also quite famous, even more than Fridays. The mobility activity distribution during the week reinforces this perception.

When analyzing urban mobility behavior in great metropolitan areas over time, an important characteristic in many studies is that some locations are more frequently visited than others, and as a consequence, some paths are more frequently used than others. The urban mobility distribution is nonlinear, and often follows a power law, which represents a functional relation between two variables, where one of them varies as a power of another. Generally, the number of cities having a certain population size varies as a power of the size of the population. This means that few cities have a huge population size, while most of the cities have a small population. Similar behavior is found in the movements. The power law distribution in the urban mobility behavior contributes highly to the migration trend patterns and frequent paths in great cities, states, and even countries.

For example, consider the number of subscribers' visits to the cell towers, which ultimately represents the subscribers' displacements. We can observe that some locations are visited more than others. Based on the sample data, approximately 8% of the towers receive less than 500 visits per week and another 7% receive between 500 and 1000 visits. Almost 25% of the towers receive 1000 to 5000 visits a week, 11% 5–10 K and 15% 10–25 K. Those figures represent 66% of the locations, receiving only 15% of the traffic. On the other hand, 16% of the towers receive over 20% of the traffic, 9% almost 30%, 5% almost 15%, and 3% more than 11% of the traffic. That means one third of the locations represent over three quarters of the traffic.

The average level of mobility in the city of Rio de Janeiro is substantial. Most users visit a reasonable number of distinct locations during their frequently used paths. The average number of distinct locations visited on a weekly basis is around 30. However, a crucial metric in urban mobility is the average distance traveled. Rio de Janeiro is the second largest metropolitan area in Brazil, the third in South America, the sixth in America, and the 26th in the world. Because of that, it is reasonable that people need to travel midi to long distances when moving around the city, for both work and leisure. The average distance traveled by the subscribers during the frequent paths is around 60 km daily. As an example, the distance between the most populated neighborhood, Campo Grande, and the neighborhood with more workplaces, Centro, is 53.3 km.

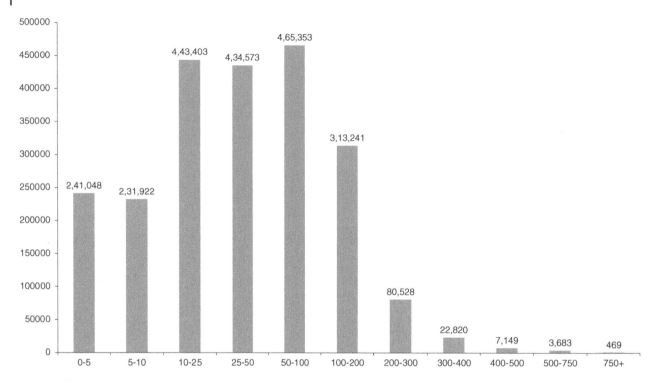

Figure 5.39 Distance traveled by subscribers.

Everyone who lives in Campo Grande and works in Centro needs to travel at least 106 km daily. The distribution of the distance traveled by subscribers along their frequent paths is presented in Figure 5.39.

We can observe that only 11% of the users travel on average less than 5 km daily. As a big metropolitan area, it is expected that people have to travel more than this. For example, 70% of the users travel between 5 and 100 km daily, 14% between 100 and 200 km, and 5% travel more than 200 km.

There is an interesting difference between the urban mobility patterns between weekdays and weekends. The mobility behavior in both cases is similar when looking at the movements over time, except by the traffic and the peaks. The number of movements on weekdays is around 75% higher than weekends. The second difference is the peak of the movements. During weekdays, people start moving earlier than during weekends. Here, the first movement is calculated from the presumed domicile to the first visited location. Figure 5.40 shows the movements along the day on both weekdays and weekends. The first movement starts around 5 a.m. and the traffic on weekdays is significantly higher than on weekends. Also, there is a traffic peak at 8 a.m. during weekdays, which means most of subscribers start moving pretty early. On the weekends, there is no such peak. There is actually a plateau between 11 and 17. This is expectable as during weekdays people have to work and during weekends people rest or are not in a rush.

The analysis of the urban mobility has multiple applications in business and government agencies. Carriers can better plan the communications network, taking into account the peaks of the urban mobility, the travels between locations, and then optimize the network resources to accommodate subscribers' needs over time. Government agencies can better plan public transportation resources, improve traffic routes, and even understand how spread diseases disseminate over metropolitan areas. For example, this study actually started by analyzing the vectors for the spread of the Dengue disease. Public health departments can use the vector of movements over time to better understand how some virus spreads out over particular geographic areas. Dengue is a recurrent spread disease in Rio de Janeiro. It is assessed over time by tagging the places where the diagnoses were made. However, the average paths of infected people may disclose more relevant information about vector-borne diseases like Dengue and possibly more accurate disease control measures. Figure 5.41 shows some focus areas of Dengue in the city of Rio de Janeiro.

Transportation agencies can evaluate the overall movements over time and the current existing routes in the city and better understand possible areas for improvement. For example, the downtown area in the city of Rio de Janeiro hosts most of the workplaces in the region, like the headquarters for large and medium companies, business offices, and corporate

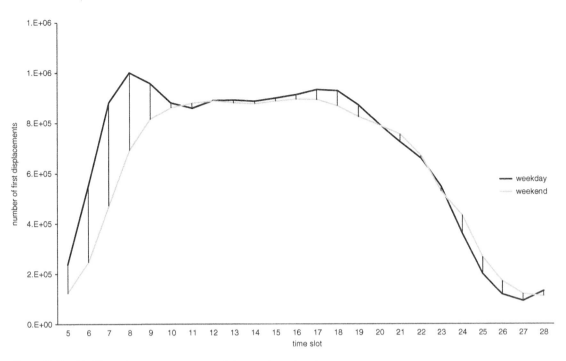

Figure 5.40 Traffic and movements on weekdays and weekends.

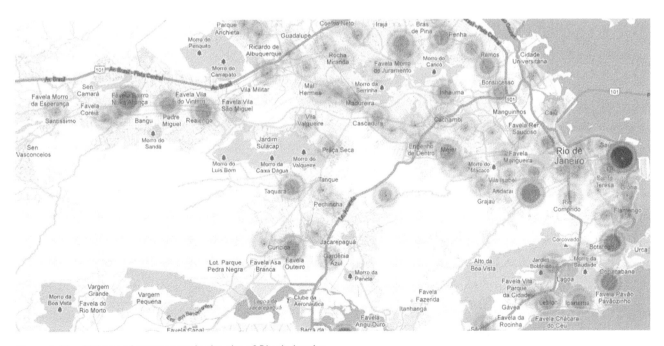

Figure 5.41 Volume of movements in the city of Rio de Janeiro.

buildings. Due to this, the majority of the other neighborhoods in the city have a significant traffic heading downtown. However, the existing routes are not exactly the same as the vector of movements identified by the analysis of the mobility behavior, as shown in the top left corner of Figure 5.42 by the thicker arrow crossing the bay. This vector represents a substantial traffic between downtown and the region of the airport, which is an island. There is a small bridge connecting this neighborhood to the shore. The existing route to reach for example the airport from downtown is shown on the right side of

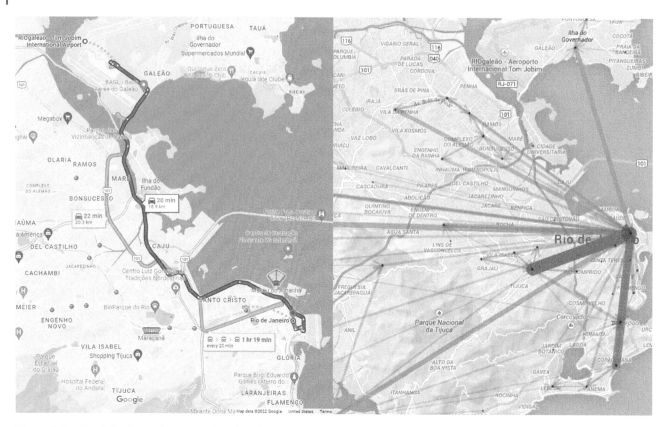

Figure 5.42 Possible alternative routes based on the overall population movements.

Figure 5.42. Eventually, it would be more effective to have another bridge connecting downtown straight to the island or a ferry line serving as a route. There are already both a ferry and a bridge connecting downtown to a nearby city called Niteroi. The solution could be an extension of the existing routes.

Urban mobility behavior is a hot topic in network science and smart cities. This subject comprises many possibilities to better understand patterns in urban mobility, like frequent trajectories, overall paths, commuting behavior, and migration trends. Such analyses may be applied on different geographic scales, lower as to cell towers and greater as to cities, states, or countries. One of the best sources of data to explore and understand the urban mobility behavior is the mobile records from carriers. Due to the high penetration of cell phones in great metropolitan cities, the analysis of subscribers' movements is closely related to the population's movements.

5.6 Fraud Detection in Auto Insurance Based on Network Analysis

Insurance companies have evolved into a highly competitive market that requires companies to put in place an effective customer behavior monitoring system. Unusual behavior in insurance can represent different business meanings, such as users properly using the services or suspicious transactions on the claims. Network analysis can be used to reveal important knowledge on the relationships between all the different actors, disclosing customers' behavior and possibly unexpected usage. Frequent or heavy usage of insurance services substantially impacts the operational cost, either by claims expenses or possible fraud or exaggeration events. Network analysis can be used to evaluate the separate roles assigned to the multiple actors within the claims, and their relationships, allowing companies to implement more effective processes to understand and monitor the commonly great volume of transactions.

The main goal of this case study is a deeper look into the claims, identify all the actors from the transactions, their relationships, and then to highlight suspicious correlations and occurrences. It shows the benefits of using a network analysis approach for fraud detection, a distinct method that focuses more on the transaction relationships than on the individual transactions themselves. This case study is based on data assigned to an auto insurance company. It contains information about the claims in a real auto insurance operation. The main goal of the network analysis in this particular case is to highlight unusual behavior from the participants assigned to the claims. These participants may have multiple roles, like policy holders, suppliers, repairers, specialists, witnesses, and so on. The idea of the network analysis is to identify unexpected relations among participants and identify suspicious groups or individuals. This approach can reveal possible fraudulent connections inside the network structure, indicating from the participants what claims may have a high likelihood of being fraudulent or exaggerated.

The identification of the suspicious claims is mostly based on the values of the network metrics, or on the combination of them when looking into the claims and the network structures created based on the transactions. The steps assigned to create the network, identify structures and groups, and compute network metrics will be briefly described later in this chapter.

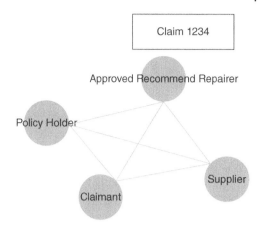

Figure 5.43 Nodes and links assigned to an individual claim.

The network structure in this case study is defined upon the claim's records. From these records, the participants, and the distinct roles they can possibly play within the claim, are selected, and then their relationships to other participants are defined. All participants in one particular claim are connected to each other in order to create all possible links between them. For example, claim 1234 shown in Figure 5.43 has one policy holder A, one claimant B, one supplier C, and one approved recommended repairer D. Based on this approach, a combination of 3 by 2 should be performed to produce a small social structure consisting of four nodes and six links. That gives us the nodes A, B, C, D, and Links A-B, A-C, A-D, B-C, B-D, C-D.

For this particular case study, the network structure was created based on 22 815 claims. All transactions were aggregated to consider distinct claims with multiple occurrences of participants. Also, the participants were considered based on a smooth match code of their names, addresses, dates of birth, and other individual descriptions, avoiding possible occurrences of the same individual in multiple transactions as a different person. From the claims used, 41 885 distinct participants were identified, along 74 503 relationships. In this network analysis approach, a participant is defined as a node and a relationship is defined as a link.

By connecting all participants to each other, a considerable number of connections among them can be reached. For example, in the network presented in Figure 5.44, there are two participants who appear simultaneously in two distinct

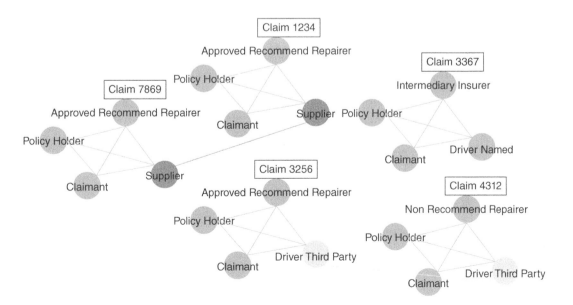

Figure 5.44 Relationships between claims due to same participants.

subnetworks. Both claims 1234 and 7869 have the same supplier, and both claims 3256 and 4312 have the same driver third party. Based on that, all participants in the claims 1234 and 7869 are connected. Similarly, all participants in the claims 3256 and 4312 are also connected. Both network structures created by these pair of claims sharing the same participant are conserved as complete graphs.

When analyzing the entire network structure and all correlations among the nodes, the network approach may consider all types of roles within the claims. Specific analyses can be performed upon different types of participants. For instance, the network metrics are computed considering the entire network, accounting for all nodes and links in the graph, regardless of the participants' roles. The computed network metrics, assigned to all individual nodes or participants are assessed. However, distinct assessments can be performed for individual participants, organization participants, and so on. This approach allows us to compare the average behavior of individuals, organizations, but still accounting for the entire network structure. The same method of analysis can be conducted on the different roles of the participants, such as policy holders, third party drivers, approved repairers, and so forth. These more specific analyses tend to better highlight, or more accurately identify, unexpected behaviors on the relationships throughout the claims by the distinct types of participants. For example, a particular approved repairer could be an outlier based on its network metrics when just the category of approved repairers is considered, but are not considered an outlier when all types of participants are accounted for. Figure 5.45 shows the network structure with nodes as individuals colored and nodes as organizations faded.

The overall analytical method used in this case study combines network analysis to build the network and compute centrality metrics, and outlier analysis to subsequently explore the network metrics distributions seeking unexpected behavior on the transactions, either in terms of nodes or links. That means unusual transactions on the participants' levels or on the relationship's occurrences. The outlier detection is mostly based on the univariate analysis, principal component analysis, and clustering analysis upon the network metrics.

To perform an outlier analysis on the individual participants, a categorization was defined based on the names of the participants. Again, the network metrics were computed considering the entire network structure, but the exploratory and outlier analyses were conducted based on the several types of participants, as individual or organization, and also considering the distinct roles. Figure 5.45 shows graphically this particular approach on the individuals and organizations analyses. All links in the network and all participants classified as individual are highlighted, and all participants classified as organization are faded. When computing the network metrics, all links and nodes are counted, but when performing the outlier analysis, just the individuals are considered.

The traditional outcome of outlier analysis and anomaly detection on transactions can be described in terms of a set of rules and thresholds, considering all the information within the transactions, or the claims. On the other hand, in the network analysis approach, the identification of unusual behavior and the presence of outliers are based on the relationships

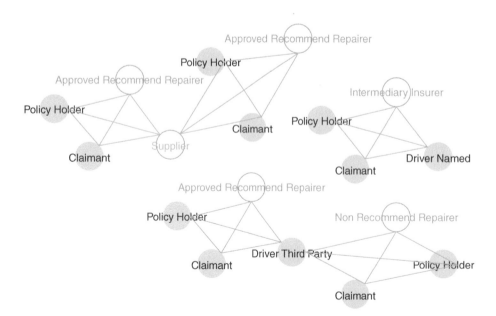

Figure 5.45 Network structure highlighting different types of participants.

between the participants rather than on their individual attributes. The values within the claims of a particular participant are less important than the combined values considering all relationships of that participant in multiple claims and eventually the separate roles this participant has throughout the network structure. For instance, a high number of similar addresses or participants involved in different claims can trigger an alert about the participants, and therefore, about the claims involved. Or a switch in participants' roles can raise an alert. For example, in one claim participant A is the driver and participant B is the witness. In a different claim, the same participant A is the witness, and the previous participant B is now the driver. As network analysis looks into frequency and types of relations as well as attributes and types of nodes, this particular case would be easily identified by network outcomes. Another recurrent important outcome from network analysis is how frequently the nodes occur in some relationships throughout the network structure, the size of their subnetworks, the strength of their relationships, or simply the number of distinct connections. A common example is a particular individual within the network with multiple connections, along separate roles over time. Or a particular individual with very few connections but all of them strong, or with high frequency. All these cases can raise some alerts for further investigation. And this is an important aspect of fraud or anomaly detection. Further investigation, usually preceded by business analysts, are always required. Outlier occurrences can be the result of transactions of a heavy user (good) or of an illegal user (bad). The manual analysis on these cases is part of the continuous learning process. Based on that approach, different analytical methods can highlight distinct sets of rules and thresholds upon the network metrics and subnetwork identifications. A combined approach to catch suspicious transactions therefore can consider the outcomes from all those approaches, combining the rules and the thresholds into a more holistic perspective of the cases.

The network analysis developed for this particular case study focused on highlighting suspicious participants, or participants with unusual or outlier relationship behavior, considering the network connections as a result of the claims. Traditionally, if several claims have average values for their attributes, such as total amount, number of participants, number of roles, addresses, and so on, neither of those claims would be raised as suspicious, or flagged to be monitored and investigated by using traditional approaches. However, if a particular participant, such as a supplier or a claimant, is involved in all those claims, its relationship level could be considered too high in comparison to the same type of participants, and an alert would be triggered for further monitoring or investigation. The individual attributes values assigned to the claims in a transaction approach, may all be normal or expected. However, the strength of the connections for the participants in those claims may be significantly higher than the normal values when a network structure is considered. Based on that, alarms can be triggered for manual actions, like case analysis and investigation, or automatic actions can be performed, such as user suspension on temporary cancelations. The definition of the manual or automatic actions can be based on the threshold values, which often come from the outlier percentiles. For example, all cases in between the outlier percentiles 90 and 95 are triggered for further analysis, and all cases in the percentiles 95 are automatically blocked.

This approach does not discard any traditional method. Business rules based on previous experience, individual attributes thresholds upon significant values, or any other straightforward strategy can certainly be effective and should be put in place throughout. The main gain in the network analysis approach is the possibility to highlight unusual behavior based on the existing relations rather than on the individual attributes. Often, the most suitable approach to monitoring transactions in a fraud perspective is by combining multiple methodologies. Each analytical method is more or less suitable to catch a particular type of suspicious event, and therefore, a combination of distinct approaches allows a wider perspective on the fraud scenarios.

One of the most important outcomes from the network analysis are the measures in relation to the participants' subnetworks, which shows how the participants relate to each other, in what frequency, and how relevant their relationships are. From the network analysis outcomes, some suspicious relationships can be highlighted, and as a consequence, also some participants. The network measures are computed upon the overall network, considering all relations of each node and some individual attributes. The following measures describe the structure and the topology of the network. From the 41 885 nodes within the social structure analyzed, 30 428 are considered individual nodes, and 4423 are considered as organization nodes. In addition to the distinct analyses according to the role of the participants, individual or organization, a set of exploratory analyses are performed aiming to highlight unusual groups of participants, either in terms of nodes or links. There are some different types of group analysis that can be performed within the network structure defined, such as community detection, clustering, connected components, and bi-connected components, among others.

There are some relevant concepts assigned to network analysis that describes the importance of some particular nodes and links, and also the overall structure of the network. These concepts can be used to determine suspicious groups of nodes, or participants, and the claims associated with them. Some of the methods to identify groups of nodes are based on similarity, distance, or the internal paths inside the subnetworks.

The concept of connected components is the first one. When there is a group of nodes, where each node can reach any other node in the group, even going through multiple steps or passing by several other nodes, this group of nodes is defined as a connected component. The connected component can be understood as a close group of nodes, separated from the rest of the network, but each node being reachable by any other node within that group. The network assigned to this particular case comprises 6263 connected components. It is important to notice that during the data cleansing some particular entities were removed from the network. There were some participants within the claims that are way too recurrent, such as government agencies and some insurers, among others. If we include these participants in the study, the entire network will be considered as a one single connected component as those participants will virtually connect all other entities within the claims.

The second approach used the concept of the bi-connected components. When there is a connected component that comprises a particular node where, if it is removed from the group, the original component would be split into two distinct connected components, this original subnetwork can be considered as a bi-connected component. Most important in our case study, that particular node that, if removed, splits the original component into two disjoint groups is called as articulation point. The articulation point plays an important role in the network, keeping possibly disjoint groups all together. The articulation point plays as a bridge connecting two disjoint lands, like a shore and an island. Without a bridge, there is no connection, and the island would be isolated. Without the articulation point, the connected component would be possibly split into multiple groups. In the network there are 15 734 bi-connected components and 2307 articulation points. Analogously to the connected components, the articulation point can be highlighted to be further investigated as they are not exactly expected in the network or at least they have a special role in terms of network structure. Also, from those 2307 articulation points, 984 of them are participants defined as individual, which can be considered even more suspicious.

The third approach is the community detection. There are some techniques, which are put in place in order to identify communities inside the entire network. Different from the connected components, which are isolated from the rest of the network, a community can hold some nodes, which have connections outside the community. In this way, a community can be connected to other communities through one or more nodes. Communities inside networks can be understood as clusters inside populations. The study of the communities can highlight some important knowledge assigned to the structure of the network and the behavior of the nodes, allowing some specific analyses to point out suspicious individuals. For this particular network, 13 279 communities were identified, considering all nodes, no matter if they were classified as individual or organization.

The next phase in this analytical approach based on network analysis is to consider the overall centrality metrics for all individuals. These centralities reveal important characteristics of the individual components within the social structure. These centrality metrics are computed to each and every node within the network. Then, the second step is to compare the nodes centralities to the average metrics for the network as a whole, or at least, to some particular types or categories of individuals, such as policy holders, repairers, or third parties, etc., all unusual or unexpected difference associated with the nodes and links can be considered a possible case for further investigation. There are several centralities to be computed. For this particular case, only six measures were considered: degree, eigenvector, closeness, betweenness, influence 1, and influence 2. All these centralities were described in Chapter 3. To briefly summarize them, the degree represents the number of connections, the eigenvector represents a measure of importance of the node within the network, the closeness represents the average shortest paths for that particular node to all other nodes in the network, the betweenness represents how many shortest paths the node partakes, the influence 1 represents the weighted first order centrality, and finally the influence 2 represents the weighted second order centrality. All these measures were considered to highlight the outliers within the social structure created by the claims.

In order to identify the unexpected transactions, the outlier analysis was performed over the centrality metrics. The outlier analysis considers the average metrics considering the entire network and centralities for each individual node and link. Distinct approaches to identify occurrences of interest were combined, such as univariate analysis, principal component analysis, and clustering analysis. The combination of these techniques allows the definition of a set of rules based on thresholds to eventually determine the likelihood of the suspicious transactions, or the identification of suspicious actors within the network structure. A similar comparison approach was performed on each subnetwork identified, considering all different types of subgroups, such as the connected components, the bi-connected components, and the communities.

The univariate analysis evaluates network centralities and defines a set of ranges of observations according to predefined percentiles. All observations that, in conjunction, satisfied the defined rules assigned to the outlier percentile should trigger a case for further investigation. For example, considering the individual centralities, when the degree is greater than 9, the eigenvector is greater than 6.30E-03, the closeness is greater than 1.46E-01, the betweenness is greater than 1.12E-04, the influence 1 is greater than 2.32E-04, and the influence 2 is greater than 1.44E-01, then the node is considered an outlier and should be flagged for further analysis. Based on the univariate analysis, there are 331 nodes considered as outliers. The average behavior of these 331 nodes creates the thresholds previously presented. If applied in a production environment, this set of rules would flag 21 participants for further investigation.

The principal component analysis evaluates the network centralities by reducing the dimensionality of the variables. All measures in relation to the network are comprised then into a single attribute, which can represent the original characteristics of the nodes. The outlier analysis points out observations according to the top values among all the transactions. Considering the individual centralities, when the degree is greater than 13, the eigenvector is greater than 3.65E-03, the closeness is greater than 1.42E-01, the betweenness is greater than 9.10E-05, the influence 1 is greater than 3.22E-04, and the influence 2 is greater than 8.48E-02, then the node is considered an outlier and should be flagged for further investigation. Based on the principal component analysis, there are 305 nodes considered as outliers. The average behavior of these 305 nodes creates the thresholds previously presented. If applied in a production environment, this set of rules would flag 23 participants for further investigation.

The clustering analysis evaluates the network centralities creating distinct group of nodes according to their similarities, or the distribution of their centralities. These similarities are based on the network metrics degree, eigenvector, closeness, betweenness, influence 1 and influence 2. The clustering analysis highlights observations not based on buckets or percentiles but instead according to unusual or low frequent clusters. The vast majority of the clusters identified within the network usually consists of several nodes. Based on that, when some particular clusters consist of a few number of members, they are defined as uncommon, and therefore as outliers in terms of groups of nodes. As a consequence, all nodes comprised in those unexpected clusters would be considered as outliers as well. The averages centralities for the outliers' clusters define the thresholds to identify which nodes within the network would be considered as outliers. According to the clusters' thresholds, when a degree is greater than 36, the eigenvector is greater than 1.04E-02, the closeness is greater than 1.45E-01, the betweenness is greater than 6.50E-04, the influence 1 is greater than 9.35E-04 and influence 2 is greater than 1.49E-01, then the node is considered as outlier and should be flagged for further investigation. Based on the clustering analysis, three clusters were considered as outliers, comprising 18, 2, and 3 nodes, respectively. In summary, these three outliers' clusters contain 23 nodes as outliers. The average behavior of these 23 nodes creates the thresholds previously presented. If applied in a production environment, this set of rules would raise only two participants for further investigation.

All previous rules and thresholds were defined based on the analysis of outliers, considering the individual nodes centralities. Although the clustering analysis takes into consideration a group of nodes, the individual nodes centralities were used to determine the thresholds. However, in terms of network analysis is also possible to analyze the behavior of the links among the nodes. The link analysis may highlight the outliers' links and therefore identifying the nodes assigned to these particular relationships. Similar to the clustering analysis where an uncommon group of nodes are considered outlier, and then its individuals, the link analysis considers uncommon connections among the nodes and then points out the uncommon individuals. In relation to the characteristics of the links, there is a measure that identifies the geodesic distance among the nodes, or the shortest paths associated with them. The link betweenness represents how many shortest paths a particular links partakes. For example, links partaking multiple shortest paths might have higher link betweenness than those do not. The nodes comprised in those outliers' links are also considered as uncommon, and then also set as outliers' nodes. The averages network centralities for the outliers' links determine the thresholds to highlight which nodes within the network would be considered as outliers according to their uncommon links. Based on the links' thresholds, when the degree is greater than 9, the eigenvector is greater than 4.30E-03, the closeness is greater than 1.41E-01, the betweenness is greater than 2.32E-04, the influence 1 is greater than 2.26E-04 and the influence 2 is greater than 9.37E-02, then the node is considered as outlier and should be flagged for further investigation. Based on the link analysis, 300 links were identified as outliers' connections, comprising 199 nodes. The average behavior of these 199 nodes creates the thresholds previously presented. If applied in a production environment, this set of rules would flag 18 participants for further investigation.

The social structure in this case contains 15 734 bi-connected components. Interconnecting these bi-connected components there are 2307 articulation points, which 984 are participants classified as individual. The claims in insurance supposed to have not too many individual participants connecting them into a major subnetwork. Companies can appear several times in the claims, once they are suppliers, repairers, or even other insurers. However, individuals appearing several times, connecting different claims, should be more suspicious and eventually be sent for further analysis. There are two different ways to use the articulation points approach. The first one is by directly highlighting them. The second one is to collect their average behavior, as done in the other approaches, and identify all nodes which match to this criteria. The averages network centralities for the articulation points would therefore determine the thresholds to highlight the outliers' nodes. Based on this, when the degree is greater than 20, the eigenvector is greater than 5.05E-03, the closeness is greater than 1.43E-01, the betweenness is greater than 4.27E-04, the influence 1 is greater than 5.00E-04, and the influence 2 is greater than 1.03E-01, then the node is considered as outlier and should be flagged for further investigation. Based on the bi-connected component analysis, 99 articulation points were identified as outliers' observations. The average behavior of these 99 nodes creates the thresholds previously presented. If applied in a production environment, this set of rules would flag 15 participants for further investigation.

Connected components hold a strong concept of relationship among the nodes. All nodes can reach each other, no matter the path required to do that. A connected component analysis was performed over the entire network. Differently than the previous analyses, where the evaluation considered just participants classified as individual, the connected components approach to identify outliers considered all nodes, both individuals and organizations. Additional information in terms of number of nodes, number of individuals and organizations, as well as the total amount of the claims where the connected components nodes are were considered for this analysis. A principal component analysis was performed considering the network centralities, subgroup characteristics, and business measures, like the average values for degree, eigenvector, closeness, betweenness, influence 1 and 2, the number of nodes for each connected component, and the total ledger, the amount claimed, the number of vehicles involved, and the number of participants, among others. The averages network centralities for the nodes comprised into the outliers' connected components would therefore determine the thresholds to highlight the overall outlier nodes. Based on this, when the degree is greater than 7, the eigenvector is greater than 6.97E-17, the closeness is greater than 8.33E-02, the betweenness is equal to 0.00E-00, the influence 1 is greater than 1.89E-04, and the influence 2 is greater than 1.50E-03, then the node is considered as outlier and should be flagged for further investigation. Based on the connected component analysis, three connected components were identified as outliers, comprising 30 nodes. The average behavior of these 30 nodes creates the thresholds previously presented. If applied in a production environment, this set of rules would flag 555 participants for further analysis. Eventually, due to this high number of participants raised by the connected components analysis, this approach could be considered not appropriated to be deployed in production, which would flag a large number of participants to be further investigated. This analysis will then be discarded.

Analogously to the connected components, the most relevant characteristic of the communities is the relationships between the nodes. As performed previously, additional information in terms of number of nodes, like individuals or organizations, and also, business information such as the total amount of the claims, or the total ledge, where the community's nodes are involved were taken into consideration to perform this analysis. Once again, a principal component analysis was performed over the network and business measures for the communities, considering the same attributes described in the connected component analysis. The averages network's measures for the nodes comprised into the outliers' communities would therefore determine the thresholds to highlight the overall outlier nodes. Based on this, when the degree is greater than 13, the eigenvector is greater than 3.20E-03, the closeness is greater than 1.42E-01, the betweenness is greater than 7.10E-05, the influence 1 is greater than 3.26E-04, and the influence 2 is greater than 8.02E-02, then the node is considered as outlier and should be flagged for further investigation. Based on the community analysis, 13 communities were identified as outliers, comprising 68 nodes. The average behavior of these 68 nodes creates the thresholds previously presented. If applied in a production environment, this set of rules would flag 24 participants for further investigation.

The entire process to identify outliers based on the different types of analyses produces the following figures:

Technique	Individual participants
Univariate analysis	21
Principal component analysis	23
Clustering analysis	2
Link analysis	18
Bi-connected component analysis	15
Connected Components Analysis	555
Community analysis	24

Notice that the connected component analysis was discarded. Outliers' nodes might be raised by distinct techniques. The total number of participants outliers classified as individual, considering all previous techniques is about 33. These participants are involved in 706 claims. They would definitely need to be sent for further investigation.

The analysis of social structures based on the computation of subgroups and centralities is one of the possible approaches to recognize the pattern associated with the relationships. In order to detect unexpected behavior on these social structures a set of exploratory analyses must be conducted, like the ones performed in this case, such as univariate analysis, principal component analysis, clustering analysis, subgroup analyses including bi-connected components, connected components and communities, and link analysis. The combination of these approaches constitutes a distinguish methodology to recognize social patterns and then unexpected events, which might be considered as high risk for a substantial number of industries.

Individual attributes associated with the claims can definitely hold useful information about the transactions in place. However, these attributes might poorly explain correlations on some particular type of event such as fraud or exaggeration. On the other hand, the multiple relationships between the claims can actually reveal significant characteristics of the participants within the claims. Network analysis can definitely reveal participants in a high number of roles, groups of participants with high average values for some specific business attributes, and strong links among participants, among many other features that traditional analysis would not raise. For example, Figure 5.46 shows a particular doctor specialist who are quite frequent within claims above a specific amount of value. He connects a high number of distinct policy holders. The star network describes a node central to holding lots of connections to different nodes. Further analyses showed completely different geographic addresses for these connected nodes. This might be quite uncommon in the auto insurance industry and hence this specialist would be flagged for further investigation.

Another approach in highlighting unusual behavior is by analyzing the identified subnetworks, such as the connected components, the bi-connected components, and the communities. Some subgroups are unexpected by their own internal average centralities or characteristics, like the number of nodes or links with respect to the average numbers of the entire social structure. Other groups might be pointed out by business attributes associated with the network metrics, like the amount of claim, the number of distinct roles, the number of cars involved, or the different addresses within the claims associated with the nodes within the subnetwork, among many others. All these attributes within the groups can also be compared to the average numbers of the entire social structure. For example, Figure 5.47 shows some subnetworks that were considered as outliers based on their own centralities or their cumulative business attributes.

Exaggeration is a recurrent problem in most insurance companies, sometimes caused by customers, sometimes by the suppliers. This can represent a substantial amount of money for the insurers, affecting the operational costs and the overall cash flow. Network analysis can be as flexible and dynamic as the market in identifying suspicious events and actors with

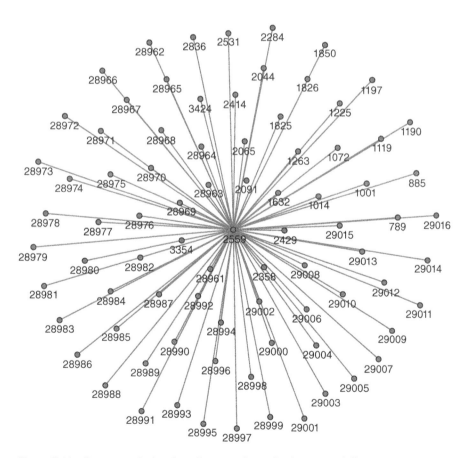

Figure 5.46 Star network showing all connections of a doctor specialist.

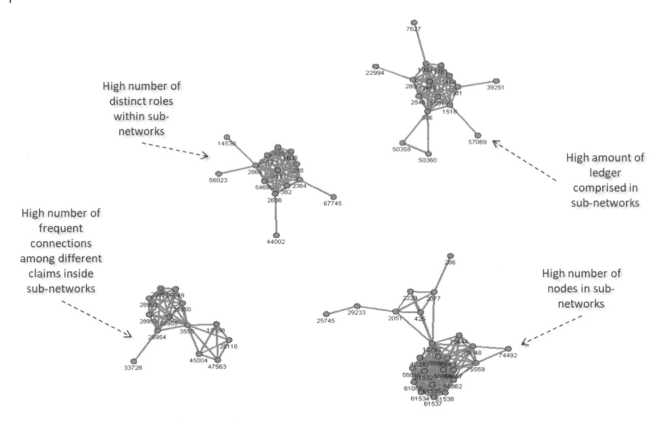

High number of distinct roles within sub-networks

High amount of ledger comprised in sub-networks

High number of frequent connections among different claims inside sub-networks

High number of nodes in sub-networks

Figure 5.47 Subnetworks considered as outliers.

social structures created by transactions. In some business scenarios, the attributes on the relations can be more important than the individual attributes, and then, reveal relevant information about the operational process.

5.7 Customer Influence to Reduce Churn and Increase Product Adoption

The telecommunications industry evolves into a highly competitive and multifaceted market, including internet, landline, mobile, content, news, streaming, etc. which demands that companies implement a holistic approach for customer relationships management, looking into multiple aspects of the customers' behavior. Social network analysis can be used to increase the knowledge about the customers' influence, highlighting relevant information on how customers interact with each other and ultimately how they may influence each other in some particular business events like churn and product adoption.

The most important characteristic of any community is the relationship between their members. Any community is established and maintained based on these relations. With all the modern technologies and devices currently available, there are new types of relationships and that characteristic that creates and holds communities gained even more relevance. From multiple possibilities of communication types, these relationships expanded in terms of flexibility and mobility, defining new boundaries for social networks. With social networks evolving and enhancing, the influence effect of some members of these communities become even more relevant to the companies.

Several events in social communities can be faced as a cascade process, where some members of the network initiate a sequence of similar events over time based on their influence in these social structures. The influential members within the social structures may decide to do something like make churn or start using a new product or service, and then some other members of the network might follow them on the same business event. The way these members are linked within the social network virtually defines how these events will occur over time. Often, the stronger the relationships between the members, the higher the probability that these events take place in a cascade fashion. Understanding the way of these relationships take place and identifying what influences community members to do something can be a great competitive advantage, especially when thinking in business events that may occur in a cascade approach, like churn, product adoption, and service consumption, among many others.

It is clear that in social structures, some people can influence others in multiple events. Recognizing these influencers and identifying their main characteristics can be crucial in defining and executing better business campaigns to exploit the influence effect on some business events like product adoption, and avoid the cascade effect on others, like churn. Identifying the central and strong nodes (subscribers) within these social networks and understanding how they behave when influencing others can be one of the most effective approaches to define a more holistic customer value. This latest information on the customers' value can be used to estimate the impact of churn and product adoption events over time in a cascade perspective rather than simply isolated. Assessing customers' connections and identifying the central nodes of each social structure may be the best method to prevent a cascade event of churn, and at the same way it can boost a broader product adoption diffusion.

A combination of community detection and centralities measures with correlation analysis on some particular business events like churn and product adoption can reveal a crucial knowledge on the impact of those events when triggered by a key node within the social structure. This approach allows companies to ultimately estimate the impact of a business event over time when it is initiated by a central and strong node or a satellite and weak node within the social structures. How the influence effect takes place in both scenarios. The influence here basically means the number of customers affected by each one of these business events, either churn or product adoption.

Monitoring social networks over time allows companies to assess the revenue impact associated business events in a cascade process. Based on the likelihood assigned to the events of churn and the level of influence assigned to the customers, companies can put in place direct actions to avoid the cascade of churn over time by retaining an influential customer who would probably lead other customers to make churn afterwards. Analogously, by knowing the likelihood of the customers to purchase some product or service, plus the influential factor assigned to these customers, it is possible to deploy an effective campaign to not just sell more products and services but also to diffuse them throughout the multiple social structures.

In order to evaluate the social structures over time and calculate the influential factor assigned to the customers, a set of mobile data is collected considering a particular timeframe and aggregated to account for the average customers' relationships. For example, the last three months of mobile data is collected to create the overall network. Once the network is created, the communities within the network are identified and the centrality metrics are computed for each community. Notice that the centrality metrics in this case are not calculated considering the entire network. Instead, they are computed considering each community identified. That means the degree centrality for instance for one particular subscriber does not represent the full connections of that subscriber, but only the connections within its community. The main reason to define this approach is due to the size of the communications network. For this particular case, 76 million mobile transactions generated 34 million links and 6 million nodes. The community detection algorithm identified 218,000 communities, with the average number of 30 members. Figure 5.48 shows graphically the steps assigned to the network analysis approach. The first step is to collect historical information in order to build the network, detect the communities, and compute the centrality metrics for each community. The timeframe for this first step is 12 weeks of mobile data. Then, we analyze the following eight weeks of churn and product adoption events. At this stage, we evaluate what happened with the other customers within the communities after customers making churn or purchasing a product during the first stage of 12 weeks. Did they make churn or purchase the same product on the following eight weeks? How many customers during the second stage of eight weeks did follow the initial events performed by the customers in the first stage of 12 weeks?

Once the customers in the first stage are identified, customers who made churn or purchased a product, then we evaluate what happened to their peers afterwards. How many customers were they able to influence in making churn or buying the product? Based on that, we move forward to the third step, which is correlating the centrality metrics to the customers who influenced other customers to make churn or purchase a product. What differentiates them from the customers who simply made churn or purchased a product in the first 12 weeks without influencing any of their peers to do the same in the following eight weeks?

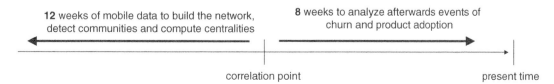

Figure 5.48 Network analysis approach.

The following numbers may better describe the process in terms of the first, second, and third stages of the network analysis process. For example, considering the first timeframe of 12 weeks, almost 40,000 subscribers made churn. From that amount, 9,000 subscribers were connected to other customers who made churn afterwards during the second timeframe of eight weeks. That means 31,000 subscribers made churn and no other subscriber they used to be connected made churn in the following eight weeks. We assume that the first 9,000 customers influenced some of their peers to make churn afterwards, while the other 31,000 subscribers did not. It is virtually impossible to know if there was really an influence effect from the first wave of churn events during the first 12 weeks to the second wave along the following eight weeks. The strength of the relationships between the customers in the first and second waves helped us to create this hypothesis. Often, they were more connected than the subscribers who did not make churn afterwards.

When we look at the top influencers, or the subscribers associated with more than one churn afterwards, the total number of customers in the first wave drops considerably, to a little less than 1,000 subscribers. When we compare these customers to the rest of the 39,000 customers, they are connected to 53% more subscribers, their communities are 60% bigger, and they influence seven times more other customers to make churn afterwards. On average, they are 140% more connected to other customers than the regular churners, they have 53% more connections off-net (subscribers in the competitors), they spent 150% more time with previous churners, and they were 106% more connected to previous churners. Those indicators describe the overall profile of the influential subscribers in the event churners. Figure 5.49 shows the comparison between the influencers and the regular churners.

Particularly to the churn event, another relevant factor of influence is the multiple occurrences of churn over time within the same community. As mentioned before, based on the historical data analyzed, 218,000 communities were detected, with 30 members on average. Recall that the timeframe to create the network was 12 weeks, and the timeframe to evaluate the cascade effect of churn and product adoption was eight weeks. Along these eight weeks, the average churn rate was around 3%, considering all communities. However, throughout this period, 141,000 communities did not observe events of churn, while 77,000 communities did. Considering only the communities with churn along these eight weeks, the churn rate increases to almost 7%. There is a quite relevant behavior associated with the members of the communities experiencing subsequent churn events. The likelihood to make churn associated with all subscribers who remain after their peers in the same community leave increases substantially. For example, all communities experiencing two subsequently weeks of churn, the churn rate increases to up 11%. For all communities experiencing four to six weeks of subsequently churn, the churn rate increases to up 14%. Finally, for all communities experiencing eight weeks of subsequent churn, the churn rate can exceed 15%. The viral effect in the first scenario, within two subsequent weeks is 28% greater than the normal rate. The second scenario, within four to six weeks, the viral effect is 37% greater than the normal rate. And for the last scenario, for the eight weeks timeframe, the viral effect exceeds 40%. Figure 5.50 shows how the churn rate evolves over time for communities experiencing subsequent churn events.

For product adoption, the viral effect is similar. For this particular product adoption event, a new product launched by this telecommunications company was selected to evaluate the impact of the viral effect on subscribers. Based on a supervised model to estimate the likelihood of customers to purchase this new product, 730,000 subscribers were targeted. In a first wave campaign, considering that initial target, a bit of less than 5,000 customers purchased the new product. In order to evaluate the impact of the viral effect throughout the social structures, two groups of subscribers were identified from those almost 5,000 buyers. For the first group, 816 customers were randomly selected. This group worked as a control group. For the second group, the leaders of each community, or the subscribers with the best centrality metrics combined were picked. They were

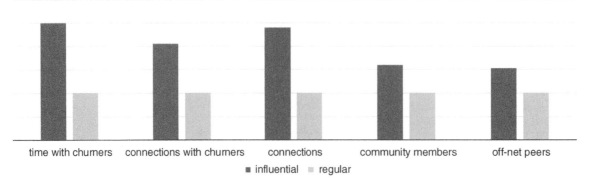

Figure 5.49 Profile of influencers and regular churners.

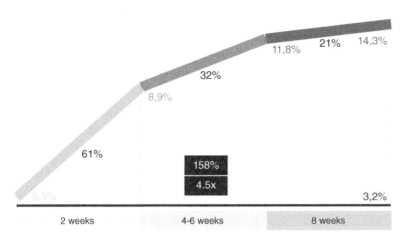

Figure 5.50   Likelihood of churn over time within the communities.

ranked by the combined centralities and the top 816 subscribers were selected. From the random 816 subscribers, after eight weeks monitoring all purchasing events, another 572 subscribers connected to those 816 initial customers also bought the same new product. The random 816 subscribers were well connected to another 8523 customers, giving an average of 10 extremely relevant relationships. For the second group, after the same eight weeks, another 4108 subscribers connected to those 816 leaders also bought the new product. The 816 leaders were well connected to another 21,013 customers, giving an average of really 26 relevant relationships. The leaders had 160% more strong connections with other customers within their social structures, or 2.6 more relevant relationships. More importantly, the leader were more than seven times more effective spreading the new product throughout their social structures. The subsequent sales assigned to the leader were 620% greater than the subsequently sales assigned to the random customers. Figure 5.51 graphically compares the viral effect in terms of product adoption considering the cascade events of purchasing initiated by the subscribers' leaders in their communities and by the random subscribers. The leaders have not only more than twice as strong connections in their social structures, but they are seven times more effective in disseminating the new production and impacting the overall sales over time.

There is an important fact along the analysis of this case study. While detecting communities and computing the network metrics to identify the leaders of each community, some of the leaders found are not customers for the company. They are subscribers for different carriers, the competitors. Eventually, they should be targeted to acquisition campaigns. For example, out of the 218,000 communities detected, almost 8,000 leaders are off-net subscribers, which means they are customers of different carriers. Even not considering all the possible connections for these subscribers, as the interactions between them and other competitors' subscribers are not in the data analyzed (each carrier has access only to transactions involving its own subscribers), they are still the most important nodes within their communities, or at least they have the highest combined centrality metrics in their community. These customers might be even more important when considering all their connections, with all subscribers, for all carriers. Based on that, these 8,000 leaders should be targeted in acquisition campaigns, as they actually can influence the current customers to eventually make churn and move to their carriers. In addition to these leaders, there are also almost 140,000 subscribers from other carriers who present a significant combination of centrality metrics and are well connected to the current customers. These subscribers belong to different carriers, but they are well connected in the social structures identified based on the company's mobile data. Similarly, these subscribers may influence the current customers to make churn and move to their originating carriers. Instead of taking the risk to lose

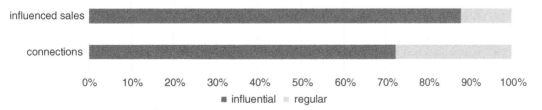

Figure 5.51   The viral effect of purchasing.

actual customers due to their connections to subscribers from different carriers, the company should target these off-net subscribes in acquisition campaigns. In addition to acquiring possibly very good subscribers, the company would protect the current customers from making churn and moving to the competitors.

5.8 Community Detection to Identify Fraud Events in Telecommunications

This case study presents how social network analysis and community detection based on telecommunications data can be a valuable tool in understanding many different aspects of subscribers' behavior, but particularly in terms of relationships. The analysis of social relationships points out distinct aspects of customers' behavior and possible viral effects on business events within the social structures. One of the most relevant social structures within networks are the communities. The viral effects associated with business events usually takes place within these social structures, following the relationships of its members. The viral effect normally works as a cascade process. A particular subscriber – often an influential one – does something and a subset of its peers follows the same process. For example, an influential customer, well connected within its social structure, decides to make churn at some point. Afterwards, some of its peers end up following this initial event and also make churn after a period of time. A similar scenario can also occur with product adoption. Both cases were described in more detail in the previous case study.

Fraud is a common and recurrent event in telecommunications. The hypothesis here is that fraud is a business event (business at least for the fraudster) that may also be impacted by the viral effect. For example, someone identifies a breach in a telecommunications process. A fraudster sees that breach as an opportunity and creates a scenario where the fraud event can occur unobserved, at least for a period of time. Regardless of the fraud aims to generate revenue to the fraudster or to simply allow people to consume communications services for free, the initial event propagates over time as more and more subscribers start using or consuming that service. Eventually, these people, the initial fraudster, and the following subscribers, are somehow connected within the social structures. Perhaps they are more connected than the other subscribers not committing fraud. If this is the case, a viral effect may occur for fraud events, as might happen for business events such as churn and product adoption.

Social network analysis often raises relevant knowledge about the customers' behaviors, particularly in relation to their relations within the social structures, which is highly correlated to the influence effect of business events over time. This particular case covers few steps on the network analysis process. Based on a specific timeframe of mobile data, which has information about subscribers' connections, the social network was created. The timeframe selected is about six months of transactions, consisting of a couple billion records. The network created from that original data contains 16 million nodes and 90 million links. Upon this network created in this first step, a community detection algorithm was performed to identify the multiple social structures within the communications network. This second step identified 700,000 communities. The average number of members within the communities was around 33 subscribers. The third step is to compute the network metrics for each subscriber considering their communities. Notice that the network metrics are not computed considering the entire network, but by the communities detected in the second step. The fourth step is to aggregate the network centralities for each community within the network. Based on the communities' measures, and the fifth step is to perform an outlier analysis, highlighting the communities with uncommon characteristics or unexpected behavior upon that timeframe defined. There are two main approaches for the outlier analysis. The first is based on the subscribers' centralities within their communities. Subscribers with way different measures compared to the rest of the community may be assigned to further investigation. The second approach is based on the communities' metrics upon the entire network. Communities with way different metrics considering the entire network may be highlighted and all members be flagged to be further investigated. Figure 5.52 shows graphically the overall process.

Most of the analytical approaches employed in fraud detection accounts for usage and demographic variables, which describe the individual behavior assigned to the subscribers. Network analysis on the other hand evaluates the relationships

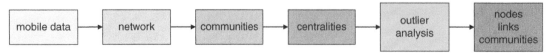

Figure 5.52 The overall process based on the social network analysis.

among subscribers, regardless of their individual information. In fact, all this individual information can be added as attributes to nodes and links, enhancing the analysis on the network, the communities, and ultimately on the relationships among customers.

The main characteristic of any community is the relationship between their members. Communities are identified over time based on that. It is important to notice how dynamic this process is, as over time, connections may come and go, changing the overall topology of the social structures. The links in the network, particularly within the communities, can be used to describe the subscribers' behavior, and therefore correlate that behavior to the business event under analysis, the fraud event. The analysis of these relationships over time allows the company to better understand how the fraud spread throughout the network and how the viral effect takes place within the social structures. For example, one important task is to identify relevant nodes and links within the network. Relevant nodes can be the central nodes or the nodes that control the information flow, the nodes playing as hubs or authorities, nodes with the higher metrics describing importance and influence, and nodes serving as bridges between different social structures. Relevant links can be straightforward and identified by the frequency and the recency between the nodes. The information about nodes and links is correlated to the occurrence of the fraud events over time, possibly allowing the identification of the sequence of those events in a cascade fashion. That allows answers to who initiated the fraud event? What is the overall characteristic of the communities experiencing frequent fraud events? What is the overall profile of the nodes initiating the fraud events over time? The overall analysis aims to detect the viral effect of fraud events within communities. What happens after the first event of fraud? What members of the social structure initiate the fraud and which ones follow? That viral effect might happen because the fraudster wants to disseminate the fraud over the social structure, or multiple structures, or simply because other subscribers get to know about the breach in the telecommunications process that allows them to use services or products for free. It can be a business fraud or an opportunistic fraud. Both have a strong component on the viral effect throughout the social structures within the network.

The most important outcome from a network analysis is the centrality metrics associated with the subscribers. These metrics describe how they relate to each other, in what frequency, and how relevant their connections are. From a network perspective, the suspicious nodes might be identified by the centralities assigned to nodes and links, as well the aggregated metrics assigned to the communities. Nodes with higher network metrics might be suspicious, particularly when compared to the other nodes within the same social structure. The nodes partaking links with high network metrics might also be suspicious. Members of communities with high network metrics might be suspicious, particularly when compared to the overall network profile.

A community usually contains nodes with connections outside the social structure, as bridges linking that community to the rest of the network. In this way, a community may be connected to other communities throughout one or more nodes. Communities within networks can be understood as clusters inside populations. The study of communities can reveal some relevant knowledge assigned to the network topology and therefore to the members of the overall structure.

In large networks such as in telecommunications, it is common to find large communities. The outcomes of the community detection process follow a power law distribution, with few communities consisting of a large number of nodes and the majority of the communities consisting of a few number of nodes. The first step of the community detection was based on the Louvain algorithm. In order to reduce the impact of large communities, we evaluated multiple resolutions while searching for the subgroups, producing several different members' distribution for the communities detected. Recall that the resolution defines how the network is divided into communities. Resolution is a reference measure used by the algorithm to decide whether or not to merge two communities according to the intercommunity links. In practice, larger resolution values produce more communities with a smaller number of members within them. When producing large communities, the algorithm considers more links between the nodes, even the weak connections that link them. Bigger communities imply in a significant number of members, and therefore, weaker connections between them. On the other hand, smaller communities imply fewer members, but with stronger connections between them. Based on the multiple resolutions, we evaluated the modularity for each level of community detection. The modularity indicates a possible best resolution for the community detection. Higher values for modularity indicate better distribution for the communities. Modularity usually varies from 0 to 1. However, the decision about the number of communities and the average number of members within each community can also be a business requirement. Identifying fraud events might require strong links between the members of the social structures. Then, a combination of the mathematical approach by using multiple resolutions and the modality, and the business requirements, is often a good approach. Finally, to diminish the effects of large communities, we executed a second community detection algorithm. This second pass was based on the label propagation algorithm, with the *recursive* option to limit the size and the diameter of the communities, as well as with the *linkremovalratio*, to eliminate small-weight links between the members of the communities.

The next phase of the network analysis approach is to execute a multivariate analysis and search for the occurrences of outliers in terms of nodes within communities and communities within the network. The outlier analysis was based on the centrality metrics for all nodes within the communities, and the aggregated metrics for all communities within the network. The centrality metrics computed were degree-in, degree-out, degree, influence 1, influence 2, PageRank, closeness-in, closeness-out, closeness, betweenness (for nodes and links), authority, hub, and eigenvector. These metrics were calculated using the by clause, which forces the calculation to inside each community.

The detection of outliers within the communities identifies unusual behavior, not in relation to individual attributes but mostly with respect to groups of subscribers and their behavior when consuming communications services and products all together. Isolated transactions may hold expected values for some particular attributes. However, the relationship among subscribers based on their connections may reveal unusual characteristics.

Fraud events have a starting point, which can be identified when the fraudster creates the proper environment to diffuse an illegal use of the telecommunications network. As long as this environment is ready to use, the fraud events start spanning throughout the network, spreading like a virus within social structures. Often, the fraud event is highly concentrated among small groups that are aware of the breach, and hence, will present a substantial distinct type of behavior than the regular subscriber groups. The outlier analysis upon the nodes as individuals may not indicate any suspicious behavior. Instead, the outlier analysis upon groups of subscribers or communities can lead the investigation toward suspicious clusters, pointing out nodes comprised within these groups.

After computing the network metrics for all communities, considering all measures previously described, the outlier analysis is performed. The univariate analysis classifies all observations into specific percentiles and points out the high values. For instance, taking the percentile 95% or 99% is possible to highlight X number of nodes, comprised into Y number of communities. These X number of nodes hold an average value for all network measures, such as degree-in d_i, degree-out d_o, closeness-in c_i, closeness-out c_o, closeness c, betweenness b, influence-1 i_1, influence-2 i_2, hub h, authority a, eigenvector e, and PageRank p. Also, we might consider the number of members comprised in the groups identified and the average of the network metrics assigned to each community. By applying these thresholds for the whole network, a Z number of nodes will be flagged, indicating possible observations to be sent for further investigation in terms of suspicious transactions of fraud.

The overall fraud detection approach consists of a sequence of steps, from identifying the set of communities within the network, to computing the network metrics for each community, identifying the outlier percentiles (95% and 99% for instance), computing the average metrics values for the outliers' observations and then applying these rules and thresholds to the entire network. The outcome from this sequence is a set of nodes, which possibly require additional investigation to validate the involvement in a fraud event. The outlier percentiles, which will determine the number of nodes flagged as suspicious is a matter of the distribution values for all network metrics but also the company capacity to investigate the cases raised by the process. Lower cutoffs for the outliers imply that more cases are raised, and higher cutoffs for the outlier percentile imply that fewer cases are raised to be investigated. Figure 5.53 shows graphically the process assigned to the rules and thresholds to identify possible cases of fraud for further investigation.

The following two figures present some suspicious behavior when the exploratory analysis considers the network measures. For example, the big node in red in the middle of the social structure showed in Figure 5.54 below makes more than 16,000 calls to the node close to it. The thicker arrow represents how strong this connection is particularly in comparison to the others. A combination of the link weight and the hub was used to highlight this particular case.

Another example, the red node with thicker arrows in the social structure showed in Figure 5.55 receives more than 20,000 calls from just 26 distinct phone numbers, represented by the nodes surrounded it. Analogously, the thicker arrows

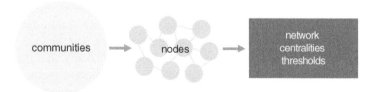

Figure 5.53 Rules and thresholds produced by the outlier analysis.

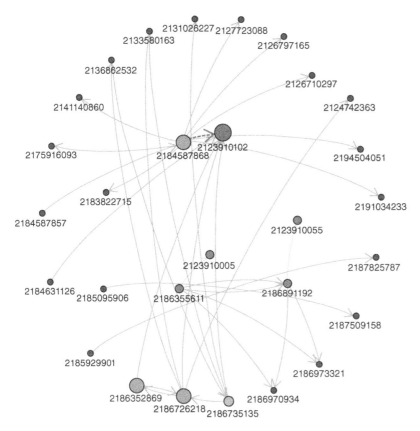

Figure 5.54 Outlier on the link weight and the hub centrality.

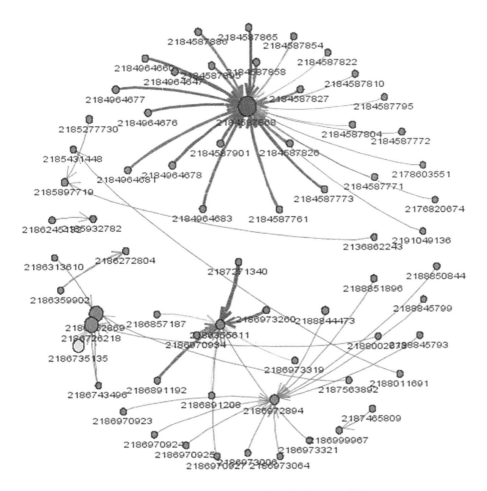

Figure 5.55 Outlier on the link weight, and hub and authority centralities.

represent the strength of the links among this particular node and its peers. A combination of the link weight and the authority was used to highlight this particular outlier.

Often, the procedures associated with fraud detection or fraud management, for example in this approach, highlights suspicious events, which from a practical point of view, flag those events for further investigation. The outlier analysis upon the social network metrics describes possibly suspicious behavior considering relationships rather than individual measures or attributes. This average behavior can be translated into a set of rules based on thresholds to point out possible cases of fraud. This set of rules based on thresholds can even be deployed in a transactional system, for example, to raise alerts in real time about possible events of fraud in course. Alerts are not stamps of fraud, but instead, they are indications on how likely these events might be fraudulent.

5.9 Summary

This chapter wraps up the content of this book by presenting some real-world cases studies. The first case showed the utilization of the TSP to find an optimal route in Paris when visiting multiple places by using different transport options. This case was a demonstration, which later on turned out to be a real-case implementation. The second case also started as a demonstration, which turned into a real deployment. It presented the VRP, which is crucial for supply chain and scheduling and delivering. Both cases were deployed in the United States. The third case presented a real-case application developed at the beginning of the COVID-19 pandemic. It employed a series of network analysis and network optimization algorithms in combination with supervised machine learning models to understand the impact of the population movements throughout geographic locations to the spread of the virus and therefore estimating possible new outbreaks. This case was deployed in a few countries around the globe, like The Philippines and Norway. The fourth case was also a real-world application to understand the urban mobility and to apply these outcomes to improve the public transportation planning, traffic routing, and disease spread. This case was initially developed in 2012 to understand the vectors for the spread of diseases like Dengue in Rio de Janeiro, Brazil. Later on, some of the algorithms were used during the World Cup in 2014 and the Olympic Games in 2016. The fifth case was also a real-world application developed in Dublin, Ireland, to identify exaggeration cases in auto insurance. The sixth case was a real-world application deployment in Turkey to identify the impact of the viral effect in business events like churn and product adoption. Finally, the seventh case was a real-world application to detect fraud in telecommunications. This case was deployed in different carriers in Brazil.

Index

Network Science: Analysis and Optimization Algorithms for Real-World Applications, First Edition. Carlos Andre Reis Pinheiro.
© 2023 John Wiley & Sons, Inc. Published 2023 by John Wiley & Sons, Inc.